Studies in History and Philosophy of Science

Volume 50

Studies in History and Philosophy of Science is a peer-reviewed book series, dedicated to the history of science and historically informed philosophy of science. The series publishes original scholarship in various related areas, including new directions in epistemology and the history of knowledge within global and colonial contexts. It includes monographs, edited collections, and translations of primary sources in the English language. These cover a broad temporal spectrum, from antiquity to modernity, and all regions of the world.

More information about this series at http://www.springer.com/series/5671

Saulo de Freitas Araujo
Thiago Constâncio Ribeiro Pereira
Thomas Sturm
Editors

The Force of an Idea

New Essays on Christian Wolff's Psychology

Springer

Editors
Saulo de Freitas Araujo
Department of Psychology
Federal University of Juiz de Fora
Juiz de Fora, Minas Gerais, Brazil

Thiago Constâncio Ribeiro Pereira
Department of Psychology
Fluminense Federal University
Volta Redonda, Rio de Janeiro, Brazil

Thomas Sturm
ICREA & Universitat Autònoma de
Barcelona
Department of Philosophy
Bellaterra (Barcelona), Spain

ISSN 0929-6425 ISSN 2215-1958 (electronic)
Studies in History and Philosophy of Science
ISBN 978-3-030-74434-2 ISBN 978-3-030-74435-9 (eBook)
https://doi.org/10.1007/978-3-030-74435-9

This Springer imprint is published by the registered company Springer Nature Switzerland AG
The registered company address is: Gewerbestrasse 11, 6330 Cham, Switzerland

Preface

This book is the result of a long-term dialogue on the history and philosophy of psychology, which began in 2009 at the Max Planck Institute for the History of Science in Berlin. In 2012, the third editor was invited by the first to visit the Federal University of Juiz de Fora (Brazil), teach a master class in the new Graduate Program in History and Philosophy of Psychology, and start conversations with his master's and doctoral students. In 2013, the first editor went to Spain as visiting scholar in the Department of Philosophy at the Universitat Autònoma de Barcelona, where the third editor was teaching and doing research. Then, in 2015, the second editor also went to Barcelona to develop part of his PhD training under the supervision of the third editor.

The common question that has provoked and sustained our conversations and collaborations is the role of eighteenth-century German philosophy and psychology in the development of psychological science. Doubtlessly, Christian Wolff (1679–1754) has a prominent place in this discussion. However, although his work is often referenced, it has not received enough attention and analysis in the history and philosophy of psychology. Wolff's psychology was prominent especially (though not exclusively) in eighteenth-century German philosophy and psychology, as witnessed by the fact that it led to a genuine school of psychology and was critically discussed until well into the nineteenth century, before it became neglected and forgotten.

In 2018, we decided that an up-to-date, comprehensive collection of essays on Wolff's psychology, its contexts, contents, and consequences was both worthwhile and a long overdue enterprise. We wanted to make Wolff's psychology more visible not only for scholars of early modern thought but also for a wider audience in history and philosophy of the human sciences, interested in the development and fundamentals of psychological science. The year 2020, in which most of the work on this volume was carried out, marks also the tercentennial of Wolff's *Deutsche Metaphysik* (The German Metaphysics), the first systematic presentation of his psychology—a good occasion for a reassessment of that psychology.

We have invited prominent international scholars, according to their respective expertise both in Wolff's work and in the topic to be addressed. They are among the best representatives of Wolff-scholarship. Their contributions focus on the major guiding ideas of Wolff's psychology, both empirical and rational, its structure, its main innovations, its relation to other areas of Wolff's work, as well as its impact on later authors, schools, and research practices. As editors, our aim is to do justice to Wolff and his psychological insights—some of which, albeit often in different disguises, are still alive today, though their founder is no longer credited. This is the force of an idea: it develops a life of its own.

Juiz de Fora, Minas Gerais, Brazil — Saulo de Freitas Araujo
Volta Redonda, Rio de Janeiro, Brazil — Thiago Constâncio Ribeiro Pereira
Bellaterra (Barcelona), Spain — Thomas Sturm

Citations and Translations

References to Wolff in the literature are not standardized. Although most scholars tend to use Wolff's *Gesammelte Werke*, published by Georg Olms (Germany), there are many exceptions. Moreover, the way to refer to passages of his work is sometimes idiosyncratic, with abbreviations and rules created for a particular purpose. For the nonspecialist reader, this can generate confusion.

In order to make the references and citations homogeneous throughout the book and more accessible to nonspecialists, we have adopted APA (American Psychological Association) style, according to the seventh edition of its *Publication Manual*. However, given the specificity of the majority of Wolff's writings, which are ordered in numbered paragraphs, we have followed the common practice of giving the corresponding paragraph number(s) for both indirect and direct citations. In the latter case, the corresponding page number(s) is also given. The complete references to Wolff's work are always given at the end of each chapter.

As for the translations, we have adopted the general rule of rendering all original passages in English. However, all authors are responsible for their own translations as well as for style.

Acknowledgments

We would like to thank, first, all contributors for sharing the central idea of our project and accepting our invitation. This project could not have been carried out without their participation.

Saulo Araujo would like to thank the National Council for Scientific and Technological Development in Brazil (CNPq) for financial support.

Thiago Pereira would like to thank the National Council for Scientific and Technological Development in Brazil (CNPq) for financial support during his PhD (2015–2016), as well as Marcos Pereira, Rogéria Ribeiro, and Fabíola Brandão for their personal support during the preparation of this volume. He also thanks his colleagues from the psychology department at the Fluminense Federal University, especially Gustavo Ferraz and Roberto Preu, for the continuous intellectual stimulation.

We are equally grateful to Bruno Fiuza at Springer for his constant encouragement and support.

Contents

About the Editors

Saulo de Freitas Araujo is professor of history and philosophy of psychology at the Federal University of Juiz de Fora in the State of Minas Gerais, Brazil. He is also director of the Wilhelm Wundt Center for the History and Philosophy of Psychology (NUHFIP) at the same institution. His work focuses on the historical development and philosophical foundations of scientific psychology, especially in the German tradition. Among numerous publications, he is author of *Wundt and the Philosophical Foundations of Psychology: A Reappraisal* (Springer, 2016).

Thiago Constâncio Ribeiro Pereira is professor at the Fluminense Federal University (Psychology Department, Volta Redonda-RJ, Brazil), where he also leads the FHIPSI (The Philosophy and History of Psychology Research Group). His work deals with the history and philosophy of psychology, focusing on Christian Wolff's psychology (on which he developed his PhD thesis entitled "The Unity of Psychology in Christian Wolff's Thought"), eighteenth-century psychology, and the mind–body relationship.

Thomas Sturm is ICREA research professor at the Autonomous University of Barcelona (Spain). His work centers on three areas: Kant's philosophy, theories of rationality at the intersection of philosophy and psychology, and the relations between philosophy and history of science. His main publications include *Kant und die Wissenschaften vom Menschen* (2009) and *How Reason Almost Lost Its Mind: The Strange Career of Cold War Rationality* (coauthored with L. Daston, M. Gordin, P. Erickson, J. Klein, and R. Lemov, 2013). He has published numerous articles in journals such as *Erkenntnis, Kant-Studien*, *Studies in History and Philosophy of Science*, and *Synthese*.

About the Contributors

Matteo Favaretti Camposampiero PhD, is a research fellow and lecturer in philosophy (philosophy of language, history of philosophy) at the Ca' Foscari University of Venice (Italy). His publications include two books (*Filum cogitandi: Leibniz e la conoscenza simbolica*, Milan 2007; *Conoscenza simbolica: Pensiero e linguaggio in Christian Wolff e nella prima età moderna*, Hildesheim, 2009) and several articles primarily focused on Leibniz, Wolff, their sources, and their influence. His research is devoted to early modern philosophy from Descartes to Kant with a special interest in metaphysics, the theories of knowledge, language, and modalities, mereology, the mind–body problem, rational theology, and the philosophy of biology and medicine.

Sonia Carboncini did her doctorate in Florence with Claudio Cesa and later in Trier with Norbert *Hinske*. She worked at the *Leibniz Archiv* in Hannover and at the *Leibniz Forschungsstelle* in Münster. She participated in the edition of the *Gesammelte Werke* by Christian Wolff. She has published numerous works on Wollf's metaphysics, on the influence of Leibniz upon it, and on the relations between Wolff and the environment of the French Encyclopedists.

Corey W. Dyck is professor of philosophy and faculty scholar for arts and humanities at Western University. He is the author of *Kant and Rational Psychology* (Oxford 2014), coeditor (with Falk Wunderlich) of *Kant and his German Contemporaries* (Cambridge 2018), and is the translator and editor of the collection *Early Modern German Philosophy: 1690–1750* (Oxford 2019). In addition, he has published a number of articles on Kant's philosophy and its history in various academic journals including *Journal of the History of Philosophy, Kant-Studien*, and *British Journal for the History of Philosophy*. He has held visiting positions at Oxford University, Johannes-Gutenberg-Universität Mainz, and at the Martin-Luther-Universität Halle-Wittenberg, where he was also recently an Alexander von Humboldt research fellow.

Werner Ludwig Euler is a visiting professor of philosophy in the Latin American Institute of Economy, Society, and Politics (ILAESP) at the Federal University for Latin American Integration in Foz do Iguaçu, Brazil. Between 2012 and 2016, he was a visiting professor in the Department of Philosophy at the Federal University of Santa Catarina (UFSC) in Florianópolis, Brazil. His work focuses on ancient, modern, and contemporary philosophy, covering the fields of theoretical and practical philosophy, the mind–body relationship, philosophy of medicine, logic, and metaphysics (ontology), aesthetics, and philosophy of art. He has participated in many international conferences on philosophy. Among numerous publications, he is author of a voluminous commentary on Kant: *Natur und Freiheit. Kommentar zu den Einleitungen in Kants "Kritik der Urteilskraft"* (Meiner, 2018).

Ursula Goldenbaum is professor at the Department of Philosophy at Emory University. She held academic positions at research institutions in Berlin and Potsdam in the GDR, then in Germany, before joining Emory in 2004. Her subject of research is the history of early modern philosophy. She edited writings and letters of Leibniz (1991), Mendelssohn's translation of Rousseau's Treatise on the Origin of Inequality (2000), and the Wertheim Bible (2011) and published monographs on Spinoza (1993), on the controversy about the principle of least action (2002), and two volumes on the public debates of German enlightenment (2004). She published about 100 articles about seventeenth- and eighteenth-century philosophers, especially on Leibniz and Spinoza. In 2007–2008, she was a member of the Institute of Advanced Study at Princeton. She is currently the president of the North American Leibniz Society and serves since 2012 as a board member of the *Journal of the History of Philosophy*.

Jean-François Goubet is professor of philosophy of education at the *Institut National Supérieur du Professorat et de l'Education* of the University of Lille, France. He wrote several articles and book chapters on Wolff's psychology and was the editor of two volumes, *Die Psychologie Christian Wolffs. Systematische und Historische Untersuchungen* (with Oliver-Pierre Rudolph, 2004) and *Psychologie et Métaphysique. Autour de Christian Wolff* (2003). His work has the classical German philosophy for object, with emphasis on psychology, logic, metaphysics, aesthetics, and also education. He signed contributions on Meier, Reinhold, Kant, Fichte, Herbart, and Natorp.

Stefan Heßbrüggen-Walter studied philosophy, musicology, and sociology at the Westfälische Wilhelmsuniversität Münster, graduating with a thesis on Kant's concept of a faculty and its precursors in seventeenth- and eighteenth-century German thought. He now teaches philosophy at the National Research University Higher School of Economics in Moscow (Russian Federation). He is mainly interested in the history of German philosophy between Melanchthon and Kant.

Dieter Hüning has studied at Philipps-Universität Marburg and at Freie Universität Berlin. He obtained his PhD with a dissertation on "Freiheit und Herrschaft in der

Rechtsphilosophie des Thomas Hobbes" (*Liberty and dominion in Thomas Hobbes' philosophy of law*) in 1996. In 2009, he completed his Habilitation on "Die Begründung des Strafrechts in der neuzeitlichen Naturrechtslehre" (*The Different Foundations of the Right to Punish in Early Modern Natural Law Doctrines*). Since 2010, he has been assistant professor at the Kant-Forschungsstelle at the University of Trier.

Ferdinando Luigi Marcolungo is currently a full professor at the University of Verona. He has conducted research in the field of theoretical philosophy, publishing both books and essays on the issue of knowledge in the works of Giuseppe Zamboni (1875–1950). Also, he has carried out researches in moral philosophy, publishing several essays on the theme. Alongside, he has worked regularly on the thought of Christian Wolff, with essays on: *Wolff e il possibile* (1982) Padova: Antenore; *Christian Wolff tra psicologia empirica e psicologia razionale* (Ed.) (2007) Hildesheim-New York: Olms; *Christian Wolff e l'ermeneutica dell'Illuminismo* (Ed.) (2017) Hildesheim-New York: Olms. In this context, he has devoted many papers to the careful examination of different aspects of Wolff's rationalism and its reception in eighteenth-century Italian philosophy.

Michael Bennett McNulty is assistant professor of philosophy at the University of Minnesota, Minneapolis-St. Paul. His research focuses on the history of philosophy of science, and he has written extensively on Kant's views on the sciences, especially his conception of chemistry. McNulty's research has appeared in journals such as *Synthese*, *Kant-Studien*, *Kantian Review*, and *Studies in History and Philosophy of Science*.

Manuela Mei studied philosophy and history at the University of L'Aquila and received her degree in philosophy in the AY 1998–2002. After a scholarship at the University of Düsseldorf in 2004, in autumn 2005 she was admitted to the PhD course at the University of Florence (2005–2008) and on April 2009 received her PhD about the Christian Wolff's theory of the *pars inferior animae* faculties. During her research activity in Trier, she analyzed the birth of Wolff's concept of *Psychometria*. In 2010, she received a scholarship from the Herzog August Bibliothek of Wolfenbüttel and focused her attention on the source of Wolff's psychological theory of the soul. She took part in national and international conventions about Enlightenment and published scientific articles, essays, and in 2014 the translation of the Georg Friedrich Meier's *Versuch einer allgemeinen Auslegungskunst* of 1757. She has been a history and philosophy high school teacher since 2012.

Andreas Rydberg is a researcher at the Department of History of Science and Ideas at Uppsala University, Sweden. In 2017, Rydberg defended the dissertation *Inner Experience: An Analysis of Scientific Experience in Early Modern Germany*. This work charted scientific experience in the intersection between philosophy, theology, and medicine by looking at a number of cases (Wolff among others) in the Prussian town of Halle, in early and mid-eighteenth-century Germany. Since 2017, he has been working on various shorter projects with a special focus on identity formation, *cultura animi*, and therapy of the soul. In spring 2019, he began a larger 3-year project on self-knowledge and objectivity in eighteenth-century Germany.

Mario Spezzapria is professor at the Department of Philosophy of the Federal University of Mato Grosso, and post-doc researcher at the University of São Paulo with a CNPq grant. At present, he deals with the history of German philosophy in the eighteenth century, in particular with questions of aesthetics and anthropology as developments of empirical psychology in *popular philosophy* and *Wolffianism*. He published *Credenza e ragione scettica: L'influenza di Hume nel pensiero di J. G. Hamann*. Prefazione di Márcio Suzuki. Torino: Nuova Trauben, 2017.

Márcio Suzuki is professor of aesthetics at the Department of Philosophy of the University of São Paulo (Brazil) and CNPq researcher. He is the author of *O gênio romântico. Crítica e história da filosofia em Friedrich Schlegel* (São Paulo: Iluminuras, 1998) and *A forma e o sentimento do mundo* (São Paulo: 34, 2014). He was visiting professor at the University Blaise Pascal—Clermont-Ferrand (France) in 2008, and at the Federal University of Paraná (Brazil) in 2009–2011.

Falk Wunderlich is Wissenschaftlicher Mitarbeiter at the Department of Philosophy at Martin-Luther-Universität Halle-Wittenberg, having previously worked at the Johannes Gutenberg-Universität Mainz, the University of Notre Dame, and the Max Planck Institute for the History of Science (Berlin). He works on early modern philosophy with a focus on Kant, Hume, and materialism, especially Priestley and German materialism. His main publications are: *Kant und die Bewußtseinstheorien des 18. Jahrhunderts* (Berlin, 2005); as coeditor: *British Journal for the History of Philosophy* 24/5 *Special Issue: Varieties of Early Modern Materialism*, with Patricia Springborg (2016); *Michael Hißmann: Briefwechsel*, with Hans-Peter Nowitzki et al. (Berlin, De Gruyter, 2016); and *Kant and his German Contemporaries*, with Corey W. Dyck (Cambridge University Press, 2018).

Chapter 1
Introduction: Reevaluating Christian Wolff's Psychology

Saulo de Freitas Araujo, Thiago Constâncio Ribeiro Pereira, and Thomas Sturm

Christian Wolff (1679–1754) is one of the leading figures in eighteenth-century Western thought, usually counted as the most eminent German thinker between Leibniz and Kant. Wolff's works found a wide audience among European philosophers and scientists from numerous fields and his fame attracted many students from different countries to come to Germany.[1] Wolff became professor in Halle and Marburg, and later a member of the Royal Society in London and the academies of Berlin, Halle, Paris, St. Petersburg, Stockholm, and Bologna. In recognition of his service in modernizing German academic philosophy, he was called *praeceptor Germaniae* (educator of the German nation) and, in 1745, received the title of Imperial Baron (*Reichsfreiherr*) of the Holy Roman Empire.[2]

Wolff was a systematic thinker and accordingly it is important to consider his philosophical views from the perspective of the whole. His work covered not only logic, metaphysics, and ethics, but also such fields as political theory, natural law, and law of peoples, mathematics, mechanics, or economics. At the same time, he

[1] Wolff was especially well received in Catholic countries in which the scholastic tradition was still alive (e.g., Italy). For the influence of the scholastic tradition on Wolff's philosophy, see Leduc (2018).

[2] For more details about Wolff's biography, see Kertscher (2018).

S. d. F. Araujo (✉)
Department of Psychology, Federal University of Juiz de Fora,
Juiz de Fora, Minas Gerais, Brazil
e-mail: saulo.araujo@ufjf.edu.br

T. C. R. Pereira
Department of Psychology, Fluminense Federal University,
Volta Redonda, Rio de Janeiro, Brazil

T. Sturm
ICREA & Universitat Autònoma de Barcelona, Department of Philosophy, Bellaterra (Barcelona), Spain

S. d. F. Araujo et al. (eds.), *The Force of an Idea*, Studies in History and Philosophy of Science 50, https://doi.org/10.1007/978-3-030-74435-9_1

tried to order and articulate them in a new way as a system.[3] Wolff invented new philosophical and scientific disciplines, although he would not draw such a distinction himself. After Wolff, it became commonplace, at least in the German tradition, to conceive of a philosophical system in terms of a general metaphysics, or ontology, and a special metaphysics—comprising cosmology, psychology, and theology—preceded by logic and followed by ethics, politics, and other practical disciplines.[4]

In the last decades, the publication of Wolff's *Gesammelte Werke* by Jean École and his collaborators has aroused new interest in his ideas, which has led to interesting and important reappraisals in the scholarly literature. For example, it has become clear that Wolff's philosophical program was neither a mere rephrasing of Leibniz's ideas nor just a preparation for Kant's critical philosophy. On the contrary, there is a growing understanding that Wolff was an original thinker, who has to be understood in his own terms. However, notwithstanding the merits of such reevaluations, many aspects of his thought remain open to new investigations and deserve further analysis and discussion.

Reappraising Wolff's philosophy, however, goes beyond a purely historical interest. Wolff's philosophical system also matters because it poses challenges that are still alive today, such as the relationship between philosophy and psychology. In particular, the meaning, scope, and impact of Wolff's psychological program have not received sufficient attention in the literature. Although he did not coin the term *psychologia*,[5] Wolff was the first to give psychology a new status: (1) by establishing it as a proper science or discipline among the special philosophical sciences (next to ontology, theology, cosmology, moral philosophy, economics, etc.), (2) by

---

[3] The spirit of systematicity is a hallmark of Wolff's work. It was not by accident that he wrote a specific essay to establish the difference between a systematic and an unsystematic intellect. According to him, "*a systematic intellect* is one that connects universal propositions to each other" (Wolff, 1729, §.2, p. 108), thus building a system of universal truths, whereas "*an unsystematic intellect* is one that … considers particular propositions as if they had nothing to do with the others" (§.5, p. 112). In this context, Wolff mentions Euclid's *Elements* and Descartes' *Meditations* as models to be followed, which betray the influence of the mathematical method on his thought. For a detailed discussion of Wolff's concept of system, see Albrecht (2019).

[4] He was so influential in the development of German philosophy in the eighteenth century that Kant, in the *Critique of Pure Reason*, referred to him as "the famous Wolff, the greatest among all dogmatic philosophers" (Kant, 1787/1998, B xxxvi). Obviously, this characterization is a double-edged sword: Wolff is famous, indeed the most outstanding of all "dogmatic" philosophers, but he represented precisely those whose thought Kant wishes to destroy, and so Wolff might be the best, but only of those who have produced a "dogmatic" philosophy—clearly not a positive characterization. After Kant, it became increasingly unpopular to follow in Wolff's footsteps, and this surely also impacted the legacy of his psychology.

[5] Talk of "psychology" probably originated in sources that are no longer accessible: Marko Marulič (1450–1524) is said to have used the title *Psichiologia de ratione animae humanae liber I* for a piece of writing in 1520. The term definitely can be found in texts by Joannes Thomas Freigius (1543–1583), and in book titles *psychologia* appears in works by Rudolf Göckel (1547–1628) and Otto Casmann (1562–1607), among others (e.g., Goclenius, 1590; Casmann, 1594). For more details, see Krstič (1964), Lapointe (1972), Brozek (1999), and Klempe (2020).

assigning to it a key role in the foundation of moral or practical philosophy, and (3) by inaugurating a division of psychological knowledge into two main branches—*psychologia empirica* and *psychologia rationalis*—, thereby setting a new agenda for debates that ranged from enthusiastic acceptance to fervent criticism.[6] Finally, one should not forget that the so-called emergence of scientific psychology in the nineteenth century cannot be dissociated from the development of a new psychological culture in the eighteenth century, largely based on Wolff's legacy.

Considering all those aspects, a new volume on Wolff's psychological program may help to consolidate his contributions not only to philosophy, but also to the human sciences in general. The last collection of essays dedicated to Wolff's psychology was published in 2007 (Marcolungo, 2007), and the last monograph appeared in 2011 (Mei, 2011). Since then, only isolated articles and book chapters have been published (e.g., Chance, 2018; Dyck, 2014; Rumore, 2018; Goubet, 2018). The present volume offers, for the first time in English, a comprehensive anthology of essays by an international group of leading scholars on Christian Wolff's psychology and its historical impact.[7] It explores Wolff's psychology comprehensively in its various aspects. Moreover, it closes a linguistic gap in Wolff scholarship: most publications on Wolff and his psychological program have appeared in German, French, Italian, or Spanish, but so far there is not a single book dedicated to Wolff's psychology in English.

Our principal goal is to offer a broad account of Wolff's psychological program and its impacts that may contribute to the disciplinary fields of historiography, philosophy, and psychology, not to mention Wolff scholarship. To do this, we have divided the contributions into two parts. Part I covers the scope and contents of Wolff's psychology, both in its internal structure and in its relation to other parts of his philosophical system, such as logic, ontology, cosmology, theology, aesthetics, and practical philosophy. Part II deals with the reception and impact of Wolff's psychology, starting with his early disciples, then moving on to Kant and others, until reaching the nineteenth century with Hegel and Wundt.

As an antechamber to both parts, Ursula Goldenbaum offers a fresh and illuminating account of the historical context underlying the development of Wolff's psychology. She uses a wide range of primary sources to highlight biographical, religious, political, and institutional aspects that help us understand the formulation and the fate of Wolff's psychological program.

Beginning Part I, Thiago Pereira and Saulo Araujo explore, in Chap. 3, the origins of Wolff's psychology in his German writings. Bringing together historical and philosophical analysis, they present content and context of its first exposition in the

[6] Before Wolff, psychological topics appeared in discussions related to either the tradition of the *scientia de anima* (science of the soul), which largely consisted of commentaries on Aristotle's *De Anima*, or medicine. It was Wolff who unified the whole field of psychological topics into a single science or discipline. For more details, see Araujo (forthcoming), Boenke (2005), and Vidal (2011).

[7] The volume by Rudolph and Goubet (2004) also explores dimensions of Wolff's psychology, but among other things does not consider its reception and impact as comprehensively as we do here.

Deutsche Metaphysik (Wolff, 1720)[8] and its further development and clarification in the *Anmerckungen* (Wolff, 1724) and the *Ausführliche Nachricht* (Wolff, 1726).

Chapters 4 and 5 analyze in closer detail the specific contents of both psychological disciplines—empirical and rational psychology. Ferdinando Marcolungo explores the relationship between reason and experience in empirical psychology, whereas Corey Dyck unveils the meaning of Wolff's rational psychology, offering a critical response to current interpretations of Wolff's rationalism.

Manuela Mei, in Chap. 6, investigates one of the many innovations of Wolff's empirical psychology, namely, his conception of *psychometria*. She shows in which sense Wolff believed in the possibility of a quantitative knowledge of the human mind, and compares his understanding of psychometrics with that of Robert Greene (1678–1730).

Next (Chapter 7), Falk Wunderlich analyzes the mind-body problem in connection with Wolff's psychology. More specifically, he deals with some of its metaphysical aspects, such as Wolff's understanding of Leibniz's monadology and the doctrine of pre-established harmony.

In Chap. 8, Matteo Favaretti Camposampiero examines the connection between Wolff's logical and psychological doctrines. He argues that a psychologistic reading of Wolff is one-sided and obscures the foundational role logic plays with respect to psychology.

Márcio Suzuki and Mario Spezzapria, in Chap. 9, delve into the relationship between aesthetics and empirical psychology. They claim that Wolff's remarks on representation as a composition give the concept of image a new status, which will have important aesthetic consequences.

The relationship between psychology and practical philosophy is explored by Dieter Hüning in Chap. 10. He examines the psychological assumptions underlying Wolff's concept of natural obligation as well as its implications for debates surrounding the concept of natural law and the will.

In the last chapter of Part I, Jean-François Goubet discusses the relationship between psychology and the other metaphysical disciplines: ontology, cosmology, and theology. He illustrates their important connection by analysing Wolff's conception of pleasure.

Part II, then, addresses the legacy of Wolff's psychology. In Chap. 12, Sonia Carboncini shows how Wolff's disciples and followers further developed and disseminated his psychological program within and beyond the German borders.

In Chap. 13, Stefan Heßbrüggen-Walter discusses the reception and the debates associated with Wolff's conception of the faculties of the soul. More specifically, he explores the realist interpretation of the faculties by authors such as Alexander Gottlieb Baumgarten (1714–1762), Johann Christoph Gottsched (1700–1766), Georg Friedrich Meier (1718–1777), or Johann Georg Sulzer (1710–1779).

[8] Although the title page gives the date of publication as 1720, the book was actually published in December 1719, as Wolff himself recalls later (Wolff, 1726, §.4).

Andreas Rydberg offers, in Chap. 14, a new look into the beginnings of experimental psychology in the eighteenth century. He traces the idea of psychological experiments back to Wolff and, in the aftermath, to three different discourses that developed in that context, namely, the experimental-philosophical, the iatromechanical, and the ethical-metaphysical.

Next, Michael Bennett McNulty (Chap. 15) illustrates the reception of Wolff's psychology by Immanuel Kant (1724–1804). In particular, he shows that Kant not only rejected the metaphysical aspects of rational psychology but also criticized the empirical side of Wolff's program.

In Chap. 16, Werner Euler discusses the reception of Wolff's psychology by Hegel. More specifically, he shows that Hegel criticized Wolff's rational psychology for being an abstract metaphysics that is unable to apprehend the essence of its object, namely, spirit. Instead of presenting yet another theory of the mind-body relation, Hegel proposed a wholly new way of approaching the study of spirit.

In the last chapter, Saulo Araujo and Thiago Pereira explore the reception of Wolff in nineteenth-century German psychology. More specifically, they show how Wilhelm Wundt (1832–1920) adopted and interpreted Wolff's psychology by way of establishing his own conception of a psychological science, though they also point out a number of important respects in which Wundt appears to have misunderstood Wolff.

Together, these chapters show that Wolff's psychological ideas are historically and philosophically more significant and interesting than conventional wisdom admits, but also that they are subject to misinterpretation. In general, Wolff's psychology remains a challenge to historians, philosophers, and psychologists. We hope this volume will contribute to bring Wolff's psychology to a wider audience.

Finally, we wish to note that, despite our comprehensive approach to Wolff's psychology, important facets and aspects have not been addressed here. For instance, the relationship between empirical psychology and practical philosophy involves many other factors. In addition, it would be interesting to show how the idea of rational psychology was carried forth in the eighteenth century by authors such as Moses Mendelssohn (1729–1786) and Johann Nicolas Tetens (1736–1807). Wolff's reception in the nineteenth century also deserves more attention. After all, it was not only Hegel and Wundt that discussed and commented Wolff's psychology. This is of course just to say that a single volume cannot exhaust the richness, complexity, and legacy of Wolff's contributions to psychology, but we hope that the present volume will serve as a foundation for further research in these and other directions, in the future.

References

Albrecht, M. (2019). Einleitung. In C. Wolff (Ed.), *Über den Unterschied zwischen einem systematischen und einem nicht-systematischen Verstand* (pp. 7–25). Meiner.

Araujo, S. F. (forthcoming). A palavra e a coisa: Sobre as origens da psicologia como ciência. *Estudos e Pesquisas em Psicologia, 21*(3).

Boenke, M. (2005). *Körper, Spiritus, Geist. Psychologie vor Descartes*. Wilhelm Fink.

Brozek, J. (1999). From "psichiologia" to "psychologia": A graphically documented archival study across three centuries. *Journal of the History of the Behavioral Sciences, 35*(2), 177–180.
Casmann, O. (1594). *Psychologia anthropologica, sive animae humanae doctrina.* Wilhelm Anton.
Chance, B. A. (2018). Wolff's empirical psychology and the structure of the transcendental logic. In C. Dyck & F. Wunderlich (Eds.), *Kant and his German contemporaries* (Vol. 1, pp. 17–34). Cambridge University Press.
Dyck, C. W. (2014). *Kant and rational psychology.* Oxford University Press.
Goclenius, R. (1590). *Psychologia, hoc est, de hominis perfectione.* Paulus Egenolphus.
Goubet, J.-F. (2018). Rationale Psychologie. In R. Theis & A. Aichele (Eds.), *Handbuch Christian Wolff* (pp. 153–174). Springer.
Kant, I. (1998). *Kritik der reinen Vernunft.* Meiner. (Original work published 1781/1787).
Kertscher, H.-J. (2018). *"Er brachte Licht und Ordnung in die Welt": Christian Wolff—Eine Biographie.* Mitteldeutscher Verlag.
Klempe, S. H. (2020). *Tracing the emergence of psychology, 1520–1750.* Springer.
Krstič, K. (1964). Marko Marulic—The author of the term "psychology". *Acta Instituti Psychologici Universitatis Zagrabiensis, 36*, 7–13.
Lapointe, F. H. (1972). Who originated the term psychology? *Journal of the History of the Behavioral Sciences, 8*, 328–335.
Leduc, C. (2018). Sources of Wolff's philosophy: Scholastics/Leibniz. In R. Theiss & A. Aichele (Eds.), *Handbuch Christian Wolff* (pp. 35–53). Springer.
Marcolungo, F. L. (Ed.). (2007). *Wolffiana III: Christian Wolff tra psicologia empirica e psicologia razionale.* Olms.
Mei, M. (2011). *L'intuizione dei pensieri. La pars inferior animae nella psicologia cognitiva di Christian Wolff.* Aracnee.
Rudolph, O.-P., & Goubet, J.-F. (Eds.). (2004). *Die Psychologie Christian Wolffs. Systematische und historische Untersuchungen.* Niemeyer.
Rumore, P. (2018). Empirical psychology. In R. Theis & A. Aichele (Eds.), *Handbuch Christian Wolff* (pp. 175–196). Springer.
Vidal, F. (2011). *The sciences of the soul.* The University of Chicago Press.
Wolff, C. (1720). *Vernünfftige Gedancken von Gott, der Welt und der Seele des Menschen, auch allen Dingen überhaut, den Liebhabern der Wahrheit mitgetheilt.* Renger.
Wolff, C. (1724). *Anmerckungen über die vernünfftige Gedancken von Gott, der Welt und der Seele des Menschen, auch allen Dingen überhaut, zu besserem Verstande und bequemerem Gebrauch derselben heraus gegeben von Christian Wolffen.* Andrea.
Wolff, C. (1726). *Ausführliche Nachricht von seinen eigenen Schrifften, die er in deutscher Sprache von den verschiedenen Theilen der Welt-Weißheit heraus gegeben, auf Verlangen ans Licht gestellet.* Andrea und Hort.
Wolff, C. (1729). De differentia intellectus systematici et non systematici. In C. Wolff (Ed.), *Horae subsecivae Marburgenses anni MDCCXXIX quibus philosophia ad publicam privatamque utilitatem aptatur* (pp. 108–154). Renger.

Chapter 2
Who Was Afraid of Wolff's Psychology? The Historical Context

Ursula Goldenbaum

> Wolff "presented the philosophers' hypotheses about the union between body and soul in greatest clarity, enabling everybody to judge with reason about any soul."[1]

2.1 Introduction

Christian Wolff (1679–1754) has suffered an increasingly bad reputation, shaped by Hegelians. They presented Wolff as an un-original thinker, a mere ruminant of Leibniz (1646–1716), whose only achievement was the methodological education of the German youth.[2] The question arises though how such a boring thinker could cause so many controversies over more than five decades. The first wave of publications began after Wolff's notorious *China Lecture* in 1723,[3] his subsequent banishment from Prussia, and the ban of his philosophy in Prussia and other territories. It produced almost 200 writings pro and con Wolff (Zedler, 1731–1754, vol. 58, columns 546–677, columns 883–1232). A new intense controversy about Wolff began in 1735, on the occasion of a Wolffian translation of the Pentateuch by Johann

[1] This is from the newspaper's review of Wolff's *Psychologia rationalis* (Neue Gelehrte Zeitungen, April 15th, 1734, pp. 269–270, here p. 270). This is my translation, just as all other translations from German sources, if not mentioned otherwise.

[2] Hegel (1770–1831), in an extremely short presentation of Wolff, somehow shaped the final judgment about Wolff up to our canon of the history of philosophy (Hegel, 1986, pp. 136–139). Lewis White Beck simply repeats (Beck, 1969), sometimes literally, what had been said by Hegel. Neither of the two seems to have studied Wolff.

[3] The best edition of Wolff's China lecture has been produced by Michael Albrecht (Wolff, 1726/1985a), containing a rich and instructive commentary about the circumstances and the research level about China at the time. For an English translation see Ching and Oxtoby (1992).

U. Goldenbaum (✉)
Emory University, Atlanta, GA, USA
e-mail: ugolden@emory.edu

S. d. F. Araujo et al. (eds.), *The Force of an Idea*, Studies in History and Philosophy of Science 50, https://doi.org/10.1007/978-3-030-74435-9_2

Lorenz Schmidt (1702–1749)—the so-called *Wertheim Bible*—which stirred up a theological turmoil with more than 100 writings pro and con (Zedler, 1731–1754, vol. 55, column 595–662; Goldenbaum, 2004, pp. 175–508). Moreover, this case against the Wolffian translator was used by Joachim Lange (1670–1744), Wolff's major theological opponent at the University at Halle to extend the ban of Wolff's philosophy to all states of the German Empire, by showing a *necessary* connection between the "horrible" *Wertheim Bible* and Wolff's philosophy and by organizing an Empire-wide campaign against Wolff (Lange, 1735). Winning this battle was crucial for Wolff; it would decide the fate of his philosophy. This second public debate about the *Wertheim Bible* is rarely recognized as a debate about Wolff since Carl Günther Ludovici (1707–1778), the likely author of all three articles concerning Wolff in *Zedler's. Universal-Lexicon*,[4] anxiously separated it from the first Wolff controversy in order to undermine Lange's strategy. He did this as well in his history of Wolffian Philosophy where he reports those writings caused by the Wertheim Bible in an Appendix only (Ludovici, 1737/1977, §§.516–521).

In the following, I will (1) discuss and answer the question how theologians got so obsessed with the German philosopher, especially during the first decades of the eighteenth century. It was mostly about Pre-established harmony, so important for Wolff's rational psychology. I will then (2) lay out the restrictive political conditions under which Wolff worked and published, including the procedures of censorship in the German Empire after the Peace of Westphalia. To challenge the theologians as Wolff did, much caution in formulation was needed, great diplomacy, outstanding teaching skills (Hartmann, 1737/1973, pp. 381–383; Ludovici, 1737/1977, §§.343–349, pp. 287–291), and the ability to win supporters and build a network, including influential allies at the courts. But besides all that it also needed much courage and commitment to stand up for his ideas.[5] And (3), I will present the means that Wolff and his partisans had available under these conditions as well as those they developed themselves to stand their ground and to increasingly take hold of the universities—against the power of their adversaries. Here, I will survey the rich landscape of German journals and Learned Newspapers which developed in early eighteenth century[6] as well as of the Wolffian Societies.

[4] He authored the articles "Christian Wolff" (Zedler, 1731–1754, vol. 58, columns 546–677), "Wertheimische Bibel" (Zedler, 1731–1754, vol. 55, columns 595–662), and "Wolfische Philosophie" (Zedler, 1731–1754, vol. 58, columns 883–1232).

[5] This shall be said against Jonathan Israel's superficial subsumption of Wolff as a moderate enlightener (Goldenbaum, 2014).

[6] Martin Welke estimates about 250,000 regular readers of newspapers in Germany at the beginning of the eighteenth century and half a million readers in the middle of the century. In 1808, the 8000 copies of *The Times* were topped by 56,000 copies of the *Hamburgische Correspondent* (Welke, 1981). For Hamburg, see also Böning & Moepps, 1996.

2.2 Pre-established Harmony as the Core of Wolff's Psychology

Wolff's Psychology appeared in two parts in 1732 and 1734, divided into empirical and rational psychology (Wolff, 1738/1968, 1740/1972). He wrote and published them during a rather calm period: he had already settled at his new position in Marburg, after his escape from Prussia, and wrapped up his defenses against the most serious official accounts from German universities that came out in the aftermath of his banishment from Halle (Ludovici, 1737/1976, pp. 155–185). Joachim Lange, also aware of the significance of official University Reviews about Wolff's philosophy, wrote an extended review of nine such Reviews using them as ammunition to raise further concerns about Wolff's philosophy (Lange, 1725/2000). Wolff could not yet anticipate that huge public debate about the *Wertheim Bible* that was soon to come, in 1735. With his *Psychologia empirica* (Empirical psychology) and *Psychologia rationalis* (Rational psychology), however, Wolff turned to the very core of the differences between him and his theological opponents, spelling out the implications of the much-attacked Pre-established Harmony for a new understanding of the soul. In contrast to our time, the term "psychology" included everything related to our soul, i.e., epistemology, theory of emotions, and theory of action.

The major points of theological criticism of Wolff's philosophy, leveled by Pietists and Orthodox theologians alike, were the following: (1) the application of the mathematical method beyond mathematics, (2) the thesis that this world is the best possible, (3) the alleged "mechanism" of this philosophy, and (4) above all, Pre-established Harmony. It was the same criticism Leibniz faced after publishing his *Theodicée* (theodicy) (Lorenz, 1997, pp. 99–150). Why would these highly abstract metaphysical ideas become such a stumbling block for theologians? How could they provoke hundreds of writings against Wolff and his disciples? Why would theologians even mobilize political authorities against him, even the Emperor? What was at stake between Wolffians and anti-Wolffians (not only Pietists), during the first half of the eighteenth century and beyond, was the question of free will. Wolff, still a young university professor, was attacked more boldly than Leibniz, the European celebrity, but Leibniz was criticized for the very same reasons as Wolff.

The mechanical explanation of natural phenomena, the latent determinism of the mathematical method with its necessary a-priori conclusions, and above all, Pre-established Harmony, were seen as so many threats to free will (*liberum arbitrium*). Allegedly, Leibniz-Wolffian Determinism would take away moral responsibility and thus provide an excuse to sinners and/or criminals. The deep gap between Leibniz, Wolff, and the Wolffians on one hand and their opponents on the other, was caused by their radically different stances toward modern science, i.e., mechanics. Whereas Leibniz and Wolff embraced it unconditionally when explaining natural phenomena, their opponents wanted to restrict the mathematical method to mathematics, including perhaps applied mathematics. They denounced mechanical philosophy as "mechanical absolutism" (Löscher, 1735, p. 239) that would lead to

Spinozism. But it was especially the explanation of the soul according to Pre-established Harmony, and of the will as determined by our intellect to act, thereby always striving for the best that caused panicking reactions. We would no longer be seen as free to determine ourselves by *mere will*, if the will were determined by our intellect. The head of the Anti-Wolffian camp, the Pietist theologian Joachim Lange gets to the heart of the theological concerns with pre-established harmony when he cries out against the Wolffian Johann Lorenz Schmidt, the author of the *Wertheim Bible*:

> I only say this [...] that the author deduces the stubbornness [of Pharaoh in Exodus 7, 13 and following] from the nexus or the fatal connection of all things, and in this way ascribes it to God according to his pre-established harmony. This nexus is the very soul of the whole system of mechanical philosophy. (Lange, 1735, p. 25)

The causal nexus of everything with everything was seen as a hidden version of Spinozistic determinism and fatalism, just as—50 years later—Friedrich Jacobi (1743–1819) would claim that all rationalism led necessarily to Spinozism, i.e., fatalism (Altmann, 1977, pp. 142–144). Lange as well as Jacobi had to ignore the careful distinction of Leibniz and Wolff between the absolute mathematical necessity of abstract things and the contingency of concrete things that allowed them to make modern science compatible with theological intentions and to overcome Spinoza's absolute determinism.

Already in his *Vernünftige Gedanken von Gott, der Welt und der Seele des Menschen* (Rational thoughts concerning God, the world, and the human soul) in 1720 (Wolff, 1751/1983a, §§.744–747), and then in his *Psychologia rationalis*, Wolff conceived the soul as *one* single force (*einige Kraft/vis unica*) (Wolff, 1740/1972, §.57). In his empirical psychology, it is distinguished into an upper and lower faculty to know, to perceive, and to reason (*pars superior/pars inferior facultatis cognoscendi*) (Wolff, 1738/1968, §§.54–55). The upper faculty of knowing was the intellect while the lower included representation, memory, and an imaginative sensing force (*Einbildungskraft/facultas imaginandi*) (Wolff, 1738/1968, §.92; 1751/1983a, §.235). The *one* single force of the soul was endued with an appetite toward the best which, if joined by rational ideas, would produce conscious volitions and thus allow for free choices of the will (Wolff, 1740/1972, §§.480–529, pp. 396–450). Most appetites though were led by the lower faculty of cognition—as in animals—and thus remained determined by external objects rather than by the soul itself. Rationally informed choices would lead to more perfection and thus to joy, while choices according to the senses could lead to less perfection and thus to sadness (Wolff, 1738/1968, §§.616–617, §§.621f., pp. 464–465). Not unlike Spinoza (1632–1677) (and Leibniz), Wolff deduces all, or at least all major human affects from this simple foundation. Human freedom, for Wolff, is not the traditional idea of free choice of the will, i.e., choosing by the mere power of our will, out of nowhere. Rather, every choice we make is determined by a *Bewegungsgrund* (motive) to choose the best. If we are determined by a rational judgment of our intellect, our choice will be free. We may well be in error about what is objectively best for us, and thus lack freedom, but we cannot desire anything than what appears best to us.

This is perfectly in agreement with Leibniz, and moreover, with Spinoza who both rejected *liberum arbitrium* as free choice of the will, as a mere act by a power of the will, independent of any cognitive power. Spinoza openly mocked free will as a human illusion while Leibniz and with him Wolff rejected it as "indifferentism". They held on to the *term* "free will," but used it for what they understood as freedom. Thus, Joachim Lange was quite right to smell Spinoza in Leibniz-Wolffian philosophy. But it was not by chance that these philosophers all ended up at freedom as choice led by reason. They tried to develop a concept of freedom that was compatible with modern deterministic science.[7] In contrast, Lange and almost all contemporary theologians and many philosophers, insisted on the separation of the soul into two forces, the intellect and the will, to vindicate absolute responsibility of individuals for their deeds to their free choice of the will.

2.3 The Political Constellation During the Rise of Wolffian Philosophy

As mentioned above, the theological battle against Wolffianism began in 1721 although tensions between theologians and Wolff had come up before.[8] It continued beyond his death but had two peaks, in terms of publications. The first wave of attacks was due to Wolff's China lecture, the second peak, less recognized was reached after the publication of the *Wertheim Bible* in 1735, turning into a public debate well beyond the walls of academia. This careful translation of the Pentateuch into then-modern German was produced in great awareness of the hermeneutical problems that beset such a project. Schmidt commented on his solutions of these problems in more than 1600 footnotes—in the spirit of Wolff's logic (Goldenbaum, 2004, pp. 195–209; Wolff, 1713/1978, ch. 10, §§.1–23; ch. 11, §§.1–8; ch. 12, §§.1–12; 1740/1983b, §§.902–981). He aimed to translate the text according to the understanding of its original audience, i.e., the ancient Jews. As a result, no allusion to the savior remained which alarmed Orthodox and Pietist theologians alike. It is from this time that the old opponents, the orthodox Lutheran Valentin Ernst Löscher (1673–1749) and the Pietist Joachim Lange made peace and united against the Wolffians.[9] Obviously, it was to his greatest dismay that, in spite of Lange's successful intervention against Wolff at the Prussian court in 1723, the philosopher continued to thrive—at the University of Marburg. In 1735 though, Lange hoped

[7] The discussion of free will is getting momentum again, due to new results of neuroscience. There appeared already an *Oxford Handbook of Free Will* (Kane, 2002). The controversial positions are presented less aggressively today but the discussion is quite heated too.

[8] Lange warned students already to attend Wolff's lectures on mathematics. He planned a refutation of Wolff's *German Metaphysics* right after its publication (Hartmann, 1737/1973, pp. 401–402).

[9] Emmanuel Hirsch speaks of an "Empire-wide General Mobilization" (Hirsch, 1951, p. 432). Even the long-lasting battles between orthodox and Pietist theologians came now to a stop (Goldenbaum, 2004, pp. 265–266).

again. Arguing that the *Wertheim Bible* was a *necessary* product of Wolff's philosophy, he aimed to defeat Wolff once and for all.

Although, at first glance, the controversies look like any scholarly controversy, the opponents did not fight equally. Theologians had the church and its administration available and were in close connection with state authorities (Hinrichs, 1971, p. 175). In Prussia and Saxony, theologians of the Lutheran church had their own journals available. These were supported and distributed by the churches, and pastors were supposed to buy them on a regular basis. In addition, theologians could use their pulpits, lecture at universities, and publish in their journals. Thus, theologians had a wide range of options to attack opponents and to spread their judgments through all levels of the Christian church. Above all, they could easily connect with state authorities even before it came to formal censorship (Wotschke, 1932, p. 54; Goldenbaum, 2004, pp. 226–232). Wolff, in contrast, began as a simple professor of philosophy, teaching mathematics at first. Thus, controversies between theologians and Wolff have never been authority-free discourses, i.e., free exchanges of arguments about philosophical differences. Theology had the monopoly in defining what the truth was and Joachim Lange still openly asked philosophy to be the maiden of the mistress theology (Lange, 1703). In these controversies, theologians did not even have to come up with novel arguments but simply nail their colors to the mast to show they belonged to the right side.[10]

There existed, however, certain accepted rules in controversies. Scholars, including theologians, had to follow the *polemical method*, i.e., to present objectively the argument of the opponent before offering one's counter-arguments (Zedler, 1731–1754, vol. 20, columns 13–37). Also, one had to respond to criticism to show one's willingness to listen and one's ability to respond. In theological controversies though, additional rules were in play. Since in the Lutheran church, all participants were considered as (equal) members capable of reading the Holy Scriptures on their own, with theologians as advisers rather than exclusive interpreters, all sides were obliged to talk to and to listen to each other. As long as deviating members were listening and considering the arguments of the community (represented by theological leaders), they remained part of the community. Only a refusal to listen, unwillingness to consider counter-arguments, and improper moral behavior could provide a case for persecution as, e.g., arrogance, vanity, or the intention to undermine religion. Only then, the church could ask the political authorities to take action. This procedure, called the *Elenchus* (Gierl, 1997, pp. 60–212), is the reason why theological attacks more often than not include *moral* blame of the dissidents in addition to arguments. It is a pity that such complaints against Wolff are still uncritically taken seriously by historians today (e.g., Schrader, 1985, pp. 180–181; Bianco, 1989, p. 112; Watkins, 1998, p. 146).

Formal censorship was, since the Westphalian Peace, no longer in the power of the Emperor. Since all three Christian denominations were now tolerated within the

[10] Historians who wonder about such redundancy are unaware of the pressure to express one's agreement with the official position of the church (Watkins, 1998, p. 148).

Empire, it was the rulers of single territories who were in charge. They decided about the "state religion", which could worship freely and publicly, but they had to tolerate their subjects who belonged to other denominations. Such decentralization had great advantages for modern philosophers of that period. It is well-known how Christian Thomasius (1655–1728) escaped persecution initiated by Saxony's theologians at the University of Leipzig by quickly moving to the nearby University of Halle in Prussia (Hinrichs, 1971, p. 353, 369). Likewise, when Wolff had been banned in 1723 by the Prussian king and had to leave the country within 48 h or be hung, he easily found another position at the University at Marburg.[11] Both philosophers could thus continue to lecture and publish within their new states, but above all, they could communicate with their colleagues and students in German language.

Since the rulers of the territories were in charge of formal censorship, they could handle it according to their own ideas. If a university existed, censorship was usually given to its theologians. But in the case of the University of Leipzig, already the city with the largest book market of at least the Protestant area of the Empire, theologians could not manage to censor all books and journals. Thus, censorship was split according to disciplines (Kobuch, 1988, pp. 18–43). Little territories with few publications decided from case to case. It was always the duty of the pastors, on any given level, to admonish the members of their communities to stay in the limits of true faith, to enter in a discourse with prospective dissidents as long as these were willing, and to ask political authorities for bans of publications and persecution of the authors if they found them unwilling.

Of course, toleration happened in different degrees in different states, and was sometimes denied altogether as, e.g., to the protestants in Salzburg. But such violations of the Westphalian Treaty did not go unnoticed. This was due to the new institution of the *Corpus evangelicorum* at the Diet, i.e., the Protestant estates; this institution could and did send protests to the Emperor on behalf of persecuted Protestants and he was supposed to take action according to the Westphalian Peace. Besides, a journal with the title *Reichs-Fama* (Fame of the Empire) regularly published the complaints of subjects who suffered intolerance (Goldenbaum, 2004, pp. 85–86). Of course, the instrument of the *Corpus evangelicorum* was cumbersome and inefficient. That it nonetheless worked can be seen in the case of the *Wertheim Bible* which was widely discussed at the Diet and in related correspondences among Protestant courts. It was likewise negotiated between them and the Emperor's institutions (Goldenbaum, 2004, pp. 289–294, 354–355, 386–400, 410–420, 443–451), always referring to the *Protestant freedom* to interpret the Bible.

These rules and procedures of the Westphalian Peace are rarely noticed in intellectual history, although European contemporaries were well aware of the legally backed religious tolerance within the Empire, much in contrast to their own countries. Among these were Locke (1632–1704), Voltaire (1694–1778), Rousseau

---

[11] Wolff had received the offer from Marburg before the ban due to the intention of Landgraf Carl I of Hessen to thoroughgoingly improve his University (Kertscher, 2018, pp. 142–147, 146–147). The number of students grew from 60 to 70 before Wolff's arrival up to 174 in 1727. Wolff got another offer from the University Leipzig right at his arrival in Marburg.

(1712–1778), and Penn (1644–1718) (Voltaire, 1764, p. 46; Raumer, 1953, pp. 326–352; Specht, 1989, p. 12). With every territory executing its own censorship, including the Free cities of the Empire (*Freie Reichsstädte*) which were directly answering to the Emperor, with lots of cities (even small cities) with printing shops, and a wide network of publishing houses distributing books and journals throughout the Empire (Goldfriedrich, 1908), the responsibilities were not always obvious. Usually, the authorities tried to get first the printer who could be found more easily (Schrader, 1985, pp. 64–69).

The Prussian theologians tried very hard to achieve a ban of Wolff-related works by showing the authors' unwillingness to listen or pointing out their moral flaws, and, moreover, by announcing their whereabouts. But their complaints at the Prussian court were of little effect if the authors lived outside of Prussia. The success against the Wolffians of the University at Jena remained limited due to the reluctance of the court at Weimar. But the tiny territory of Wertheim answered directly to the Emperor and its dukes supported Schmidt. That is why Joachim Lange, when fighting against the *Wertheim Bible* and Wolff together in the late 1730s, went out of his Protestant way and denounced Schmidt at the Emperor's court in Vienna. Arguing that Schmidt undermined all three denominations he asked for an Empire-wide ban and prosecution of the author (Goldenbaum, 2004, pp. 330, 397–400), and of Wolff's philosophy as the cause. He even agreed, clearly against the Westphalian Treaty, to hold the trial against the Lutheran Schmidt on Catholic territory to make sure the dissident would stay forever in chains in the casemates of the Bamberg castle. Although he finally succeeded with the ban of the *Wertheim Bible*, it took him years. And during these years, Schmidt was permitted to respond to all theological criticism fueling an ever-increasing public debate about his translation, the principles of translating and about the freedom of thinking (Schmidt, 1736a, b, c, d, 1738). Moreover, he formally approached the *Corpus evangelicorum* to defend, against his enemies, his Protestant freedom to translate the Bible according to his best knowledge (Schmidt, 1736e).

After a period of increasing support for Wolff at the courts and thus by state authorities during the late 1720s and early 1730s, when Lange had even been silenced by the Prussian king and needed to find colleagues outside of Prussia to write against Wolff, Lange used the publication of the *Wertheim Bible* in 1735 as a welcome opportunity to eventually secure an Empire-wide ban on Leibniz-Wolffian philosophy.[12] In this extremely dangerous situation for Wolff, he and his partisans made the greatest efforts to avoid a final Empire-wide ban. Therefore, they had to deny such a *necessary* connection between Wolffianism and the *Wertheim Bible* although they were well aware that there was one, in terms of method as well as metaphysics. This is clear from some letters Wolff as well as Mosheim (1693–1755) and Reinbeck (1683–1741) exchanged with Schmidt and with his mentor at the court of Wertheim, Johann Wilhelm Höflein (1689–1739). They expressed their

[12] An outstanding example is Johann Friedrich Bertram (1699–1741), a former student of Lange who published against Wolff on behalf of Lange to show Wolff's connection with the Wertheimer, simply because Lange was forbidden to continue his polemics (Goldenbaum, 2004, pp. 337–344).

general sympathy with the project but also uttered criticism. When Wolff was asked for support of Schmidt, he carefully excused himself, explaining that—in this situation—it was the very legality of his philosophy that was at stake. But he assured him and Höflein, that he would never join the attacks against the Wertheimer to save his skin (Goldenbaum, 2004, p. 349).

Not only was the new wave of anti-Wolffian publications due to the *Wertheim Bible*. So was the well-known Royal Commission which was installed by the Prussian king in 1736 to investigate the relation of Wolff's philosophy to the Wertheim Bible (Ludovici, 1737/1976, pp. 2–154).[13] Within a few days in June 1736, the court at Berlin suggested the ban of the *Wertheim Bible* in Prussia but cleared Wolff's philosophy from Lange's accusations (Goldenbaum, 2004, pp. 319–323). Thus, this victory of Wolffian philosophy had a price.

Clearly, Wolff was not the only victim of political persecution for holding philosophical views in the first half of the eighteenth century. The translator of the *Wertheim Bible* was sued and threatened with lifelong imprisonment in the casemates of the Bamberg castle, before he could escape to Altona (belonging to Denmark and no part of the Empire), living the rest of his life as an outlaw (Spalding, 1998). The well-known Wolffian philosopher Christoph Gottsched (1700–1766), who had still taught Pre-established Harmony after his arrival in Leipzig in 1724, turned to an agnostic position after aggressive attacks from theologians in Leipzig. He was nonetheless brought to trial in 1737, at the court in Dresden, and had, as a professor, to stand an entire day in front of a tribunal of theologians who interrogated him about single sentences of his books he was not allowed to check for the context. At the end, he was given the ultimatum—to either stop teaching Wolffian philosophy or be fired (Goldenbaum, 2004, pp. 366–367; Döring, 1999, pp. 141–152). The editor of the *Gelehrte Zeitungen* (Learned Newspapers) in Leipzig, the Wolffian Wolf Balthasar Adolph von Steinwehr (1704–1771), had to face such a tribunal in the same year (Goldenbaum, 2004, pp. 367–368). This was even true for the chief editor Friedrich Otto Mencke (1708–1754) although he, belonging to the most honorable citizens of Leipzig, had to appear before a tribunal at the University of Leipzig only (Goldenbaum, 2004, pp. 362–368). And the

[13] The Royal order about a ban of the *Wertheim Bible* was given by the king on May 27, 1736. At the end of May, the statements of Lange, Reinbeck, and Wolff pro and con Wolff's philosophy appeared as *Nouvelles pieces sur les erreurs prétendues de la philosophie de Mons. Wolff* (New Pieces about the Alleged Errors of Mr. Wolff's Philosophy) (Wolff, 1736/1985b). Reinbeck, in the third part of his *Betrachtungen zur Augsburger Konfession* (Consideration about the Confession of Augsburg) (Reinbeck, 1736), distances himself from the *Wertheim Bible*. The Royal ban against the Wertheim Bible goes out to all governments in Prussia and to all book sellers on June 2nd. On the very same day, the king sent a request to the *Reichshofrat* (Councilor of the Empire) in Vienna to ban the *Wertheim Bible* within the Empire. On June 5th, 1736, the Royal Commission gathered and lifted the ban against Wolff's philosophy, based on the statements mentioned above. They concluded their case by June 17th respectively 22nd, judging that Lange's accusations were baseless (Ludovici, 1737/1976, pp. 126–154). Also, in 1736, Reinbeck's third volume of the *Betrachtungen* (Reinbeck, 1736) came out. The preface contains a rejection of the Wertheim Bible in the paragraphs 7–10.

leading Wolffians Bernhard Bilfinger (1693–1750), Jacob Hermann (1678–1733), and Samuel König (1712–1757) found positions in Russia, Venice, or the Netherlands rather than within the German Empire; Wolff was able to pave the way for some younger partisans though, e.g., for Ludwig Philipp Thümmig (1697–1728) and Johann Ulrich Cramer (1706–1772) in Marburg.

Given the strength of theological power, the question arises how Wolffian philosophy could overcome at all? First of all, the Wolffians knew and used the political and juridical means as described above quite well and used them carefully. They interpreted "Protestant freedom" in a wide sense and referred to the Westphalian Peace; Wolff proudly stated that his philosophy is in agreement with all three denominations (Wolff, 1736/1985b, pp. 101–102). Also, what is often considered as a harsh *personal* reaction of Wolff—his dismissal of August Hermann Francke's (1663–1727) request in 1723 to turn in his China lecture to the faculty of theology—is in fact a formally correct protest according to the existing procedures according to the University's Constitution.[14] Similarly, Wolff denounced the violation of academic rules by his former student Daniel Strähler (1690–1750) to the Academic Council and to the court: Strähler had publicly criticized a colleague without following the procedure within the university (Hartmann, 1737/1973, pp. 682–687). Likewise, whenever Wolff *publicly* referred to violations of the rules or a false presentation of facts (not arguments), he published in newspapers rather than in a journal or book (Ludovici, 1737/1977, §.658, p. 574). Further, Wolff responded to all "official" University Reviews to show his willingness to the "Teaching Elenchus" and thus to avoid the opportunity for state authorities to ban his philosophy (Hartmann, 1737/1973, pp. 729–829; see also Wolff, 1724/1980a, 1724/1980b, 1725/1980c, 1724/1983c). Thus, Wolff knew the rules that limited his freedom but was also capable of using them to protect himself and his philosophy.

In addition, Wolff and his partisans approached their rulers or influential persons at the courts to obtain a fair consideration. This became easier in the course of Wolff's increasing success. The reason for getting a position at the University at Marburg, where Lutherans were formally not even permitted, and for offers from other states to come, was clearly the expectation that Wolff would draw students which was not only an educational but economic argument in his favor.[15] This is already clear from early second thoughts at the court at Berlin after Wolff's ban was followed by a remarkable loss of students.[16] It is explicitly formulated by the

[14] While §.2 of the Constitution asked for agreement among all professors about Christian doctrine, nobody was permitted to attack another in public. Rather, possible disagreement should be reported to the provost who was supposed to gather all professors to *talks* (instead of publishing) (Hinrichs, 1971, pp. 403–404).

[15] The number of students in Marburg increased such that the University's lecture halls did not suffice to place them (Kertscher, 2018, pp. 159–160).

[16] There exists a report by the supervising government at Magdeburg, on the request of the Royal court at Berlin from 1730 about the Decline of the University of Halle (GStA PK, Rep. 52. 159. N. 1), pointing to a number of only 722 students in contrast to 1000 in the decade before. In comparison, Frankfurt a.O. had 190 in 1716, Königsberg 400, and Duisburg 163 (GstA PK, Rep. 51.34). In contrast, the University of Marburg had never more than 200 students and hardly from

minister in charge of the foundation of the University at Göttingen, Gerlach Adolph Freiherr von Münchhausen (1688–1770), who wrote, ironically to Lange, on April 16, 1733: "Since, however, this Wolffian philosophy has so many partisans everywhere, and is so much applauded, it would infallibly do tort to the new University if one would not allow them to teach." (Rössler, 1855, p. 36–37). That was also the reason for a State Commission for the University of Jena in 1733 to reconsider the teaching of Wolff's philosophy (Ludovici, 1737/1977, §.648, pp. 570–571). In Berlin, the general Grumkow (1678–1739), Manteuffel (1676–1749), and finally Reinbeck became allies of Wolff and would change the mind of the Prussian king. At the end of 1733, Wolff got the first request from Prussia, through chancellor Samuel Cocceji (1679–1755), to take a position at Frankfurt an der Oder (Gottsched, 1755/1980, pp. 46–49), years before the Royal Commission in 1736 cleared Wolffian philosophy from the accusation that it would destroy the Christian religion, and his final return to Prussia in 1740.

Propst Reinbeck, however, Wolff's new protector at the Prussian court, although he did not approve of Pre-established Harmony, generally supported Wolffian philosophy, (Reinbeck, 1737, §§.xxx–xxxv). Facing the extremely precarious political situation after the publication of the *Wertheim Bible* and the theological attempts to get his philosophy banned in the Empire, Wolff needed any possible political support to keep his philosophy legal, and especially that of the court in Berlin against his archenemy Lange.[17] Therefore, in order to maintain the crucial ally Reinbeck, Wolff, and his disciples were ready to play down Pre-established Harmony—as has often been noticed, though without an understanding of the political causes.[18] From 1736 on, Wolff and his disciples emphasized even more that Pre-established Harmony could not be demonstrated and that the Wolffian system could stand without it (holding on to it though).[19] Many other Wolffians were agnostic and abstained from a decision between *influxus physicus* (physical influence), occasionalism, or

---

other territories than Hessen before 1724, but had more than 300 in 1727, among them many foreign students (Heer, 1927, p. 9).

[17] Theologians had not only the power to decide who was a heretic but also the political network, and at times even the ear of the king. For Joachim Lange's extended network to the Prussian court see (Goldenbaum, 2004, pp. 222–233, 270–279). On the other hand, Friedrich Wilhelm I (1688–1740) saw Pietist theologians as a political tool to enforce his absolutist power against the local estates (Hinrichs, 1971, pp. 216–300, p. 432).

[18] Watkins emphasizes the rapidity of the exchange of polemic writings, their length, and redundant character (Watkins, 1998, p. 149). This is easily explained though, first, by the then-common *methodus polemica* according to which one had to present the argument of the opponent before refuting it exhaustively (Zedler, 1731–1754, vol. 20, columns 13–37). The redundancy follows from the duty of all theologians to *declare* their position in the battle—no matter whether one had new arguments.

[19] Already in his *German Metaphysics*, Wolff (1751/1983a) points out that, at the beginning, he wanted to leave out the question of pre-established harmony since it could not be demonstrated, neither a priori nor a posteriori. In his *Gründliche Antwort* (Thoroughgoing Response) from 1724 though, i.e., in his response to his accuser written from Marburg, Wolff de-emphasized the importance of pre-established harmony for his philosophical system (Wolff, 1724/1980a, ad §.12). I cannot understand how Watkins would see this as Wolff moving away from Leibniz. He must

Pre-established Harmony to explain the union of body and soul as had been permitted by Wolff and suggested by political pressure.[20] To be sure, even after the victory of Wolff against Lange in 1736, it remained difficult for any career in Prussia or the Empire, in church, university or state administration, to stick with the authentic hardcore version of Pre-established Harmony.[21] But I would certainly not count those philosophers as Wolffians who indeed united *influxus physicus* with Wolffian philosophy.[22]

To sum up, while not much freedom of thinking existed in the Empire in the first half of the eighteenth century and theologians had almost unlimited authority to determine what counted as truth, some rules and procedures existed that hindered theologians, in cooperation with political authorities, from an immediate proceeding to the persecution of authors. Even if the violation of these rules and procedures happened frequently, the public denunciation in newspapers or journals not only pointed out these violations to the public, but also came with an explanation of the rules and procedures which became thus commonly known. Also, such public denunciations could possibly, although rarely, force the *Corpus evangelicorum* to urge the Emperor to intervene, according to the Peace of Westphalia.

2.4 The Public Space as a Means to Resist Theological Hegemony

Obviously, neither these limits of political authorities nor the rules and procedures in controversies can explain the extraordinary rise of Wolffianism during the first decades of the eighteenth century—a rise that led to a thorough going modernization of universities in the Protestant areas of the Empire and beyond, in terms of

completely ignore the political circumstances for such softening of Wolff's language (Watkins, 1998, pp. 140–142).

[20] Gottsched argued: "None of the three [explanations of the relation between body and soul] is completely explained or demonstrated; each of them still has its difficulties: Thus each person can maintain whichever one is most pleasing" (Gottsched, 1762/1983, §.1077). I do not see here any support in favor of the doctrine of *influxus physicus*; rather it is the attempt to escape the pressure to embrace *influxus physicus*. Watkins muses about the inconsisteny of Gottscheds positions on pre-established harmony in the course of those years, completely ignoring the changing "political weather" which made the term more or less of a theological taboo, even after Wolff's victory in 1736 (Watkins, 1998, pp. 170–174).

[21] Gottsched reported to Reinbeck that—in contrast to students of law—theology students did not dare to attend his lectures because they were afraid of not finding a position in church (Döring, 1999, p. 72).

[22] Since Hegel, Alexander Baumgarten (1714–1762) and Meier (1718–1777) are usually considered as Wolffians although they were not even in contact with Wolff when they lived in the same city. Both keep the term "pre-established harmony" but hold on to the possibility of the soul's influence on the body and vice versa—to save free will as *liberum arbitrium*. Watkins though uncritically follows Beck and Hegel (Watkins, 1998, pp. 183–191).

mathematics, sciences, modern philosophy, law, and even theology. In spite of Wolff's persecution in Prussia, and that of his disciples in other states; in spite of the dangers for a professional career of straight Wolffians—the appeal of Wolffianism steadily increased. Already in 1723, when the political authorities of Sachsen-Weimar requested a review of Wolff's philosophy to decide about its teaching at the University of Jena, two single professors refrained from the majority vote against Wolff and defended its teaching (Ludovici, 1737/1976, pp. 170–177). This won them the applause of students who protested the majority vote. Another statement requested from the teaching *magistri legentes* (i.e., teaching graduate students), had never been published (Hartmann, 1737/1973, pp. 798–813). In fact, the University of Jena, with as many students as the University of Leipzig (about 1000) although from lower income households, would develop into a Wolffian stronghold during the 1720s and early 1730s.[23] While only very few professors supported Wolff's philosophy,[24] the graduate students taught Wolffian philosophy—and also mathematics as propaedeutics for philosophy. That is where Schmidt was trained as a theology student, taking mathematics and philosophy courses with Wolffians, while studying theology with Johann Franz Budde (1667–1729) (Goldenbaum, 2004, pp. 214–217). The frequent complain of theologians about these Wolffian "magistri legentes" can be heard from the universities at Leipzig and Halle too (Goldenbaum, 2004, p. 259).

The growing enthusiasm for Wolff among the academic youth can be seen from reports about student protests against the ban of his philosophy and by the exodus of large numbers of students from Halle to Wolff's new university at Marburg which was reported in newspapers. These reports regularly mentioned the full titles of Wolff, i.e., *Hofrat* (Councilor), *Geheimer Kgl. Hofrat* (Privy Royal Councilor), member of the various academies of Prussia, France, Russia, in order to emphasize how much Wolff was honored and admired in all Europe. But an especially remarkable and expressive document for the increasing influence of Wolffian philosophy among the academic youth is the urgent call of the leader of the orthodox Lutheran church of Saxony, Valentin Ernst Löscher to the students: *Quo ruitis?* That is, where are you heading? Between 1735 and 1742, he published this extensive series of articles in his influential theological journal *Frühaufgelesene Früchte* (Early picked fruits), dedicating it to his "beloved sons", "the noble students of philosophy at the Evangelical-Lutheran universities" (Löscher, 1735, p. 71). He aimed to show the enormous damage to Christian faith done by Leibniz and Wolff's ideas and urged the students to return to Christian faith. Just as for Lange, his major concerns about this philosophy were the principle of sufficient reason (Löscher, 1735, pp. 119–128),

[23] Such resistance from the "magistri legentes" happened even in Halle, see (Hartmann, 1737/1973, pp. 822–825). That is why Lange would try to destroy the career of two such graduate students in Jena, Darjes (1714–1791) and Carpov (1699–1768), denouncing them as Wolffians and immoral people, in the second edition of his *Religionsspötter* (The Mocker of Religion) (Lange, 1736; Goldenbaum, 2004, p. 179, pp. 236–237, 477–478).

[24] Hammerstein, focusing on professors only, falsely concludes that Wolff's influence in the first half of the eighteenth century had been exaggerated (Hammerstein, 1983, pp. 266–277).

pre-established harmony (Löscher, 1740, pp. 236–252), the geometrical method (Löscher, 1735, pp. 128–131, 234–236), and the mechanism as the explanatory model in science (Löscher, 1735, pp. 227–244). Interestingly, he did not reject modern philosophy or science altogether. It was mechanical philosophy, especially that of Leibniz and Wolff which was presented as hostile to Christian faith, while Newton (1642–1726) and Boyle's (1627–1692) philosophy was praised as in best agreement with Christian religion. They, just as anti-Wolffian theologians, had rejected the *universal* demand for mechanical explanations of natural phenomena and defended final causes within science. They also accepted free will as *liberum arbitrium* (Löscher, 1740, p. 247). Löscher's enormous efforts during 7 years to win back the academic youth clearly show Wolff's increasing success among them.

Besides the enthusiasm of the students, Wolff received great support from the new media: Wolffian independent societies, newspapers—in particular the Learned Newspapers—as well as scholarly journals, all run by independent private individuals, i.e., without the backing from church or state institutions. It does not come as a surprise that the *Nova Acta Eruditorum* (New Acts of the Erudites) sided with Wolff, even after the ban of his philosophy in Prussia. After all, he had written hundreds of reviews for this scholarly journal since his arrival at Leipzig in 1702 and had a good relation to the editor Johann Burkhard Mencke (1674–1732) as well as to the latter's son and successor Friedrich Otto Mencke (1708–1754). The positive review of his *Psychologia empirica* appeared already in the December issue of 1732 (*Nova Acta Eruditorum*, 1732, pp. 575–581), and that of the *Psychologia rationalis* in October 1736, in spite of the ongoing public debate about the Wertheim Bible (Nova Acta Eruditorum, 1736, pp. 460–470). Since the journal was written for scholars, it rarely became a target for theologians.[25]

A much more influential support came from the so-called *Gelehrte Zeitungen* (Learned Newspapers). This was a new type of a newspaper after the model of political newspapers. Like them, it was based on their editors' extensive correspondence to as many places in the German Empire and Europe they could connect with to collect academic news. Just as in political newspapers, the academic news had been listed according to the name of the city where they happened. And like common newspapers, they were not distributed by bookshops but offered twice a week at post expedition offices (Neue Zeitungen von gelehrten Sachen, February 1st, 1720, p. 72). Most important, they were written in German, thus appealing to the new increasing social class of learned but not necessarily erudite people in the cities.

The first such Learned Newspaper, the *Neue Zeitungen von Gelehrten Sachen* (New Newspapers about Learned Things) was founded in Leipzig in 1715, by Johann Burckardt Mencke. He recognized the book sellers' desire to advertise their books and journals to the public and to their colleagues beyond academics, especially for the Leipzig Book Fair. The Leipzig University easily provided capable authors. The *Neue Zeitungen* at Leipzig was indeed written by adjuncts or "assistant

[25] An exception was a review of the *Wertheim Bible*. About the difficulties of the editor of the journal to solve the problem: to publish a rejection of the book due to political pressure without losing one's pro-Wolffian high reputation (Goldenbaum, 2004, pp. 277–279).

professors" who could well use an additional income. From 1715 to 1732, they were edited by Johann Gottlieb Krause (1684–1736), who had studied theology in Leipzig while taking courses in Mathematics with Wolff. He was replaced by Friedrich Wilhelm Stübner (1710–1736), a Wolffian and Mathematician who would engage in the controversy about the measure of force, taking Leibniz's side against the Cartesians. Due to his early death, he continued only till 1736. But it was he who authored the first positive review of the *Wertheim Bible* setting the stage for the public debate to come (Goldenbaum, 2004, pp. 191–194). His successor was another Wolffian, Wolf Balthasar Adolph von Steinwehr, who continued till the end of the 1730s when he became one of the Wolffian Professors at the new University at Göttingen.

All three editors of the *Neue Zeitungen von gelehrten Sachen* followed their own scholarly careers. Stübner and Steinwehr became members of the Royal Society at Berlin; Krause and Steinwehr became university professors and Stübner received offers for a professorship in Russia and Greifswald which he could not accept due to his illness. All three were members of the *Deutsche Gesellschaft* (German Society) in Leipzig.[26] This is one of the early Wolffian societies before the Aletophiles (founded only in 1736) (Döring, 2000a, pp. 95–150) which created a solid network of Wolffians with branches in many other Protestant cities of the Empire. While they are usually considered as boringly occupied with German language and poetry (Habermas, 1993, p. 95; cf. though Döring, 2000b, pp. 81–84), most of its members held Wolffian positions, and were supportive of Leibniz and Wolff against their enemies. In fact, they even supported the translator of the *Wertheim Bible*, called the Wertheimer, when he escaped (with the support of the dukes of Wertheim) from his prison at Wertheim, and headed to Altona in Denmark. Surprisingly, he who was already wanted by the Emperor did not take the shortest path to his goal but traveled through those cities where he could rely on the support of members of the *Deutsche Gesellschaften* (Spalding, 1997).

The first Learned Newspaper at Leipzig became a great success and appeared from 1715 to the end of the century (Goldenbaum, 2004, pp. 93–98). We have no numbers about the early years, but at the end of the century it still sold about 1000 copies although the competition from other newspapers had much increased. Naturally, it became a model for other *Gelehrte Zeitungen* as, e.g., the *Hamburgische Berichte von neuen gelehrten Sachen* (Hamburg Reports about new learned things), appearing since 1732, followed by the *Freye Urtheile* (Free Judgments) of Hamburg (Goldenbaum, 2004, pp. 243–253, 284–285, 294–299), and also the better-known *Göttingische Zeitungen von gelehrten Sachen* (Göttingian Newspapers about Learned Things), appearing since 1739. The *Hamburgische Berichte* was edited by the independently wealthy intellectual Johann Peter Kohl (1698–1778), who openly supported Wolff and later Schmidt. He tried very hard to keep an objective presentation of the controversies and responded to threatening letters from Lange by a

[26] See a list of members of the German Society in Leipzig (Goldenbaum, 2004, p. 381). Since this list stems from 1737, Stübner is no longer mentioned. He had passed away in August 1736.

courageous public statement in his newspaper—declaring his absolute commitment to truth, no matter what (Hamburgische Berichte, March 16th, 1736, pp. 188–189). *The Göttingische Zeitungen* was edited first by the Wolffian Wolf Balthasar Adolph von Steinwehr, the former editor of the *Neue Zeitungen* at Leipzig, until he moved to the University at Frankfurt an der Oder in 1742.

What is most interesting about this new type of media in the Protestant area of the Empire is their economic independence from institutions. While most theological and political journals got some support from church or state institutions, if only by the expectation that their employees were supposed to buy them, the Learned Newspapers could only survive on the free market. That the *Gelehrte Zeitungen* at Leipzig made it from 1715 to the end of the century and the *Hamburgische Berichte* from 1732 to 1757, although both openly supported Wolff can tell us a lot about the increasing interest for Wolff.[27]

These Learned newspapers did not so much publish reviews, although sometimes providing very short summaries of new books. Mostly, however, they reported about all kinds of academic events and publications, Europe-wide, from literature to science, and thus quite naturally about the fate of Wolffian philosophy. All the writings pro and con Wolff were reported. They informed the public about Wolff's plans for the future and reported his aim to accomplish an entire system of philosophy. Besides metaphysics and logic, the system would include psychology, ethics, physics, natural and civil law, politics, and theology. Wolff's works were announced in advance, their delayed completion was justified by Wolff's heavy burden as a Vice Dean, and newly published works were reported. These newspapers trumpeted every honor given to Wolff by his rulers in Marburg and Sweden, the offer from Russia, the Italian edition printed at Verona, the new reception of Wolff in Switzerland (Hamburgische Berichte, March 3rd, 1733, pp. 148–151; March 24, 1733, pp. 206–207), where even a Wolff medal was produced in 1733 (Neue Gelehrte Zeitungen, October 12th, 1733, p. 714), the memberships in European academies (Hamburgische Berichte, January 4th, 1734, p. 45; Neue Gelehrte Zeitungen, October 22nd, 1733, p. 754), offers from other universities, and they addressed him commonly as "our famous Prof. Wolff". The two volumes of Wolff's psychology were mentioned in the very month of their appearances, and their upcoming reviews in the *Nova Acta eruditorum* were announced in advance. Just from the increasing page numbers in the yearly index referring to Wolff, one can see the growing attention for his work as much as his productivity. Likewise, from 1735 to 1740, these Learned Newspapers kept everybody informed about the course of the public debate of the *Wertheim Bible*, first about the work itself, then about all the writings pro and con, and, of course, about all the steps of political-juridical persecution and the resistance at the *Corpus evangelicorum* (Goldenbaum, 2004, pp. 451–482). Thereby they fueled the public debate about the *Wertheim Bible* as much as about Wolff and the freedom of thinking.

---

[27] Joachim Kirchner estimated the minimal number of copies of a newspaper or journal at 500 copies to survive on the market (Kirchner, 1928, p. 54).

Sometimes, the Learned Newspapers published entire letters sent from Halle or Marburg, from Berlin or Jena, containing more political details about the state of Wolffian affairs. In January 1733, the *Hamburgische Berichte* reported such a letter from December 27, 1732, about the Commission at Jena, installed by the ruler of Saxony-Weimar. It was supposed to reconsider the teaching of Wolffian philosophy and the hiring of Wolffians (Hamburgische Berichte, January 2nd, 1733, pp. 7–8). When the Commission decided in Wolff's favor, it was immediately announced (*Hamburgische Berichte*, January 23, 1733, p. 64). The Learned Newspapers were also well informed about the earliest negotiations between the court at Berlin and Wolff in 1733, about a possible return to Prussia, emphasizing their information's reliability (Hamburgische Berichte, January 6th, 1734, p. 45; April 23rd, 1734, p. 283). Even the rumors from Göttingen that Wolff might be offered a professorship, were reported, although judged to be unlikely (Hamburgische Berichte, January 18th, 1734, p. 68). The Learned newspapers closely cooperated with the Wolffian Societies, the *Deutsche Gesellschaften,* and the *Aletophile Gesellschaft* (Society of the Lovers of Truth), especially in Leipzig.

As soon as political and legal decisions came into the game though, common political newspapers would also report the events, and the editors of all the here mentioned newspapers were pro-Wolffian. Among the large number of German political newspapers that existed since the early eighteenth century, I am only referring to the mighty and influential *Hamburgische Unpartheyische Correspondent*, read throughout the German Empire, with about 1650 copies in 1730 (Goldenbaum, 2004, pp. 93, 437–439), showing the large public audience at the time. This newspaper would quite regularly report about the most important political-juridical publications and actions taken by the different camps in the battles about Wolff and about Schmidt.[28] Its editors were Wolffians in a wide sense, i.e., not trained in Wolffian philosophy but very supportive of Wolffian ideas in general, and they directly offered their support to the Wolffian Gottsched (Goldenbaum, 2004, p. 248).[29]

Since the Newspapers were private enterprises and brought taxes to their rulers, their editors enjoyed much more leeway than authors of journals or books;[30] and since they were available at a low price, they reached a much wider audience than people at universities. These new media created the countervailing power of a public discourse that could compete with the prevailing power of the church and undermine the latter's close relation to state power.

[28] The *Hamburgische Correspondent* announced more than 30 titles of pro-Wolffian literature during the second peak of the battle against Wolff (Goldenbaum, 2004, pp. 437).

[29] The *Hamburgische Berichte* from June 9, 1733 published an appraisal of Gottsched's *Anfangsgründe der Weltweisheit* (The First Elements of Philosophy) by a Hamburgian Society. Its members asked for more such writings.

[30] About the passive resistance of booksellers against censorship, see the exchange between state authorities and the supervisors of the booksellers in Leipzig and Frankfurt a. M. (Goldenbaum, 2004, pp. 260–261, 387–388).

2.5 Conclusion

The widespread picture of Wolff as easily prevailing German eighteenth century philosophy is drafted by the victors of philosophical history who did not like him and did not hold rationalism in high regard. In fact, he had to work his way up through constant battles with theologians and traditional philosophers. While it is true that they showed some respect to Leibniz, the European genius who was celebrated everywhere, they strongly dismissed his Pre-established Harmony. Wolff's *Psychologia rationalis*, and accordingly his *Psychologia empirica*, rest on this very same metaphysical fundament. He admitted that it could not be demonstrated but argued that it was, in contrast to other hypotheses, free of contradiction. While Wolff deviated in certain respects from Leibniz, both philosophers were convinced that neither bodies can influence the souls nor souls the bodies although both acted in perfect harmony. Likewise, both philosophers saw freedom as acting according to reason and rejected indifferentism, i.e., acting without a reason, the common view about free will. It is not by chance that pre-established harmony was blamed to be Spinozism. Especially theologians were alarmed that sinners might be excused from their evil actions if their evil deeds were determined by certain reasons rather than by the mere power of their will. How could their actions then be imputed to the sinners?

The battles about free will still cause heated and emotional controversies in our days. In eighteenth century though, theologians who were up in arms had the power to silence the philosophers. In case of Christian Wolff and his partisans, they tried very hard to reach a ban of Wolffian philosophy in the entire German Empire. Theologians were almost everywhere in charge of censorship and thus able to hinder publications of books as well as of journals. They held influential positions at the courts of states and of the Empire. In case of Wolff, the German Pietist theologian Lange even succeeded in convincing his ruler that Wolff's philosophy justified deserting soldiers besides destroying religion which led the king to eventually expel Wolff from Prussia, to ban his philosophy in Prussia and to influence other rulers and the Emperor to do likewise. The theological faculties were—at all universities and higher schools—fortresses to defend the students' orthodoxy and to provide positions to those who remained orthodox.

Under such harsh circumstances, it is as amazing as interesting to see how Wolff's influence grew nonetheless, beginning in 1710. In the first place, this is due to his enormous success as a teacher. He attracted large numbers of students wherever he taught. The wit and the clarity of his lectures were widely praised. Such drawing of students was an economic advantage in itself and gave second thoughts to those rulers who hosted universities in their states. Besides, Wolff was a prolific writer delivering, volume after volume, an entire system of rational thinking which covered every single subject one could think of. His work was praised for the clarity and conclusiveness, its applicability not only to all sciences and technology but also to law, political, and economic theory, not to mention natural theology. Besides, Wolff was an effective university administrator and moreover, able to produce a reliable network with journals, publishers, city administrations, and with quite a number of rulers. He held even good relations with the Russian Court for whom he taught Russian students in private courses. Former students who became his partisans founded societies to promote his philosophy, enthusiastic graduate students taught the next generation. Further, Wolff had the leading scholarly journal *Acta eruditorum* available, for reviews of his and his partisans' books. Above all, however, it was the Learned Newspapers that forged Wolff's success. Their steady reports built and reassured his audience beyond any particular university or city. Their Wolffian editors helped to attract a wide readership for Wolff, by emphasizing the clarity of his thinking, the practicability of his philosophical thinking for the sciences, law and politics, and even for theology, laying bar the prejudices of his critics, and celebrating his European fame. They could dare to question theological accusations against Wolff, e.g., of atheism or immoral behavior, although they too published under the restricted conditions as presented above.

That is how Wolff could overcome and eventually win the battle against his powerful theological enemies and become the most influential philosopher in Germany throughout the eighteenth century—who was well received even beyond the Empire's borders. Thus, Wolff well deserves to be addressed as the *Philosophus Germanicus*[31] (The German Philosopher) of his century in spite of the fact that his powerful opponents got never tired of fighting him.

Unpublished Sources GStA PK = Geheimes Staatsarchiv Preußischer Kulturbesitz Berlin-Dahlem. Acta wegen der in Verfall geratenen Universität Halle: Rep. 51 and 52. (https://gsta.preussischer-kulturbesitz.de).

References

Altmann, A. (Ed.). (1977). *Moses Mendelssohn. Gesammelte Schriften. Jubiläumsaugabe* (Vol. 13). Frommann-Holzboog Verlag.

Beck, L. W. (1969). *Early German philosophy*. Harvard University Press.

Bianco, B. (1989). Freiheit gegen Fatalismus: Zu Joachim Langes Kritik an Wolff. In N. Hinske (Ed.), *Halle: Aufklärung und Pietismus* (pp. 111–156). Schneider.

Böning, H., & Moepps, E. (1996). *Hamburg. Kommentierte Bibliographie der Zeitungen, Zeitschriften, Intelligenzblätter, etc. sowie biographische Hinweise zu Herausgebern, Verlegern und Druckern periodischer Schriften. Von den Anfängen bis 1765*. Frommann-Holzboog.

Ching, J., & Oxtoby, W. (Eds.). (1992). *Moral enlightenment. Leibniz and Wolff on China*. Routledge.

Döring, D. (1999). *Die Philosophie Leibniz' und die Leipziger Aufklärung in der ersten Hälfte des 18. Jahrhunderts*. Hirzel.

Döring, D. (2000a). *Johann Christoph Gottsched in Leipzig*. Verlag der Sächsischen Akademie.

[31] This grand title was given to Wolff by Herm. Adolph. Le Fevre (1708–1745), a lawyer at Lubeck and a former law student from Jena, whose Wolffian thesis (defended in Strasbourg on July 17, 1733) was reviewed and cited with aplomb by the *Hamburgische Berichte* on December 15, 1733.

Döring, D. (2000b). Beiträge zur Geschichte der Aletophilen in Leipzig. In D. Döring & K. Nowak (Eds.), *Gelehrte Gesellschaften im mitteldeutschen Raum (1650–1820)* (pp. 95–150). Verlag der Sächsischen Akademie der Wissenschaften zu Leipzig.

Gierl, M. (1997). *Pietismus und Aufklärung. Theologische Polemik und die Kommunikationsreform der Wissenschaft am Ende des 17. Jahrhunderts*. Vandenhoek & Ruprecht.

Goldenbaum, U. (2004). Der Skandal der Wertheimer Bibel. Die philosophisch-theologische Entscheidungsschlacht zwischen Pietisten und Wolffianern. In U. Goldenbaum, *Appell an das Publikum. Die öffentliche Debatte in der deutschen Aufklärung 1687–1796* (pp. 175–508). Akademie-Verlag.

Goldenbaum, U. (2014). Some doubts about Jonathan Israel's new compartmentalization of enlightenment. A plea for Hobbes, Voltaire, and Mendelssohn. In F. Grunert (Ed.), *Concepts of (radical) enlightenment. Jonathan Israel in discussion* (pp. 52–80). Mitteldeutscher Verlag.

Goldfriedrich, J. (1908). *Geschichte des deutschen Buchhandels* (Vol. 2). Verlag des Deutschen Börsenvereins.

Gottsched, J. C. (1980). Historische Lobschrift des Herrn Christians Wolff. In H. W. Arndt (Ed.), *Christian Wolff: Gesammelte Werke* (Biographie) (Vol. 10, Pt. II). Olms. (Original work published 1755).

Gottsched, J. C. (1983). Erste Gründe der Weltweisheit. In J. École (Ed.), *Christian Wolff: Gesammelte Werke* (Erste Gründe der gesamten Weltweisheit) (Vol. 20.1, Pt. III). Olms. (Original work published 1762).

Habermas, J. (1993). *Strukturwandel der Öffentlichkeit*. Suhrkamp.

Hamburgischer Unparteyischer Correspondent. (1712–1851). Hamburg, Germany.

Hammerstein, N. (1983). Christian Wolff und die Universitäten. Zur Wirkungsgeschichte des Wolffianismus im 18. Jahrhundert. In W. Schneider (Ed.), *Christian Wolff (1679–1754). Interpretationen zu seiner Philosophie und deren Wirkung* (pp. 266–277). Meiner.

Hartmann, G. V. (1973). Anleitung zur Historie der Leibnitzisch-Wolffischen Philosophie. Part 2. In J. École (Ed.), *Christian Wolff: Gesammelte Werke* (Anleitung zur Historie der Leibnitzisch-Wolffischen Philosophie) (Vol. 4, Pt. III). Olms. (Original work published 1737).

Heer, G. (1927). *Marburger Studentenleben 1527–1927. Eine Festgabe zur 400jährigen Jubelfeier der Universität Marburg*. Elwert.

Hegel, G. W. F. (1986). *Vorlesungen über die Geschichte der Philosophie*. Meiner.

Hinrichs, C. (1971). *Preußentum und Pietismus. Der Pietismus in Brandenburg-Preußen als religiös-soziale Reformbewegung*. Vandenhoek & Ruprecht.

Hirsch, E. (1951). *Geschichte der neuern evangelischen Theologie* (Vol. 2). Gütersloh.

Kane, R. (2002). *The Oxford handbook of free will*. Oxford University Press.

Kertscher, H.-J. (2018). *"Er brachte Licht und Ordnung in die Welt". Christian Wolff—Eine Biographie*. Mitteldeutscher Verlag.

Kirchner, J. (1928). *Die Grundlagen des deutschen Zeitungschriftenwesens mit einer Gesamtbibliographie der deutschen Zeitschriften bis zum Jahr 1790 (1. Teil)*. Hiersemann.

Kobuch, A. (1988). *Zensur und Aufklärung in Kursachsen*. Böhlau.

Kohl, J. P. (Ed.). (1732–1759). *Hamburgische Berichte von neuen Gelehrten Sachen*. Ort.

Lange, J. (1703). *Libri de medicina mentis*. Without Publishing House.

Lange, J. (1735). *Der philosophische Religionsspötter*. Without Publishing House.

Lange, J. (1736). *Der philosophische Religionsspötter* (2nd. ed.). Walther.

Lange, J. (2000). Ausführliche Rezension der wider die Wolfianische Metaphysic auf 9. Universitäten […] edirten 26. Schriften. In J. École (Ed.), *Christian Wolff: Gesammelte Werke* (Schriften über Joachim Langes und Johann Franz Buddes Kontroverse mit Christian Wolff) (Vol. 64.1, Pt. III). Olms. (Original work published 1725).

Lorenz, S. (1997). *De mundo optimo. Studien zu Leibniz' Theodizee und ihrer Rezeption in Deutschland (1710–11)*. Steiner.

Löscher, E. V. (1735). Quo ruitis? Treuhertzige Anrede eines bejahrten Lehrers an die den Philosophischen Studiis ergebene Jugend, wegen der zur Herrschaft sich dringenden neuen

Philosophie (16 Pensa). In E. V. Löscher (Ed.), *Frühaufgelesene Früchte der Theologischen Sammlung von Alten und Neuen Sachen* (pp. 1735–1742). Braun.

Löscher, E. V. (1740). Quo ruitis? Treuhertzige Anrede eines bejahrten Lehrers an die den Philosophischen Studiis ergebene Jugend, wegen der zur Herrschaft sich dringenden neuen Philosophie (16 Pensa). In E. V. Löscher (Ed.), *Frühaufgelesene Früchte der Theologischen Sammlung von Alten und Neuen Sachen* (pp. 1735–1742). Braun.

Ludovici, C. G. (1976). Sammlung und Auszüge der sämmtlichen Streitschriften wegen der Wolffischen Philosophie, 1. Theil. In J. École (Ed.), *Christian Wolff: Gesammelte Werke* (Vol. 2, Pt. III). Olms. (Original work published 1737).

Ludovici, C. G. (1977). Ausführlicher Entwurf einer vollständigen Historie der Wolffischen Philosophie, Anderer Theil. In J. École (Ed.), *Christian Wolff: Gesammelte Werke* (Psychologia empirica) (Vol. 1.2, Pt. III). Olms. (Original work published 1737).

Friedrich Otto Mencke. *Nova Acta Eruditorum* (1732-1776).

Neue Gelehrte Zeitungen. (1715–1784). Leipzig, Germany.

Reinbeck, J. G. (1736). *Betrachtungen über die in der Augsburgischen Confession enthaltene und damit verknüpfte Göttliche Wahrheiten.* Haude und Spener.

Reinbeck, J. G. (1737). *Erörterung der Philosophischen Meynung von der sogenannten HARMONIA PRAESTABILITA.* Haude.

Rössler, E. F. (1855). *Die Gründung der Universität Göttingen. Entwürfe, Berichte und Briefe der Zeitgenossen.* Vandenhoeck & Ruprecht.

Schmidt, J. L. (1736a). *Die göttlichen Schriften vor den Zeiten des Messie Jesus.* Nehr.

Schmidt, J. L. (1736b). *Die vestgegründete Wahrheit der Vernunft und Religion.* Nehr.

Schmidt, J. L. (1736c). *Vertheidigung der freyen Übersetzung.* Nehr.

Schmidt, J. L. (1736d). *Beantwortung verschiedener Einwürfe.* Nehr.

Schmidt, J. L. (1736e). *Öffentliche Erklärung vor der ganzen evangelischen Kirche, die freye Übersetzung der göttlichen Schriften betreffend.* Nehr.

Schmidt, J. L. (1738). *Sammlung derjenigen Schriften welche bey Gelegenheit des wertheimischen Bibelwercks für oder gegen dasselbe herausgekommen sind.*

Schrader, H.-J. (1985). Pietistisches Publizieren unter Heterodoxieverdacht. Der Zensurfall 'Berleburger Bibel'. In H. Göpfert & E. Weyrauch (Eds.), *"Unmoralisch an sich ...". Zensur im 18. Und 19. Jahrhundert* (pp. 61–88). Harrassowitz.

Spalding, P. S. (1997). Im Untergrund der Aufklärung: Johann Lorenz Schmidt auf der Flucht. In E. Donnert (Ed.), *Europa in der Frühen Neuzeit* (Vol. 4, pp. 135–154). Böhlau.

Spalding, P. S. (1998). *Seize the book, jail the author. Johann Lorenz Schmidt and censorship in eighteenth-century Germany.* Purdue University Press.

Specht, R. (1989). *John Locke.* Beck.

Voltaire. (1764). *Abhandlung über die Religionsduldung.* Fritsch.

von Raumer, K. (1953). *Ewiger Friede.* Alber.

Watkins, E. (1998). From pre-established harmony to physical influx: Leibniz's reception in early 18th-century Germany. *Perspectives on Science, 6*, 136–203.

Welke, M. (1981). Die Legende vom 'unpolitischen Deutschen.' Zeitungslesen im 18. Jahrhundert als Spiegel des politischen Interesses. *Jahrbuch der Wittheit zu Bremen, 25*, 161–188.

Wolff, C. (1968). Psychologia Empirica, methodo scientifica pertractata, qua ea, quae de anima humana indubia experientiae fide constant, continentur et ad solidam universae philosophiae practicae ac theologiae naturalis tractationem via sternitur. In J. École (Ed.), *Christian Wolff: Gesammelte Werke* (Psychologia Empirica) (Vol. 5, Pt. II). Olms. (Original work published 1738).

Wolff, C. (1972). Psychologia rationalis, methodo scientifica pertractata, qua ea, quae de anima humana indubia experientiae fide innotescunt, per essentiam et naturam animae explicantur, et ad intimiorem naturae ejusque autoris cognitionem profutura proponuntur. In J. École (Ed.), *Christian Wolff: Gesammelte Werke* (Psychologia rationalis) (Vol. 6, Pt. II, 2nd ed.). Olms. (Original work published 1740).

Wolff, C. (1978). Von den Kräften des menschlichen Verstandes und ihrem richtigen Gebrauche in Erkenntnis der Wahrheit. In H. W. Arndt (Ed.), *Christian Wolff: Gesammelte Werke* (Deutsche Logik) (Vol. 1, Pt. I). Olms. (Original work published 1713).

Wolff, C. (1980a). Des Herrn Doct. und Prof. Joachim Langens oder: Der Theologischen Facultæt zu Halle Anmerckungen über des Herrn Hof-Raths und Professor Christian Wolffens Metaphysicam, [...], nebst beygefügter [...] Gründlicher Antwort. In J. École (Ed.), *Christian Wolff: Gesammelte Werke* (Kleine Kontroversschriften mit Joachim Lange und Johann Franz Budde) (Vol. 17, Pt. I). Olms. (Original work published 1724).

Wolff, C. (1980b). Nötige Zugabe über Herrn Buddens Bedencken von der Wolffischen Philosophie. In J. École (Ed.), *Christian Wolff: Gesammelte Werke* (Kleine Kontroversschriften mit Joachim Lange und Johann Franz Budde) (Vol. 18, Pt. I). Olms. (Original work published 1724).

Wolff, C. (1980c). Klarer Beweis, daß Herr D. Budde die ihm gemachten Vorwürfe einräumen und gestehen muß. In J. École (Ed.), *Christian Wolff: Gesammelte Werke* (Kleine Kontroversschriften mit Joachim Lange und Johann Franz Budde) (Vol. 18, Pt. I). Olms. (Original work published 1725).

Wolff, C. (1983a). Vernünftige Gedanken von Gott, der Welt und der Seele des Menschen, auch allen Dingen überhaupt. In C. Corr (Ed.), *Christian Wolff: Gesammelte Werke* (Deutsche Metaphysik) (Vol. 2, Pt. I, 11th ed.). Olms. (Original work published 1751).

Wolff, C. (1983b). Philosophia rationalis sive Logica. In J. École (Ed.), *Christian Wolff: Gesammelte Werke* (Philosophia rationalis sive logica) (Vol. 1.3, Pt. II). Olms. (Original work published 1740).

Wolff, C. (1983c). De differentia nexus rerum sapientis et fatalis necessitatis [...] luculenta commentatio. In J. École (Ed.), *Christian Wolff: Gesammelte Werke* (Opuscula Metaphysica) (Vol. 9, Pt. II). Olms. (Original work published 1724).

Wolff, C. (1985a). Oratio de Sinarum philosophia practica. In M. Albrecht (Ed.), *Rede über die praktische Philosophie der Chinesen*. Meiner. (Original work published 1726).

Wolff, C. (1985b). Nouvelles pièces sur les erreurs prétendues de la philosophie de Mons. Wolff. In J. École (Ed.), *Gesammelte Werke* (Nouvelles pieces) (Vol. 22, Abt. III). Olms. (Original work published 1736).

Wotschke, T. (1932). Wolffs Briefe über seinen Streit mit den halleschen Pietisten. *Thüringisch-sächsische Zeitschrift für Geschichte und Kunst, 21*, 51–74.

Zedler, J. H. (1731–1754). *Grosses vollständiges Universal-Lexicon aller Wissenschaften und Künste* (64 vols.). Zedler.

Part I
The Scope and Contents of Wolff's Psychology

Chapter 3
The Origins and Development of Wolff's Psychology in His German Writings

Thiago Constâncio Ribeiro Pereira and Saulo de Freitas Araujo

3.1 Introduction

Despite the fact that general textbooks on the history of psychology usually ignore Christian Wolff's contributions, there can be no doubt that his psychological project played an important role in the development of German scientific psychology (e.g., Araujo, 2012; Feuerhahn, 2002; Gundlach, 2006; Hatfield, 1995; Smith, 1997; Sturm, 2009; Vidal, 2006). However, studies on his psychological ideas traditionally focus on his two famous Latin works—*Psychologia Empirica* (1732) and *Psychologia Rationalis* (1734)—overlooking his German psychological writings.[1]

The problem lies in the fact that, although Wolff's Latin works represent his mature positions on psychological matters, the idea of a psychological doctrine as part of his philosophical system was not a later result of his intellectual development. Instead, his German writings already contain the essence and the fundamentals of his mature psychology (e.g., Wolff, 1720, 1724, 1726).

It is not sufficient, however, just to consider Wolff's psychological ideas as they appear in his German writings. His writing style and aims as well as the particularities of his professional life, are of the utmost importance to understanding the force and impact of his ideas at that time. This is one of the reasons why both Wolff

[1] There are exceptions, however. Historians of philosophy have emphasized the importance of Wolff's German writings to an adequate understanding of his psychological ideas (e.g., Campo, 1939/1980; Dyck, 2014; Goubet, 2018; Marcolungo, 2007a; Rumore, 2018).

T. C. R. Pereira (✉)
Department of Psychology, Fluminense Federal University,
Volta Redonda, Rio de Janeiro, Brazil
e-mail: tcrpereira@id.uff.br

S. d. F. Araujo
Department of Psychology, Federal University of Juiz de Fora,
Juiz de Fora, Minas Gerais, Brazil

S. d. F. Araujo et al. (eds.), *The Force of an Idea*, Studies in History and Philosophy of Science 50, https://doi.org/10.1007/978-3-030-74435-9_3

scholarship and the historiography of psychology have not yet recognized the full originality, consistency, and influence of Wolff's psychological ideas or their significance within the eighteenth-century intellectual context, particularly for the development of psychology as a science.

The goal of this chapter is to offer a general historical and philosophical account of Wolff's early psychology, based on his German writings. We will consider his most relevant works in the period between 1720 and 1726, which include his *Deutsche Metaphysik* (The German metaphysics, 1720),[2] *Anmerckungen zur Deutschen Metaphysik* (Annotations to the German metaphysics, 1724), and *Ausführliche Nachricht* (Detailed report, 1726).[3] In addition, we will explore the relationship between those works and Wolff's professional career as a professor in Halle and with the debates following their reception. In so doing, we hope to present a novel perspective on the topic, thus contributing to further debates in the historiography of psychology and Wolff scholarship.[4]

3.2 Emergence of Wolff's *Deutsche Metaphysik*

In 1707, Wolff delivered his first lectures as a professor of mathematics and natural philosophy at the University of Halle, which had been recently founded (1694). Before that appointment, his work, which had mostly been related to the field of mathematics, had gained him the sympathy and support of G. W. Leibniz (1646–1716), whose prestige helped Wolff achieve his professorship. However, Wolff's interests went well beyond mathematics. His writings and lectures soon covered practically every field of knowledge, making him a recognized *Gelehrter* (scholar) within Prussian territory (Hettche & Dyck, 2019).

In contrast to the more traditional German universities, the University of Halle established German as one of its official languages. This was very innovative, since Latin was still the current language in European academic circles. Wolff was quite an enthusiast of this new trend and developed a series of German textbooks aimed at his students with a primarily didactic purpose, namely, to help them follow his classes without the distraction of having to take notes. The works were also intended to be didactic from a

[2] Actually, it was published in December 1719 (see footnote 8 in Chap. 1).

[3] Except when otherwise indicated, all the translations from the original German sources were made by the authors of this chapter.

[4] This chapter developed out of our previous work on the topic (Araujo, 2012; Araujo & Pereira, 2014; Pereira & Araujo, 2015; Pereira, 2017). However, it contains ideas and clarifications that we could not present in those publications.

logical point of view: they were to be organized in such a way that the simplest subject would always come first, and with every antecedent containing the ground of its consequent, thus resulting in an interconnected demonstrative system.[5]

Wolff's major German treatises are known as his *Vernünfftige Gedancken* (rational thoughts).[6] In addition, to differentiate them from their Latin counterparts, they are usually referred to in the literature with the adjective *deutsch* (German), covering the fields of logic (*Deutsche Logik*, 1713), metaphysics (*Deutsche Metaphysik*, 1720), ethics (*Deutsche Ethik*, 1720), politics (*Deutsche Politik*, 1721), physics (*Deutsche Physik*, 1723), teleology (*Deutsche Teleologie*, 1724), and physiology (*Deutsche Physiologie*, 1725).

In fact, Wolff's *Vernünfftige Gedancken* inaugurates a new phase in German philosophy. Before that, philosophical curricula at German universities were based on the old Lutheran-scholastic treatises of P. Melanchthon (1497–1560) and others (Hammerstein, 1983; Hettche & Dyck, 2019). Wolff's contributions extended far beyond the pedagogic dimension, however, to include linguistic and epistemic innovations, since he created not only the German philosophical vocabulary but also many of the philosophical disciplines that would be established as autonomous fields of knowledge taught at German universities (such as psychology, general cosmology, and ontology). One can say that he made a triple synthesis: early-eighteenth-century German thought, the new ideas of the *Aufklärung*, and the Scholastic tradition (Schwaiger, 2010).[7]

Wolff's psychology emerges in this context, although it is true that he had already published some psychological writings before 1720. Two examples should be mentioned here. In his 1707 paper *Auflösung einiger Schwierigkeiten, welche bei der menschlichen Seele vorkommen* (Dissolution of some difficulties concerning the human soul), Wolff argues that the absence of an appropriate way to philosophize about the soul was an obstacle to the progress of metaphysical knowledge. Here, he discusses how the understanding (*der Verstand*) develops, and why children, deaf-mutes, and persons during sleep do not make adequate use of reason (Wolff, 1737/1981). Most importantly, in his *Ratio Praelectionum* (Wolff, 1718), a general introduction to his courses in Halle, Wolff presents a condensed version of his psychology, commenting on the main

[5] Wolff mentions this pedagogical role of his German writings in the first prologue of his *Deutsche Metaphysik* (Wolff, 1720) and in the second chapter of his *Ausführliche Nachricht* (Wolff, 1726), where he comments on his writing style.

[6] Each title starts with the phrase "Rational thoughts on."—The "rational" aspect of these writings is to be found in the way they are organized (as a demonstrative chain), as well as in their general appeal to reason rather than tradition, authority, or faith.

[7] For a more detailed presentation of Wolff's impact on German universities, see Albrecht (2018) and Hammerstein (1983).

questions concerning the knowledge of the soul to be addressed in his subsequent metaphysics. Although both publications already indicate that psychology is central to Wolff's philosophical program, a systematic presentation appears for the first time only in his *Deutsche Metaphysik* (Wolff, 1720).

The full title of the *Deutsche Metaphysik—Rational thoughts on God, the world and the human soul, and also on all things in general*—refers to each of the four metaphysical disciplines that are its concern: natural theology, general cosmology, psychology, and ontology, respectively. However, following Wolff's pedagogic intentions, the contents are displayed in a different order, from the most evident to the most abstract. The first chapter—"On how we know that we are and what use this knowledge has for us" (§§.1–9)—serves as a general introduction and marks the beginning of empirical psychology.[8] The second chapter—"On the first principles of our knowledge and all things in general" (§§.10–190)—corresponds with ontology. The third—"On the soul in general, namely, what we actually perceive of it" (§§.191–539)—is also related to empirical psychology. The fourth—"On the world" (§§.540–726)—belongs to general cosmology. The fifth—"On the essence of the soul and of a spirit in general" (§§.727–927)—addresses questions associated with rational psychology. Finally, the sixth—"On God" (§§.928–1089)—deals with natural theology.

After this brief description, it becomes clear that psychology occupies a significant place in the *Deutsche Metaphysik*. Beginning with the empirical (psychological) knowledge that we are conscious of our own existence, Wolff is led to clarify the logical bases of all knowledge in general (e.g., the principle of sufficient reason), as well as the general notions applicable to all things, including the soul (e.g., the concept of a simple thing). This constitutes the content of ontology. Next, he investigates what we perceive of the soul in our concrete experience (empirical psychology), which leads him to search for its transcendent ground (rational psychology). Before that, his general cosmology establishes concepts (e.g., the concept of the world) that will help psychological investigation to penetrate that transcendent dimension (e.g., the essence of the soul involves representing the world). It is only after this that he concludes his work with natural theology.

[8] Wolff neither calls his psychological chapters by the name "psychology" nor offers a formal definition of psychology. Everything here falls under the general label "metaphysics." However, there is no doubt that they represent what Wolff would later call empirical and rational psychology. Those terms appear for the first time in his *Ausführliche Nachricht* (Wolff, 1726, §.79), referring to Chaps. 3 and 5 of his *Deutsche Metaphysik* (see Sect. 3.5).

3.3 The Soul as the Subject Matter of Psychology in the *Deutsche Metaphysik*

Wolff starts his metaphysics with a psychological observation.[9] According to him, "we are conscious.[10] No one can doubt this who has not been completely robbed of his senses" (Wolff, 1720, §.1, p. 1).[11] In so being, he continues, we must exist, since whoever is conscious, exists.[12]

The purpose here is twofold. First, Wolff wants to establish the most basic knowledge, which is an immediate empirical finding, an intuition that cannot be doubted. Second, he intends to guarantee certainty by revealing the real form of that knowledge—a syllogism akin to a geometrical demonstration[13]—which will serve as an ideal form for the whole subsequent investigation.[14]

In Chap. 3, Wolff announces the first part of his psychological program: "I still do not intend to show here what the soul is and how modifications occur in it; rather, my intention is solely to recount what we perceive of it by means of daily experience" (§.191, p. 89). This is a kind of descriptive exercise that any of us should be able to do, provided we pay attention to ourselves. We are, so to speak, at the common sense level of experience.[15] Wolff makes clear that what we are going to observe

[9] It is controversial as to whether this first chapter belongs to psychology, logic, or ontology (e.g., Arnaud, 2003; Paccioni, 2001). For us, it is clearly a part of psychology, since it deals with the empirical consciousness of our own existence and how it is achieved, topics that are related to the soul's activities (consciousness and inference). More specifically, it should belong to empirical psychology, because it refers to what can be apprehended through experience. This becomes obvious when one considers that Wolff's *Psychologia Empirica* begins in the same way (see Wolff, 1738/1968, §§.11–19).

[10] From the second edition of the *Deutsche Metaphysik* onward, Wolff added this: "we are conscious of our things and other things" (Wolff, 1751/2003, p. 1, §.1). For an analysis of this change, see Marcolungo (2007b).

[11] Here and in other passages, we have benefited much from Corey Dyck's partial translation of the first edition of the *German Metaphysics* (Dyck, 2019). However, we have not always followed it.

[12] Some scholars see here a kind of *Wolffian cogito*, implying Wolff's adoption of Cartesian philosophy (e.g., Arnaud, 2002; Vittadello, 1973). Blackwell (1961) and Euler (2004) have pointed out some problems with this reading. We would like to add to this debate three distinctive aspects of Wolff's approach: first, it is the existence of real conscious beings that is being inferred here, not of a purely metaphysical *res cogitans*; second, the ultimate ground of metaphysical certainty is the logical form of demonstration, not the existence of a thinking substance; third, the senses play here an important role, which is not the case in Descartes, for whom they are not to be trusted.

[13] For the syllogistic form of the phenomena occurring in the soul, see the contribution from Matteo Camposampiero to this volume.

[14] For Wolff, geometrical truths are as certain as our consciousness of ourselves because their demonstration is equivalent (a valid syllogism). In the latter case, the major premise is "Whoever is conscious of themselves, exists" whereas the minor premise is "We are conscious of ourselves." The conclusion ("Therefore, we exist") follows necessarily.

[15] It is important to note, however, that this descriptive exercise is not the only purpose of empirical psychology. In later editions, Wolff adds that it will also provide empirical rules for the soul's

in ourselves is the soul, which he defines as "that thing which is conscious" (§.192, p. 90).[16]

After the observation that the soul is conscious of itself, Wolff examines the cognitive faculties of the soul (§§.194–403). The most basic of these is sensation (*Empfindung*), which is the capacity of the soul to represent something present that causes changes in the body (§.220). However, the soul does not represent only present things. Imagination (*Einbildung*) is the faculty through which the soul represents an absent thing connected to a present one (§.235). This process happens according to the rule of imagination: "when our senses present something to us that has something in common with a sensation we had at another time, that same thing also occurs to us again" (Wolff, 1720, §.238, p. 113).[17] Memory (*Gedächtniß*) is the capacity of the soul to recognize a thing that has been already represented at a different time (§.249). The soul can also attend to a sensation or image to the detriment of others, thus increasing its consciousness of a thing. When this happens, we say that this thing becomes clearer to us. This is what Wolff calls attention (*Aufmercksamkeit*) (§.268). Understanding (*Verstand*) is the ability of the soul to represent distinctly (§.277).[18] It involves three functions: (1) the capacity to determine the general concepts of things by means of distinguishing them and finding their general similarities and differences, that is, their species and genera; (2) the capacity to represent the union and separation of concepts, which is called judgment (*Urtheil*); and (3) the special order of succession of the changes in the soul, which is called inference or syllogism (*Schluß*). Finally, the soul can also represent the interconnection of truths, which Wolff calls reason (*Vernunft*) (§.368). For Wolff, both reason and experience (*Erfahrung*) lead to truth, but in different ways.[19] The

operations that will serve as the fundamentals of logic, ethics, and politics (Wolff, 1751/2003, §.191). This grounding role of psychology for other disciplines is of paramount importance to Wolff.

[16] It should be noted that empirical psychology itself already presupposes the concept of the soul as a thing, derived from ontology, even though Wolff will only develop it in his rational psychology.

[17] One might understand this to mean that only the common element reappears in the imagination. However, in the later editions, Wolff added, "when a part of the whole present sensation is part of a previous one, then the whole previous sensation reappears" (Wolff, 1751/2003, §.238, p. 132). It is thus the whole past sensation that reappears as part of our imagination.

[18] Here, it is important to keep in mind that, for Wolff, representations can achieve four degrees of clearness. For instance, if we cannot see the difference between a square and a triangle, our representation is obscure. If we can, then our representation is clear. Moreover, if we can see that this difference consists in the number of sides, our representation is not only clear, but also distinct. However, if we are unable to do this, our representation is clear but indistinct (§§.198–214). Through the senses and imagination, we can have only clear representations. Through attention and the understanding, we achieve distinctness.

[19] It is important to note that experience, for Wolff, is a general concept that includes two subordinate concepts, namely, common experiences (*gemeine Erfahrungen*) and experiments (*Versuche*). "The knowledge we acquire when we pay attention to our sensations and the modifications of the soul we call *experience.* And when sensations occur spontaneously, we call it *common experiences*; *experiments*, by contrast, when we acquire it through our efforts" (Wolff, 1720, §.325, p. 159). In principle, therefore, the possibility of an experimental psychology is given already at this early stage of Wolff's theory.

latter is based on the senses and is thus limited to contingent concepts and judgments (§.368). The former, being grounded in the understanding, establishes general truths and represents their interconnection. This is the only way to find the ultimate reasons for whatever is given in experience (§.372).[20]

This is the basic picture of the cognitive faculties of the soul. It is important to note, however, that they do not work in mutual isolation. Wolff demonstrates in each case how those faculties involve many others (e.g., attention unfolds in reflection, while memory involves remembering, reminiscence, and forgetting) and derives therefrom many other notions (e.g., intuitive and symbolic knowledge from sensation and understanding, invention from imagination, discovery from inference, and science and certainty from reason).

Continuing the examination of experience, Wolff notices that the soul is capable of more than just knowing things. Once things are known, they produce in the soul either pleasure or displeasure, and are able to make the soul more or less perfect, so that it judges them as good or bad for itself, and inclines itself to approach or deviate from them. This is the basis of what Wolff calls the volitional faculties of the soul: sensual desire and aversion (the inclination toward or deviation away from what the soul represents *intuitively* as good or bad), affects (degrees of sensual desire or aversion), will and not will (inclination toward or deviation away from what the soul *distinctly* represents as good or bad), and freedom (the soul's ability to determine itself—not by nature or anything external—in whatever direction most pleases it) (§§.404–526). Here, we have again important rules, such as the rule of the will, through which the soul only wants what it sees as good and does not want what it sees as bad (§.506). For Wolff, this is one of the psychological rules that will have high value for other disciplines, such as practical philosophy.

At the end of the chapter, Wolff presents some comments on the harmony (*Uebereinstimmung*) between the body and the soul (§§.527–539). First, he states that this harmony is an empirical finding for which we do not know the reason. For him, by observing how changes occur either in the body or in the soul, it becomes perfectly clear that there is a harmony between them; this is why people usually talk about a union between them and of the regulation of the body by the soul. However, Wolff insists, if one wants to stay at the level of experience—without adding any opinion or hypothesis to what one perceives in it—one is forced to admit that experience reveals neither bodily effects on the soul nor effects of the soul on the body nor the absence of such effects. All experience offers is the existence of a harmony. If one wants to understand the reasons for it, one must adopt another method of investigation that goes beyond experience. This is where Wolff ends his early empirical psychology, thus making room for rational psychology, which begins in Chap. 5.

At the opening of that chapter, he says:

[20] In the later editions, after explaining the difference between experience and reason (§.371), Wolff adds that "science, however, comes from reason" (Wolff, 1751/2003, §.371, p. 228). This makes clear that, for him, experience alone cannot lead to scientific knowledge (broadly understood).

> In fact, I have already dealt extensively with the soul in the third chapter, above, but only insofar as we perceive its effects and are able to achieve a distinct concept of them (§.191). Now we must investigate in what the essence of the soul and of spirit in general consists, and how what we perceive of it and have noticed before is founded in that essence. Therewith, it will be possible to deal with other things about the soul to which experience does not lead us immediately. (Wolff, 1720, §.727, pp. 401–402)[21]

The chapter concentrates on four main subjects: the essence of the soul, the relationship between the body and the soul, the grounding of the faculties of the soul in the essence of the soul, and knowledge about spirits in general. This investigation thus plays a double role: it justifies and expands empirical psychology.

First, Wolff resumes his psychological observation that the soul is conscious of itself (§§.728–734). This is only possible, he says, because the soul can represent, compare, and differentiate things both from each other and from itself. Consciousness arises not from representation alone but from comparison and differentiation. Therefore, he argues, the soul must be different from the body, for bodies cannot compare and differentiate things, although they can form material representations, which are specific dispositions that their parts assume in their general composition when they meet other bodies in space (§§.738–741). The soul, then, is not a compound thing (as the body is), but a simple thing.[22] Hence, it must possess a force from which all its changes derive, and in which its essence consists (§§.742–752): the force of "representing the world according to the position of its body in the world" (§.753, p. 415). This definition leads Wolff to the second theme of the chapter: the harmony between the body and the soul (§.760).

What Wolff presents, in fact, is not an empirical investigation, but an examination of existing theories (§§.761–780). For, since experience shows only the existence of the harmony between the body and the soul, but not the reason for this harmony, all we can do is to speculate on the basis of that empirical finding. Wolff considers three theories: natural influx, God's immediate interventions (occasional causes), and pre-established harmony.

The natural influx theory holds that *experience shows* the effect of the force of the body on the production of the soul's thoughts and, inversely, the effect of the force of the soul in the body's movements. Occasionalism states that God produces thoughts in the soul on every occasion there is a change in the body, but also movements in the body whenever a will occurs in the soul. Wolff rejects both theories for the same two reasons. First, they do not correspond to what experience actually reveals or to what the concepts of body and soul express. Second, they appeal to

[21] From the third edition (1725) onward, Wolff adds the following comment to this passage: "One sees, thus, that what was mentioned before about the soul based on experience is the touchstone of what is taught here about its nature and essence as well as those effects grounded in it. However, by no means what is taught here is a touchstone of what experience teaches us" (Wolff, 1751/2003, §.727, pp. 453–454). We will explain later why Wolff made such an addition.

[22] In Chap. 2 (on ontology), after presenting the concepts of simple and compound things (§§.75–80), Wolff explains that simple things cannot have any property of compound things, such as extension or spatial movement (§§.82–91), and thus must be substances that have a force that constitutes their essence and nature (§§.112–117).

phenomena that contradict the natural laws and Wolff's conception of nature (§§.761–764).

Wolff then comes to Leibniz's theory of pre-established harmony, which maintains that both the soul and the body have their own forces, by means of which they produce all of their changes autonomously and successively. It also assumes that the two chains of parallel processes are set in harmony by God at the Creation, so that they would be forever and naturally harmonized. Insofar as this does not contradict either experience or the concepts and laws already demonstrated in his system, Wolff considers this to be not only possible but also the best available theory.

Wolff sees two difficulties in this theory, however (§.781). First, how would it be possible for the body—through its movements and without any contribution of the soul—to act in the same rational way we observe in experience, expressing universal truths, making inferences, and discovering things? Second, how would it be possible to sustain human freedom, if the body has no understanding and all of its movements are necessarily determined by other bodily movements? One can see that, whereas the first difficulty is relative to the cognitive dimension of the soul, the second refers to its volitional faculty.

In order to solve these problems, Wolff approaches the third theme of the chapter: the grounding of the faculties of the soul in its representative force (§§.782–885). That the soul's faculties derive from its force is demonstrated by the fact that they possess the same three fundamental features of this force, namely, (a) representation of the world, (b) the limitation of representations by bodily conditions, and (c) the constant impulse to change that affects them. He concludes, then, that the faculties are simply different limitations of the one and the same representative force of the soul.

In this way, Wolff achieves two goals in his rational psychology: he demonstrates the ultimate reason for the faculties of the soul, and he answers the questions raised about pre-established harmony. In short, the body can speak and infer rationally because words are sounds, that is, material phenomena reproducible by the body as a simple machine, in correspondence with the properly rational occurrences in the soul (§§.835–845). Moreover, Wolff rejects the possible interference of the body in human freedom. He claims that both the necessary movements of the body and the free acts of the soul are chains of events that run parallel to each other, that are synchronized, and that do not interfere with each other. As the Creator of the world, God could have chosen differently. However, since it was possible for Him to choose as He did, it follows that pre-established harmony is possible (§§.883–885).

Finally, Wolff arrives at the fourth topic of rational psychology, which deals with different kinds of beings with representative force and their specific properties (§§.788–927). Leibniz's monads, for example, represent the world obscurely. Animal souls can represent the world clearly but not distinctly. Spirits in general are additionally endowed with understanding and freedom. The human soul belongs here. It has a representative force that represents things clearly and distinctly, and it possesses other properties, such as wisdom, personality, and immortality.

Wolff concludes his rational psychology by making room for natural theology (§§.928–1089), for God is also a spiritual being. However, since this topic does not belong in a chapter on psychology, we will leave it here aside.

3.4 Amendments to the *Deutsche Metaphysik*

Wolff's *Deutsche Metaphysik* became a bestseller. The novelty of its subject and the fact that it was written in German and intended for teaching purposes made it very popular at German universities. Many reviews appeared in journals, and less than 2 years after its first edition, a second and expanded edition came out (Wolff, 1722). This, however, is just part of the story. Beyond this intellectual dimension, social and institutional factors are relevant to the fate not only of the *Deutsche Metaphysik* but also of Wolff's career.

3.4.1 Wolff's Banishment from Halle

Wolff's philosophical work and teaching style had been attracting admirers and disciples for some time. However, it also caused some discomfort in Halle, especially among the pietists of the Theology department. For them, Wolff's emphasis on the precedence of reason over tradition, authority, and faith had a deterministic and atheistic tone. The publication of the *Deutsche Metaphysik* and other *Vernünfftige Gedancken* served to consolidate that impression, which intensified his personal and academic disagreements with his opponents.[23] To make matters worse, in 1721 Wolff presented his *Oratio de Sinarum Philosophia Practica* (Discourse on the practical philosophy of the Chinese). Here, he praised Chinese philosophy for grounding morality in reason rather than in revelation, which mirrored his own philosophy, including his emphasis on the role of empirical psychology in politics and ethics. This was taken as a personal affront by the pietists, thus giving rise to a dispute that would last for decades (Corr, 1983; Hettche & Dyck, 2019; Schwaiger, 2010).

In 1723, various publications by Wolff's opponents began to appear. They attacked several theses from the *Deutsche Metaphysik*, especially the cosmological notion of *nexus rerum* (connection of things) and his defense of pre-established harmony. Broadly speaking, the attacks revolved around a central accusation, namely, that his philosophy was a deterministic and Spinozistic system that led to atheism and the consequent subversion of morality and religion (École, 1983). Wolff defended his system extensively in several works that came to be known as his *Schutzschriften* (defense writings). Some of his disciples took part in these

---

[23] By 1720, Wolff had already published not only his *Deutsche Metaphysik*, but also his *Deutsche Logik* (1713) and his *Deutsche Ethik* (1720).

disputes as well, publishing various texts in defense of his system.[24] However, this clash had unanticipated consequences.

In November 1723, responding to insistent requests from Halle pietists, the Emperor of Prussia, Frederick William I (1688–1740), deposed Wolff (along with some of his colleagues and disciples) from his university position and banished him from Prussian territory under the threat of the death penalty.[25] Wolff left for the University of Marburg, where he would hold a post for the next 17 years. Opposition to his writings did not cease, however. In 1724, shortly after Wolff's expulsion from Halle, J. F. Budde (1667–1729), a church counselor and former Halle professor, began a new wave of attacks (Corr, 1983; Drechsler, 1997; École, 1983; Hettche & Dyck, 2019; Watkins, 1998).

In order to respond to these criticisms of his *Deutsche Metaphysik*, Wolff decided to publish the *Anmerckungen zur Deutschen Metaphysik* (Wolff, 1724).[26] He wanted to show that his ideas fostered the progress of truth and promoted religion and the practice of virtue without implying determinism and conflicting with revelation or custom. Here, we will focus on his annotations to psychology, more specifically on what we call the epistemic restriction of pre-established harmony.[27]

---

[24] For a general presentation of Wolff's disciples, see the contribution from Sonia Carboncini to this volume.

[25] The history of the conflict leading to Wolff's expulsion from Halle is a complex subject. It has generally been regarded as a series of theoretical disputes between Wolff and the Halle pietists. Some scholars even suggest that it was a consequence of a particular interpretation of pre-established harmony offered to the emperor (himself a pietist and militarist), according to which army deserters could not be punished if their acts had been pre-established by God (e.g., Drechsler, 1997; Watkins, 1998). However, a more recent study (Holloran, 2010) claims that Wolff's personal, political, and administrative disputes with his colleagues (notably with Lange) were the actual reason for his expulsion. This interpretation challenges the more romantic narrative of a struggle between religious intolerance and enlightened reason, in which Wolff appears as a champion of academic freedom and Enlightenment principles. Be that as it may, most authors agree that the emperor's reaction was not expected or intended by Wolff's opponents and that it had the side-effect of promoting an image of Wolff as a martyr of the Enlightenment, which gained him more advantages than disadvantages over the subsequent years. For more details about the controversies surrounding Wolff's career and his banishment from Halle, see Biller (2018), Kertscher (2018), Pečar et al. (2015), and the contribution from Ursula Goldenbaum to this volume.

[26] A more detailed review of the various writings published by Wolff's opponents, and Wolff's respective answers to them, can be found in École (1983) and Watkins (1998).

[27] In general, Wolff's amendments and additions to psychology consist in a series of conceptual clarifications that serve different purposes (Pereira, 2017), the most important of which is to explain the place and role of pre-established harmony in his psychological system. It is important to note that the structural division between empirical and rational psychology is not useful to a proper understanding of Wolff's *Anmerckungen*, since his conceptual clarifications and discussions about pre-established harmony are cross-disciplinary.

3.4.2 Wolff's Epistemic Restriction of Pre-established Harmony

In the *Deutsche Metaphysik*, crucial aspects of psychology are intertwined with discussions of the body-soul relationship and pre-established harmony. We have seen that Wolff connects the demonstration of how the faculties of the soul derive from its fundamental force (Wolff, 1720, §§.782–885) to the solution of the problems related to pre-established harmony (§.781, §§.835–845, §§.883–885). Wolff also says that "since we have proven that the soul has a force to represent for itself that which causes changes in its body, we must now investigate how it is possible that the soul and the body agree" (Wolff, 1720, pp. 416–417, §.760). In this way, Wolff seems to lend support to his critics, who understood his psychology as being dependent on pre-established harmony.[28]

This caused him many problems, especially charges of determinism, materialism, and atheism. Thus, one of his main goals with the additional notes to the *Deutsche Metaphysik* was to prove that neither psychology nor any discipline that receives principles from it (logic, ethics, and politics) depends on the acceptance of pre-established harmony. For him, even if accepted, the latter does not imply determinism or atheism.

First, Wolff emphasizes that his empirical psychology is based on what is given in experience and that the empirical (perceived) harmony between the soul and the body does not depend on any hypothetical explanation (theory) at all. Thus, the main psychological concepts and ideas formulated in empirical psychology do not depend on the acceptance of pre-established harmony, which is only one hypothesis among others. As long as no empirical data are denied or contradicted, one can choose whatever hypothesis one wants (Wolff, 1724, Ad §.539). The same goes for those disciplines that are built on psychological concepts (Wolff, 1724, Ad §.191, Ad §.527ff.).[29] As Wolff says,

> If one had to learn how to write, would it not be absurd to worry first about how the hand can move the feather? One can learn how to write without knowing this. Likewise in ethics with the exercise of virtue and the avoidance of vices. [...] One asks merely what the soul necessarily has as its motive so that it resolves itself to this or that movement of the bodily

[28] In the preface to the second edition of his *Deutsche Metaphysik*, Wolff recognizes that the rational psychology chapter could be understood as if it were "nothing but an explanation of pre-established harmony between the soul and the body" (Wolff, 1751/2003, p. 16).

[29] The characterization of those theories (natural influx, occasionalism, and pre-established harmony) as hypotheses is an important amendment to the *Deutsche Metaphysik*, which Wolff employed to answer his critics. For him, no fundamental truth can be derived from hypotheses. Thus, the truths of empirical psychology and practical philosophy do not depend on pre-established harmony. However, Corey Dyck (personal communication) argues that, rather than an amendment, Wolff's response may be seen merely as a clarification, representing no significant change in Wolff's early account. We thank him for presenting us with this interpretation. For a discussion of Wolff's notion of hypothesis, see Leduc (2017).

organs, instead of worrying about how the soul must have done to start this movement. (Wolff, 1724, Ad §.191, p. 105)[30]

Second, Wolff claims that the demonstrations of rational psychology are also independent of one's acceptance of pre-established harmony. For him, neither the essence of the soul as a force nor the thesis that its faculties originate in that essence depends on any explanation of the relationship between the body and the soul. He insists that deriving the soul's faculties from its force should not be confused with explaining how that force is determined, which is precisely the subject matter of occasionalism, natural influx, and pre-established harmony. According to occasionalism, the force of the soul is constantly determined by God; in natural influx, it is determined by the force of the body; and in pre-established harmony, it is determined by its very essence and nature, given by God. The important point, here, is that in each theory both the force of the soul and the faculties that derive from it remain the same (Wolff, 1724, Ad §.753).[31]

Wolff goes on to ask, rhetorically, why we should worry about those theories if they are irrelevant to practical life and to the establishment of psychological truths (empirical and rational)? He responds that it is because the empirical (perceived) harmony between the body and the soul generates great intellectual perplexity, which leads us to search for explanations for it. Materialism (only the body exists) is one possible explanation, but that is exactly what Wolff is trying to reject. The solution is therefore to demonstrate that the empirical harmony can be reasonably explained by other means. This is precisely the role of those competing dualistic theories (natural influx, occasionalism, and pre-established harmony). In this regard, Wolff admits that they are equally useful and that one can choose whichever one wants.[32]

Given these clarifications, it is clear that Wolff restricts the meaning and weight of pre-established harmony in his system. It is but one of the useful hypotheses on the relationship between the body and the soul, and it does not constitute the basis of his psychological theory. This is what we mean by the epistemic restriction of pre-established harmony.

However, it is worth remembering that Wolff still prefers it because (1) it better explains how the soul remains active and autonomous, while still limited to

[30] In the following passage, Wolff reinforces the irrelevance of those theories for practical life: "as if people's salvation and blessedness, as well as the Roman Empire's well-being, depended on the explanation of how body and soul interact with each other" (Wolff, 1724, Ad §.527ff, p. 232).

[31] In this attempt to separate the demonstration of the faculties of the soul from the explanation of the body-soul relationship, Wolff indicates his intention to develop this idea in a future Latin work (Wolff, 1724, Ad §.876). This is confirmed by his later statements that the *Psychologia rationalis* was organized in such a way that the independence of those topics is absolutely clear (Wolff, 1740/1983, Ad §.844 and Ad §.876). This shows that the development of Wolff's German psychology is important for understanding his Latin psychology.

[32] We can understand, then, why Wolff spends so much time discussing the advantages and disadvantages of these theories in a balanced manner (e.g., Wolff, 1724, Ad §.727, Ad §.760, §§.780–782, Ad §.842, Ad §.844). It allows him not only to avoid materialism but also to escape the accusation of having defended pre-established harmony to the detriment of other theories.

(regulated by) bodily states (Ad §.815, Ad §.818); (2) it better explains how the body can act rationally without interference from either the soul or God (Ad §.836ff); and (3) it allows him to reject both materialism and atheism, thus making room for revealed religion (Wolff, 1724, Ad §.760, Ad §.765).

3.5 Wolff's Final Statement of His Early Psychology

In 1726, Wolff published his last major philosophical work in German. It consisted of a synthesis and general explanation of his previous philosophical works: *Ausführliche Nachricht von seinen eigenen Schrifften, die er in deutscher Sprache heraus gegeben* (Detailed report on his own writings published in German). Only 2 years after the appearance of the *Ausführliche Nachricht*, Wolff would change the language of his writings from German to Latin, thus inaugurating a series of new philosophical treatises aimed at the European public at large (Arndt, 1973).

Broadly speaking, in the various chapters of this work, Wolff presents the reasons for his writing style and his use of the German language, as well as the main features of his doctrine and of each philosophical discipline, with their subjects and connections. It was not his goal to present substantial innovations. Nevertheless, there are relevant clarifications concerning his psychological thinking. Here, we will focus on the division of psychology into two parts.

First, Wolff describes how the metaphysical disciplines were arranged in the *Deutsche Metaphysik* and how psychological knowledge appeared there. He says,

> I have dealt with a part of *psychology* before *cosmology*. The reason for this is that I divide *psychology* into two parts. One deals with that which is known about the human soul from experience, whereas the other explains everything on the basis of the nature and essence of the soul, and gives the reasons for that which is observed. The first part I call *Psychologia empirica*, whereas the other *Psychologia rationalis*. *Psychologia empirica* is properly a history of the soul and can be known without the help of other disciplines.[33] By contrast, *Psychologia rationalis* presupposes a knowledge of *cosmology*. Therefore, if one wants to treat the disciplines in their specific order, *cosmology* comes right after *ontology*, and *psychology* after *cosmology*. [...] But since I have not treated the disciplines that belong to the chief science [metaphysics] in their specific order, I have placed one part of psychology, namely, the *empirica*, before *cosmology*, because it is easier than the latter and more attractive for beginners, eliminating their frustration with *ontology* caused by having to pay more attention to various things than they are used to. (Wolff, 1726, §.79, pp. 231–232, original italics)

Here, Wolff makes clear that the peculiar arrangement of metaphysical disciplines in the *Deutsche Metaphysik* was driven by the same pedagogical motivation underlying it. Separating psychology into two parts and placing the empirical one before cosmology facilitates the study of metaphysics because empirical psychology is easier to grasp than cosmology. He acknowledges, however, that there are

[33] On the meaning and the problems aroused by Wolff's reference to empirical psychology as a history of the soul, see Corr (1975) and Pereira (2017).

other possible arrangements for presenting metaphysics, as two of his disciples, L. P. Thümmig (1697–1728) and G. B. Bilfinger (1693–1750)[34] did in their own works (Bilfinger, 1725; Thümmig, 1725).

Most important of all, though, is the fact that Wolff makes his first direct reference here to the separation of psychology into two major parts and uses for the first time the terms that will become the hallmark of his psychological system: empirical psychology and rational psychology.[35] Moreover, he offers another reason for that separation:

> Since what one knows of the soul through experience comprises important truths, these become highly valued: not only the rules of logic—according to which the understanding is guided to know the truth—but also the rules of ethics, after which man's will is moved toward good and away from evil, find their proofs here. One should not accept as the foundation of such important disciplines anything except truths that one can convince oneself of immediately. Now, what one knows of the soul through infallible experiences has such a nature that, by paying attention to oneself or to others, one can be immediately convinced of it. (Wolff, 1726, §.89, p. 252)

Now, the reason is not pedagogical but epistemic. The truths conveyed by empirical psychology are based on common and infallible experiences. As soon as one pays attention to oneself and to others, those truths are immediately recognizable. Thus, they are fundamental for other disciplines, such as logic and ethics. This is not the whole story, however. There is still another aspect to be considered here, as Wolff reveals:

> Although I see now that in the other part [rational psychology] I have also shown the correct reason for that which we perceive of the soul through careful attention, not everyone is used to paying so much attention required to grasp the foundation of truth. Besides, my enterprise is something new, and people are not yet used to it. However, everyone knows that truth is always contradicted when it is new. (Wolff, 1726, §.89, pp. 252–253)

It is not only that empirical psychology is easier than cosmology and facilitates the study of metaphysics, or that it serves as the foundation of logic and ethics. It is also that, for Wolff, the secure truths of empirical psychology should not be blended with these more controversial ones found in rational psychology, however trustworthy they may be. In sum, the general value of psychology should not be challenged by the novel and controversial character of rational psychology. Hence the need for the separation. We can call this the pragmatic argument for Wolff's division of psychology.

---

[34] Wolff makes several references to Thümmig and Bilfinger throughout this work. He stresses that their work exemplifies those principles presented by him in the *Deutsche Metaphysic* and *Anmerckungen* (Wolff, 1726, §§.90–103). In the Latin writings, he adopted many features displayed in the works of Thümmig and Bilfinger. It is possible that Wolff owes more to them than is generally acknowledged in the literature. See the contribution from Sonia Carboncini to this volume.

[35] It should be noted that Wolff recognizes (Wolff, 1726, §.90) the prior use of those same terms by Thümmig (1725). It seems, then, that Thümmig was the first to name those two parts of psychology.

Finally, it is important to notice that, notwithstanding his particular cautiousness with regard to rational psychology, Wolff is rather optimistic about it in the *Ausführliche Nachricht*. He stresses its originality in deriving all the faculties of the soul and its rules from a single concept of the soul, which in his view opens the way to a new and comprehensive explanation of the soul's activities and contributes to humanity's life and other theoretical demonstrations (Wolff, 1726, §.102). Even possible errors committed in this discipline, he adds, would not imply any harm to truths in general, since they would only consist of false and disposable reasons for true and lasting experiences. Moreover, its truths are also correctable, as all hypotheses are, so that nothing would prevent this science from advancing progressively toward the truth (Wolff, 1726, §.104). Thus, although rational psychology is subject to error by its very nature and evokes adverse reactions from more conservative scholars, Wolff recognizes its relevance and the value of its existence alongside empirical psychology.

3.6 Conclusion

Our analysis allows us to draw three main conclusions. First, Wolff's German writings contain not only the roots but also central ideas of his psychological program, which he would subsequently develop. A comprehensive account of Wolff's psychology (and philosophy in general) should not disregard them. Second, Wolff's intellectual development is closely related to the vagaries of his professional life. The rise of his psychology, the way it was published, its influence, its consequences, and its changes over the years represent a fine illustration of that relationship. Third, one cannot apprehend the scope and meaning of Wolff's psychology without considering both its historical and philosophical aspects.

The secondary literature has frequently overlooked these points, either by ignoring the German writings or by restricting itself to one-sided analyses (either philosophical or historical). We hope to have demonstrated that an integrated analysis of Wolff's works, particularly regarding his German phase, opens up new ways for understanding them. In addition, the historiography of psychology has largely failed to notice Wolff's relevance to the development of psychological science, to the extent that it has ignored Wolff's German writings and their impact on his disciples and opponents.

Here, we have offered but a start. Many topics remain to be addressed by future research, such as the precise relationship between Wolff's German and Latin psychological writings. Be that as it may, we hope that this chapter contributes to a better understanding of Wolff's intellectual development regarding psychology.

References

Albrecht, M. (2018). Wolff an den deutschsprachigen Universitäten. In R. Theis & A. Aichele (Eds.), *Handbuch Christian Wolff* (pp. 418–465). Springer.

Araujo, S. F. (2012). O lugar de Christian Wolff na história da psicologia. *Universitas Psychologica, 11*(3), 1013–1024.

Araujo, S. F., & Pereira, T. C. R. (2014). La idea de psicología racional en la *Metafísica Alemana* (1720) de Christian Wolff. *Universitas Psychologica, 13*(5), 15–26.

Arnaud, T. (2002). Le critère du métaphysique chez Wolff: Pourquoi une Psychologie empirique au sein de la métaphysique? *Archives de Philosophie, 65*(1), 35–46.

Arnaud, T. (2003). Dans quelle mesure l'ontologie est-elle fondamentale dans la métaphysique allemande de Wolff? *Revue Philosophique de la France et de L'étranger, 128*(3), 323–336.

Arndt, H. W. (1973). Einleitung [Ausführliche Nachricht]. In H. W. Arndt (Ed.), *Christian Wolff: Gesammelte Werke*. (I. Abt., Bd. 9) (pp. v–xxii). Olms.

Bilfinger, G. B. (1725). *Dilucidationes Philosophicae de Deo, Anima Humana, Mundo, et Generalibus Rerum Affectionibus*. Cotta.

Biller, G. (2018). Biographie und Bibliographie. In R. Theis & A. Aichele (Eds.), *Handbuch Christian Wolff* (pp. 5–31). Springer.

Blackwell, R. J. (1961). Christian Wolff's doctrine of the soul. *Journal of the History of Ideas, 22*(3), 339–354.

Campo, M. (1980). Cristiano Wolff e il razionalismo precritico. In J. École et al. (Eds.), *Christian Wolff: Gesammelte Werke* (III. Abt., Bd. 9). Olms. (Original work published 1939).

Corr, C. A. (1975). Christian Wolff's distinction between empirical and rational psychology. *Studia Leibnitiana Supplementa, 14*, 195–215.

Corr, C. A. (1983). Introduction [Anmerckungen zur Deutschen Metaphysik]. In C. A. Corr (Ed.), *Christian Wolff: Gesammelte Werke*. (I. Abt., Bd. 3) (pp. 1–25). Olms.

Drechsler, W. (1997). Christian Wolff: A biographical essay. *European Journal of Law and Economics, 4*, 111–128.

Dyck, C. W. (2014). *Kant and rational psychology*. Oxford University Press.

Dyck, C. (2019). *Early modern German philosophy (1690–1750)*. Oxford University Press.

École, J. (1983). Introduction de l'editeur [Opuscula Metaphysica]. In J. École (Ed.), *Christian Wolff: Gesammelte Werke*. (II. Abt., Bd. 9) (pp. V–LXXXIV). Olms.

Euler, W. (2004). Bewuβtsein—Seele—Geist. Untersuchungen zur Transformation des Cartesischen "Cogito" in der Psychologie Christian Wolffs. In O.-P. Rudolph & J.-F. Goubet (Eds.), *Die Psychologie Christian Wolffs. Systematische und historische Untersuchungen* (pp. 11–50). Max Niemeyer.

Feuerhahn, W. (2002). Comment la psychologie empirique est-elle née? *Archives de Philosophie, 65*(1), 47–64.

Goubet, J.-F. (2018). Rationale Psychologie. In R. Theis & A. Aichele (Eds.), *Handbuch Christian Wolff* (pp. 153–174). Springer.

Gundlach, H. (2006). Psychology as science and as discipline: The case of Germany. *Physis, 43*(1–2), 61–89.

Hammerstein, N. (1983). Christian Wolff und die Universitäten. Zur Wirkungsgeschichte des Wolffianismus im 18. Jahrhundert. In W. Schneiders (Ed.), *Christian Wolff 1679–1754. Interpretationen zu seiner Philosophie und deren Wirkung* (pp. 267–277). Meiner.

Hatfield, G. (1995). Remaking the science of mind: Psychology as natural science. In C. Fox, R. Porter, & R. Wokler (Eds.), *Inventing human sciences: Eighteenth-century domains* (pp. 184–231). University of California Press.

Hettche, M., & Dyck, C. (2019). Christian Wolff. In E. N. Zalta (Ed.), *The Stanford encyclopedia of philosophy*. (Winter 2019 Edition). Retrieved July 20, 2020, from https://plato.stanford.edu/archives/win2019/entries/wolff-christian

Holloran, J. (2010). Wolff in Halle—Banishment and return. In J. Stolzenberg & O.-P. Rudolph (Eds.), *Christian Wolff: Gesammelte Werke*. (III. Abt., Bd. 105) (pp. 365–375). Olms.

Kertscher, H. J. (2018). *"Er brachte Licht und Ordnung in die Welt". Christian Wolff—Eine Biographie*. Mitteldeutscher Verlag.

Leduc, C. (2017). Wolff on hypothesis. In A. Pelletier (Ed.), *Wolffiana VI: Christian Wolff's German Logik—Sources, significance, and reception* (pp. 77–97). Olms.

Marcolungo, F. L. (Ed.). (2007a). *Wolffiana III: Christian Wolff tra psicologia empirica e psicologia razionale*. Olms.

Marcolungo, F. L. (2007b). Christian Wolff e il progetto di una psicologia filosófica. In F. Marcolungo (Ed.), *Wolffiana III: Christian Wolff tra psicologia empirica e psicologia razionale* (pp. 15–34). Olms.

Paccioni, J.-P. (2001). Wolff et la constitution d'une science psychologique. *Annales Doctorales de l'Université de Dijon, 4*, 67–85.

Pečar, A., Zaunstöck, H., & Müller-Bahlke, T. (Eds.). (2015). *Die Causa Christian Wolff. Ein epochemachender Skandal und seine Hintergründe*. Franckesche Stiftungen.

Pereira, T. C. R. (2017). *A unidade da psicologia no pensamento de Christian Wolff*. PhD Dissertaton, Universidade Federal de Juiz de Fora, Juiz de Fora.

Pereira, T. C. R., & Araujo, S. F. (2015). A psicologia empírica na *Metafísica Alemã* (1720) de Christian Wolff. In S. F. Araujo & F. Caropreso (Eds.), *Temas Atuais em História e Filosofia da Psicologia* (pp. 13–42). Ed. UFJF.

Rumore, P. (2018). Empirical psychology. In R. Theis & A. Aichele (Eds.), *Handbuch Christian Wolff* (pp. 175–196). Springer.

Schwaiger, C. (2010). Wolff, Christian (von). In H. F. Klemme & M. Kuehn (Eds.), *The dictionary of eighteenth-century German philosophers* (pp. 1290–1297). Continuum.

Smith, R. (1997). *The Norton History of the Human Sciences*. Norton & Company.

Sturm, T. (2009). *Kant und die Wissenschaften vom Menschen*. Mentis.

Thümmig, L. P. (1725). *Institutiones Philosophiae Wolfianae in Usus Academicos Adornatae*. Renger.

Vidal, F. (2006). *Les sciences de l'âme: XVIéme–XVIIIéme siècle*. Honoré Champion.

Vittadello, A. M. (1973). Expérience et raison dans la psychologie de Christian Wolff. *Revue Philosophique de Louvain, 71*, 488–511.

Watkins, E. (1998). From pre-established harmony to physical influx: Leibniz's reception in eighteenth century Germany. *Perspectives on Science, 6*(1–2), 136–203.

Wolff, C. (1718). *Ratio Praelectionum Wolfianarum in Mathesin et Philosophiam Universam*. Renger

Wolff, C. (1720). *Vernünfftige Gedancken von Gott, der Welt und der Seele des Menschen, auch allen Dingen überhaut, den Liebhabern der Wahrheit mitgetheilt*. Renger.

Wolff, C. (1722). *Vernünfftige Gedancken von Gott, der Welt und der Seele des Menschen, auch allen Dingen überhaut, den Liebhabern der Wahrheit mitgetheilt* (2nd ed.). Renger.

Wolff, C. (1724). *Anmerckungen über die vernünfftige Gedancken von Gott, der Welt und der Seele des Menschen, auch allen Dingen überhaut, zu besserem Verstande und bequemerem Gebrauch derselben heraus gegeben von Christian Wolffen*. Andrea.

Wolff, C. (1726). *Ausführliche Nachricht von seinen eigenen Schrifften, die er in deutscher Sprache von den verschiedenen Theilen der Welt-Weißheit heraus gegeben, auf Verlangen ans Licht gestellet*. Andrea und Hort.

Wolff, C. (1968). Psychologia Empirica, methodo scientifica pertractata, qua ea, quae de anima humana indubia experientiae fide constant, continentur et ad solidam universae philosophiae practicae ac theologiae naturalis tractationem via sternitur. In J. École (Ed.), *Christian Wolff: Gesammelte Werke* (Psychologia Empirica) (Vol. 5, Pt. II). Olms. (Original work published 1738).

Wolff, C. (1981). Auflösung einiger Schwierigkeiten, welche bei der menschlichen Seele vorkommen. In *Christian Wolff: Gesammelte Werke* (Gesammelte kleine philosophische Schriften) (Vol. 21.2, Pt. II, pp. 169–186). Olms. (Original work published 1737).

Wolff, C. (1983). Der vernünfftigen Gedancken von Gott, der Welt und der Seele des Menschen, auch allen Dingen überhaupt, anderer Theil, bestehend in ausführlichen Anmerckungen, und zu besserem Verstande und bequemerem Gebrauche derselben heraus gegeben. In C. A. Corr (Ed.), *Christian Wolff: Gesammelte Werke* (Anmerkungen zur Deutschen Metaphysik) (Vol. 3, Pt. I). Olms. (Original work published 1740).

Wolff, C. (2003). Vernünfftige Gedancken von Gott, der Welt und der Seele des Menschen, auch allen Dingen überhaupt. In C. Corr (Ed.), *Christian Wolff: Gesammelte Werke* (Deutsche Metaphysik) (Vol. 2, Pt. I). Olms. (Original work published 1751).

Chapter 4
Empirical Psychology: Between Reason and Experience

Ferdinando Luigi Marcolungo

4.1 Introduction

The two great Latin treatises of *Empirical Psychology* (1732) and *Rational Psychology* (1734) constitute an essential reference in understanding the articulation of Christian Wolff's entire metaphysics. The formulation itself of both titles is unanimously recognised as a precise reference to his thought, although it does not often result in a correct comprehension of it. In fact, the judgment expressed by Kant (1724–1804) in *Transcendental Dialectics* still prevails in the history of philosophy books: rational psychology, which develops from "the sole text" of the "I think" (Kant, 1787/1911, p. 264), should, in fact, from his point of view, leave aside any empirical reference which must be excluded, as a matter of principle, from an investigation that operates rigorously *a priori*. However, Kant's interpretation contains a misunderstanding that arises from the opposition between rationalism and empiricism, a contrast which is instead denied by the close link between the two great Latin treatises by Wolff; this is proof of the close connection between reason and experience that distinguishes his entire thought.

Here we would like to resume the discussion a little further, in order to better understand the purpose of the title *Empirical Psychology*, which designates the first of the two great Latin treatises dedicated to philosophical psychology. With the term "philosophical psychology", we intend to encompass both treatises, believing that they jointly constitute a unitary study able to make us better understand the perspective of the author (Marcolungo, 2005, 2007). It is no coincidence that the titles of both texts express the scientific character that unites the two volumes, which intend to follow a strictly scientific method—*methodo scientifica pertractata* (treated according to the scientific method). The distinction between the two moments is part of a unitary framework, in which the empirical reference, on the one hand, and the *a priori* deduction starting from the essence of the soul, on the other hand, contributes to the same objective differently.

F. L. Marcolungo (✉)
University of Verona, Verona, Italy
e-mail: ferdinando.marcolungo@univr.it

S. d. F. Araujo et al. (eds.), *The Force of an Idea*, Studies in History and Philosophy of Science 50, https://doi.org/10.1007/978-3-030-74435-9_4

To the lack of awareness of the mutual reference between the two Latin treatises, it should be added that many scholars share the belief of a substantial unity of the entire path of Wolffian thought, with the result of no longer being able to perceive its developments and at the same time the specificity of the several moments. The titles of the two great Latin treatises are in fact mostly used also to indicate the corresponding sections of *German Metaphysics* (1719) respectively the third and fifth chapters. This leads to neglecting the specificity of the path taken by Wolff in the decade that separates that work from the beginning of the series of Latin treatises, but also the reasons that led to the specific use of the title *Empirical Psychology*, upon which we now intend to focus firstly. Considering we are now accustomed to its connotation in today's philosophical language, we may no longer realise the once negative meaning that the term "empirical" had at the beginning of the eighteenth century. Wolff himself, as we will see, appears to reluctantly favour this term, only after criticism arose following his teaching of Halle.

Secondly, it will become necessary to discuss the relationship between "historical knowledge", a term that refers to everything we can acquire through experience, "philosophical knowledge" and "mathematical knowledge", referring to the epistemological model suggested in the *Preliminary Discourse* placed at the beginning of *Latin Logic* (1728). It is here that Wolff systematically proposes the "marriage between reason and experience", which can be particularly relevant when distinguishing between Empirical Psychology and Rational Psychology. In this regard, it will later prove useful to refer back to the two precautionary methods previously defined in *Ratio praelectionum* (Lessons plan, 1718), which recommends applying the principle of reduction and to avoid falling into the *vitium subreptionis* (surreptitious error) in order to guarantee the correct relationship between experience, philosophical explanation and the use of mathematics.

Thirdly, we will illustrate the relationship between *a posteriori* and *a priori*, referring to the great Latin logic, in order to concretely assess how the intertwining of reason and experience is presented. On the one hand, we shall remember the importance of *intuitive judgements* while, on the other hand, we shall highlight the relevance of conceptual elaboration and the use of hypothesis, necessary in deepening knowledge acquired through experience.

The example of astronomy shall programmatically return in the fourth section in order to clarify the relationship between Empirical Psychology and Rational Psychology, referring to the *Prolegomena* of the two great Latin treatises. Once again, it will become clear that the marriage between reason and experience is fundamental, also, to the mutual reference between the two disciplines, which are, according to Wolff, part of a unitary project.

4.2 The Reasons for a Title

The first point that must never be forgotten is that the term "empirical" certainly did not enjoy a good reputation during the eighteenth century, as has been witnessed since the previous century—suffice to consider the seventh objections and answers in Descartes' *Metaphysical Meditations* (1642). In the eighth volume of Zedler's *Grosses Universal-Lexicon* (Great Universal Lexicon), published in 1734, under the entry "*Empirici*", we find a significant view of the mostly negative meaning of this term amongst scholars: if one intends to reach real knowledge only through experience, one risks taking a long path and obtaining an uncertain outcome, even if it is wise to combine theory with experience, using the latter as a concrete touchstone. Under the entry, the ancient distinction between "empirical" and "dogmatic doctor" is also found and it is specified that it was not then a question of "sects" or "philosophical schools", opposed to each other, given that there were no real "dogmatic" doctors, or reference authors recognised by everyone: "Today, instead, the empirical term is used for those who in medicine boast of their discoveries and neglect the principles of Hippocrates and Galen" (Zedler, 1731–1754, vol. 8, column 1042).

Under the entry "*Erfahrung*", in the same volume (Zedler, 1731–1754, vol. 8, columns 1596–1597), the corresponding Latin term "*Experientia*" can be found and it is outlined that it represents the sum of what we obtain in an immediate way through the senses, as we say for example of any well accomplished doctor who has experienced the benefits and application of a medicine for certain diseases. Shortly after, under the entry "*Experimentum*" (Zedler, 1731–1754, vol. 8, columns 2344–2345), a neologism taken from Latin, next to which the German term "*Versuch*" is located, with a meaningful exchange between the two languages compared to what had happened for the German word "*Erfahrung*", it is specified that it is an experience about things that are produced through our concern: in this sense, the experiment is opposed to the simple observation that relates to what nature spontaneously offers. In science, we are concerned with providing the causes for which something happens in a certain way and not otherwise. We cannot know it *a priori* as well as in pure mathematics. However, through knowledge of the effects and the annotation of what happens in their production, we have to determine what produces a certain effect. In this way, Zedler's *Lexicon* clarifies the difference between bare experience and experiment, on the basis of what Wolff had already observed in *German Metaphysics* (Wolff, 1751/1983a, §.325). This difference is decisive in understanding the specific meaning of the expression "Empirical Psychology" and its link with the scientific method mentioned in the subtitle of the great Latin work of 1732.

In this regard, it is once again useful to refer to Zedler's *Lexicon* and consider the entry "*Seelen-Lehre*" (Zedler, 1731–1754, vol. 36, columns 1168–1169), a term now commonplace, to which the German neologism "*Psychologie*" is added, together with the Latin "*Psychologia*", as evidence of the increasing presence of the new term in the early eighteenth century. The word, in the 36th volume of 1743, has always referred to the Wolffian distinction between Empirical Psychology (which deals with the soul based on what we know through experience) and Rational Psychology (which instead aims to clarify everything starting from the nature and essence of the soul).

Referring to the two disciplines, however, Zedler proposes two German terms, while the Latin meanings are shown in parentheses, almost confirming their meaning. On the one hand, the term "*Seelen-Geschichte*" relates to Empirical Psychology, referencing the historical knowledge on which Wolff had focussed not only at the beginning of the *German Logic* (1713), but also, above all, in the articulated *Preliminary Discourse* in the great *Latin Logic*. On the other hand, Rational Psychology is presented as a *Seelen-Wissenschaft*, as if only the latter could be understood as a strictly scientific discipline, while the former would be limited to the level of mere experience, which, according to the conclusion of the *Preliminary Discourse*, could be called "bare knowledge of the fact [*nuda facti notitia*]" (Wolff, 1740/1983b, §.7, p. 3).[1]

However, the distinction immediately refers to the strategic role of Empirical Psychology within the overall framework of the system. In fact, from the experience of what happens in our own soul we derive important truths useful not only to logic, but also to ethics, given that what is known about the soul through certain experiences is such that one remains convinced when considering himself or anyone else. Furthermore—Zedler reminds—Wolff says that the task he pursues in Rational Psychology is "new and unusual" (Paccioni, 2004; Goubet, 2018), but at the same time he reminds us that in Empirical Psychology we can already find the correct foundation of what we perceive of the soul through careful reflection (Zedler, 1731–1754, vol. 36, column 1168; Feuerhahn, 2002; Paccioni, 2006). Significantly, the entry of Zedler's *Lexicon* then refers to the text of Thümmig's *Institutions* (Thümmig, 1725/1982), rather than to the great Latin treatises of Wolff, which were certainly well known at the time when the 36th volume of the *Lexicon* appeared in 1743. After specifying the main contents of the two parts, the conclusion is that, through these empirical researches, "the changes that take place in the soul are clarified in a rational way, similar to what happens in physics, due to the changes that occur in nature" (Zedler, 1731–1754, vol. 36, column 1169).

The reference to Thümmig's *Institutions* is not accidental. Although the term "empirical", which points to the first area of psychology, refers back to the teachings of the master—"The philosopher [Wolff] divides psychology into two parts, one of which he calls empirical, the other rational" (Thümmig, 1725/1982, p. 115)—the fact remains that the expression "Empirical Psychology" appears for the first time in his *Institutions of Wolffian Philosophy*, published in early 1725.

Shortly after being dismissed from Halle (in November 1723), at the beginning of 1724, Wolff published the *Annotations to the German Metaphysics*, in which he stressed that what he presented about the soul in the third chapter of the German work had been obtained from experience, a concept which linked to the rational considerations suggested in the fifth chapter (Wolff, 1740/1983c, §.55). What he stated in the latter chapter about the pre-established harmony was to be accepted as a philosophical hypothesis, without obscuring the results of the previous research, even if the challenge, which he had first attempted, to provide an explanation stemming from the very concept of the soul was certainly not to be overlooked for its importance (Wolff, 1740/1983c, §.261).

[1] For the translations of *Preliminary Discourse,* see Blackwell (1963).

In the *Annotations* there is no indication of the two distinct disciplines: Empirical Psychology and Rational Psychology. Two years later, in the *Ausführliche Nachricht* (Detailed Report) of 1726, Wolff could not fail to face the problem which arose from the distinction officially introduced by Thümmig's *Institutions* the previous year. Later, when examining cosmology (which, today, is presented as the second branch of metaphysics), Wolff firstly pointed out that this change of order between Cosmology and Empirical Psychology, compared to the order he followed in *German Metaphysics*, had been introduced by Thümmig (Wolff, 1733/1973, §.79); but this happened for a reason, he confessed, since, as he specified, "cosmology can make us understand the nature of the soul, in which the world ideally exists" (Thümmig, 1725/1982, p. 71). At the same time, however, he reminded us that his path had been different, given that a first chapter of psychology had been placed immediately after ontology and before cosmology. Wolff specified here, for the first time, that he used "Empirical Psychology" to refer to this first investigation, which deals with what we know about the soul through experience, while the other expression, "Rational Psychology", stands for the discipline which shows the foundation of what we observe based on the nature and essence of the soul. Significantly, again referring to the distinction between historical knowledge and philosophical knowledge, already introduced in the *Prolegomena* of the *German Logic*, he spoke here of Empirical Psychology as a real "history of the soul [*eine Historie von der Seele*]" (Wolff, 1733/1973, §.79, p. 231), which can be approached independently of any other discipline; whereas Rational Psychology presupposes cosmology and for this reason both Thümmig and Bilfinger had rightly placed the latter immediately after ontology and before psychology (Wolff, 1733/1973, §.79). And yet, in some way, Wolff intended to reaffirm his own initial view in this regard, since his purpose was not to deal with the individual disciplines separately, but rather to show a path that would progressively lead to further study of the problems. This is why Empirical Psychology in *German Metaphysics* had been placed before cosmology, precisely "because it was easier and more accessible to beginners" (Wolff, 1733/1973, §.79, p. 232).

Apart from the fact that in 1726 Wolff seems to have accepted the distinction between the two disciplines, it does not fail at the same time to stress their close connection, which corresponds to a precise epistemological model. A little further on, in fact, he returns to clarify that he has never presented in Rational Psychology anything other than statements that had previously been established through experience in Empirical Psychology, precisely because the intent was still to show *a priori*, based on principles and in a rational way, that which conforms to experience. He additionally reminds us of the difference between doctrines and hypotheses, as in the field of physics. Even in metaphysics one can employ hypotheses, which are nevertheless an indispensable tool in the gradual acquisition of the truth, but they "should not be confused with what is obtained beyond any doubt by examining experience" (Wolff, 1733/1973, §.104, pp. 291–292).

The great *Preliminary Discourse* placed at the beginning of *Latin Logic* (1728) then further clarifies that the two moments, the empirical and the rational, must be understood in close connection, as part of a twofold process. By categorising the different disciplines which hold equal weight within philosophical knowledge, we find in fact the close link between Empirical Psychology and experimental physics reaffirmed: quite differently from the mere historical knowledge that would be limited to *nuda facti notitia*, Wolff strongly claims the "philosophical" and "scientific" character of Empirical Psychology, understood as "science capable of establishing principles through experience, so as to account for what happens in the human soul" (Wolff, 1740/1983b, §.111, p. 51). Indeed, it does not just keep track of what is observed, but, based on observation, it creates the notions of the faculties and habits and formulates many other principles which can give reason to various aspects: "which is the task of philosophical knowledge and cannot refer at all to historical knowledge only" (Wolff, 1740/1983b, §.111, p. 51; see also Arnaud, 2002).

Compared to the negative meaning that the term "empirical" could still have in the language of that time (which also emerges from the distinction between historical, philosophical, and mathematical knowledge in the great *Preliminary Discourse* at the beginning of the *Latin Logic*), "Empirical Psychology" acquires now a strictly philosophical depth that underlines the role it plays in the articulation of Wolffian thought. This role confirms, as we will see, the idea of the centrality of philosophical psychology within the system, a centrality that also passes in some way from the interaction between Empirical Psychology and Rational Psychology, within the framework of an overall vision in which the so-called "experimental philosophy" occupies a leading role.

4.3 Between Historical Knowledge, Philosophical Knowledge and Mathematical Knowledge

In order to understand the epistemological framework outlined in the *Preliminary Discourse* (*Latin Logic*), it is useful to go back to the programmatic declarations proposed by Wolff in the *Ratio Praelectionum* of 1718, which, once again, give us precise proof of the centrality of Empirical Psychology. In fact, Wolff sums up the meaning of his own teaching as follows:

> There are two things that I felt I had to worry about first, that is (1) that my metaphysical doctrines were free from any school bond and (2) that could be made through them reason for the changes in the mind in the same way that one makes reason through the physical principles of the changes that occur in the bodies. (Wolff, 1735/1972a, p. 146)

The *Preface* to *Empirical Psychology* refers precisely to this statement, claiming the autonomy of the research with respect to the various metaphysical hypotheses about the relationship between soul and body:

> If so far the philosophers have not managed to give a complete picture of the virtues and customs, it is because they have not yet noticed how through the laws followed by the faculties of the soul, those same actions are subjected to our decision [...]; thus the individual human actions are explained through constant laws in an intelligible way and deduced a priori from the nature of the soul, just as in physics today we explain the actions of bodies and the changes that derive from them in the world. (Wolff, 1738/1968, *Preface*, pp. 13*–14*)

Thus, the task of Empirical Psychology becomes essential as part of an overall framework marked by a sort of parallelism between the soul and the body. Here two precise notes concerning methodological questions (which had already been developed in the *Ratio Praelectionum*) recur: the first refers to the so-called "reduction principle," while the second relates to the so-called *vitium subreptionis*, which leads to an undue transition from experience to principles.

The first principle, indicated in 1718 as "a great analytical principle", can be applied especially to mathematics, as it allows us to "reduce the unknown to the known" (Wolff, 1735/1972a, p. 23). This principle, when applied, in particular, to the field of algebra and symbolic knowledge, as well as to infinitesimal calculus, is quite effective in acquiring a deeper knowledge of the soul: operations of the mind represent a continuum of which we are conscious (Wolff, 1735/1972a, p. 99). In the *Empirical Psychology*, Wolff further highlights that this term indicates that artifice that allows us to broaden our knowledge by explaining something by means of something else, so that we can apply what we already know to the new situation by virtue of the notion common to both cases (Wolff, 1738/1968, §.472). This principle is useful in the art of discovery (*Ars inveniendi*) and finds application not only in mathematics, but also in physics and morality (Wolff, 1738/1968, §.473).

The second methodological note constitutes a precautionary rule, which is necessary when we examine experience with the aim of clarifying what it actually offers us. This is a detail that we could define phenomenological. Falling into the *vitium subreptionis*, as specified in the *Preliminary Discourse* of 1728, means not following the necessary order of demonstration, attributing to a given subject a predicate that goes beyond what had previously been demonstrated (Wolff, 1740/1983b, §.130). Later, in the *Latin Logic*, this vice is once again highlighted in reference to the relationship between the soul and the body: sometimes there is a risk of taking as a datum of experience something that is not of this nature, such as the alleged "physical influx" of the soul on the body, or the attracting force of the magnet or the feelings of love and hate in inanimate things (Wolff, 1740/1983d, §.668). Previously, in the *Ratio Praelectionum*, he stated that the study of mathematics can help us accurately evaluate our experiences, so as to avoid falling into the *vitium subreptionis*, that is, to take as a result of experience propositions with no evidence. This often occurs, causing serious damage to the sciences (Wolff, 1735/1972a, p. 7).

These two notes, the first positive (the principle of reduction) and the other negative (the *vitium subreptionis*), in accordance with Wolff, demonstrate that there is a close connection between historical knowledge, philosophical knowledge and mathematical knowledge. In the *Preliminary Discourse*, this link also plays an important role in relation to the great Latin works that were to follow shortly thereafter. If, on the one hand, the distinction seems to date back to the very beginning of *German Logic* (Wolff, 1754/1965, p. 115), it should not be forgotten that it now additionally assumes crucial importance with regard to the epistemological model that Wolff explicitly claimed a desire to pursue. Unsurprisingly, the distinction between the various branches of philosophy refers, in the third chapter of the *Preliminary Discourse*, not only to the distinction between the different areas of knowledge, mentioned in the first chapter (Wolff, 1740/1983b, §§.1–28), but also to the definition of philosophy as a science of the possible as possible, to which the second chapter is dedicated (Wolff, 1740/1983b, §§.29–54).

Here the reference to the possible plays an essential role, as specified in the definition of psychology dealing with the soul, as "science of those things which are possible through human souls" (Wolff, 1740/1983b, §.58, pp. 29–30). However, the expression does not temporarily clarify what this reference means. This becomes clearer later, relating also to the distinction between the two areas of psychology: the empirical and the rational. Nevertheless, this distinction appears only towards the end of the third chapter, after having clarified the role of physics in metaphysics and the sense of experimental philosophy (Wolff, 1740/1983b, §§.94–95, §§.106–110). It is only at this point that Wolff introduces the definition of Empirical Psychology, as "the science of experientially establishing the principles from which the reason is given for those things which occur in the human soul" (Wolff, 1740/1983b, §.111, pp. 50–51). It is precisely this criterion of the possible that has to guarantee the intertwining of experience, philosophy and mathematics, overcoming any reductionist approach and at the same time ensuring further developments in our knowledge. Experience does not consist only of mere sense, but is also enriched with further insights related to our ability to reflect.

This is evident from the beginning when one considers that sort of parallelism witnessed by a twofold contribution, that of the senses, through which "we know things which are and occur in the material world", and that of the mind, that "is conscious of the changes which occur within itself": to be convinced of it, "let one merely direct one's attention to one's self" (Wolff, 1740/1983b, §.1, p. 1). Having observed that, Wolff points out that it is not a question of establishing the limits of these sources of knowledge, since it is enough for us to know that we cannot doubt what we acquire through the senses and self-awareness (Wolff, 1740/1983b, §.2). Historical knowledge, which examines "those things which are and occur either in the material world or in immaterial substances", also shows us that "things which are or occur possess a reason from which it is understood why they are or occur", as we can understand as long as we pay the necessary attention to our experiences (Wolff, 1740/1983b, §§.3–4, p. 2). Now, the knowledge of this reason is matter of philosophical knowledge, without the need to specify its limits at that given moment: in any case, it is a question of seeking the reasons for what occurs, as when one wants to clarify, for example, the relationship between the speed of the waters and the inclination of the riverbed, or that between the tendency towards an object and the perception we have of it (Wolff, 1740/1983b, §.6).

Only after having clarified the particular features of philosophical knowledge, he focusses on historical knowledge, specifying the difference between the "bare knowledge of the fact" (Wolff, 1740/1983b, §.7, p. 3) and the search for the reason of that same fact, on the basis of the Aristotelian distinction between "know-that" and "know-why" (Aristotle, 1997, I, 1, 981a28–981a30). Shortly afterwards, however, he reaffirms the link between the two types of knowledge, since experience allows us to specify some elements from which we can find the reason for what is and occurs, or can occur, and in this way offers a foundation for philosophical knowledge, although this does not yet provide a completely clear explanation (Wolff, 1740/1983b, §.10). For this reason, historical knowledge is an indispensable premise of philosophical knowledge: the two types of knowledge must be constantly combined, to the point that, as Wolff says immediately after, in every part of philosophy, the marriage (*connubium*) between reason and experience must be kept firm (Wolff, 1740/1983b, §.12; 1738/1968, §.497).

Next to philosophical knowledge, which seeks reason for what is or occurs, it is necessary to place mathematical knowledge, based on the quantitative dimension that distinguishes every finite reality, which, as such, can increase or decrease (Wolff, 1740/1983b, §.13). In the note, in accordance with the parallelism between the world and the mind, Wolff once again suggests, beside physical examples (such as the inclination of the sun during the year, the speed of a stream or the movement of the planets), many examples drawn from the human world: "attention in different men differs by degrees", and, similarly, there are "various degrees of virtue and vices for a diversity of subjects" that we can take into consideration (Wolff, 1740/1983b, §.13, p. 6). Mathematical knowledge is intertwined with previous disciplines: historical knowledge and philosophical knowledge can both offer a foundation for mathematical knowledge, as is evident from Galilei's (1564–1642) example and from the role that experiments play in science (Wolff, 1740/1983b, §.18). This is because, in order to discover the secrets of nature, no element should be overlooked, as Newton's (1642–1726) *Optics* (1704) reminds. The relationship between the different kinds of knowledge is clear in every field of experimental philosophy, not only in astronomy, but also in psychology and moral philosophy (Wolff, 1740/1983b, §.20).[2]

As mentioned above, the distinction between Empirical Psychology and Rational Psychology is proposed in the third chapter of the *Preliminary Discourse* dedicated to the areas of philosophy, after a thorough examination of the links between logic, ontology and psychology (an item that cannot be developed here). It is no coincidence, however, that this distinction follows almost immediately the introduction of experimental philosophy within the more general field of physics, that is, of science that "gives the reason of those things which can occur through bodies" (Wolff, 1740/1983b, §.107, p. 48). Now, it is precisely the relationship between the various fields of knowledge that can explain the new discipline that is experimental physics, defined as "the science of experimentally establishing the principles from which the reason can be given for what occurs in the nature of things" (Wolff, 1740/1983b, §.107, p. 48). In a note, Wolff specifies that this epistemological model can be applied to every field of philosophical knowledge, so that he is not afraid to introduce the expression "experimental theology" (Wolff, 1740/1983b, §.107, p. 48), as well as to discuss the possibility of experiments in the field of morality and politics (Wolff, 1740/1983b, §.107).

---

[2] See the Psychometrics project Wolff, 1738/1968, §.522; see also Feuerhahn (2004) and Rumore (2018).

In a similar way, this occurs in the field of psychology, where principles of such great importance should first be established on the basis of experience and arranged in such a way that, building upon these principles, as in experimental physics, everything that follows is justified. In this sense, Empirical Psychology is defined as a science in which "experience establishes the principles from which the reason can be given for those things which occur in the human soul" (Wolff, 1740/1983b, §.111, p. 51). As Wolff expressly points out, it is not therefore merely a historical discipline, but rather a philosophical knowledge in the strictest sense, because, despite stemming from experience, the task is nevertheless to clarify the faculties and principles necessary to explain what is observed in the soul (Wolff, 1740/1983b, §.111).

4.4 A Posteriori and A Priori

For a complete understanding of the reasons behind the relationship between historical, philosophical and mathematical knowledge, it is useful to consider what Wolff wrote in the great *Latin Logic* about how, building first on our experience, we can refer back to the principles via those *intuitive judgments* that are based on particular experiences (Wolff, 1740/1983d, §§.669–709). Here the fundamental difference of mere empiricism emerges, considering the representative character of our every notion. As Wolff states from the very beginning, with regard to the three operations of the mind (Wolff, 1740/1983d, §§.30–58), through the senses we perceive external things that affect our sensory organs, but, at the same time, our mind is aware of itself and conscious, if not of everything, at least of something that passes in itself, and is therefore able to perceive itself with a sort of "internal sense" (Wolff, 1740/1983d, §§.30–31, pp. 125–126). Now, the contribution that is provided by the senses is not limited to actual perception, because we are also able to reproduce, by means of imagination, the things that are currently absent, as well as to recall something that had previously been perceived together with the content of the actual perception. In the note, Wolff states that this "law of the imagination" (Wolff, 1740/1983d, §.32, p. 126), on which he then focusses in the context of psychology, is of utmost utility also in the field of logic. When it becomes clear and familiar, we will be able to investigate the reason why we now imagine different contents. The principles that are obtained *a posteriori* must in fact be confirmed and inculcated by means of examples (Wolff, 1740/1983d, §.32).

The structure of cognitive intentionality emerges from the presence of two intrinsically connected moments: the first, simple apprehension (*simplex apprehensio*); the other, intellectual elaboration of the notion or idea. In the first case, the attention turns to the current content of the senses or to the image that represents it (Wolff, 1740/1983d, §.33); in the second, it is the representation of things in the mind that attracts our attention, whether it refers to the reality outside of us, or to what happens within ourselves (Wolff, 1740/1983d, §.34). The two moments are linked to each other, since if we did not have a notion or an idea, we would not even be able to notice a content. Once again, our self-awareness (namely, the ability to be aware not only of the contents present, but also of ourselves) plays a key role: in this sense, the notion consists in the act itself of representing the content offered by apprehension (Wolff, 1740/1983d, §.35).

Later, in the second section of *Latin Logic* (dedicated to the application of logic in the different fields of knowledge) and, in particular, in the second section, Wolff focusses on the usefulness of logic in the search for truth. After the first chapter, dedicated to experience in general, in the second chapter he addresses the possibility of forming intuitive judgments and *a posteriori* notions, before concluding in the third chapter with discursive judgments and *a priori* definitions. Experience is acquired by way of knowledge that can be obtained from our perceptions: if we are aware of ourselves, we know that we are unable to give consent to things that are contradictory to each other, such as, for example, accepting at the same time that it is raining and it is not raining (Wolff, 1740/1983d, §.664). Now, while it is true that there is no experience if not in particular cases, we can derive knowledge of the universal through our experience; it is therefore possible to prove *a posteriori* knowledge of the universal by obtaining it from experience (Wolff, 1740/1983d, §.665). At the same time, paying attention to the content of the experience protects us from the *vitium subreptionis* in which we can fall by claiming to derive from experience what the experience is actually unable to offer (Wolff, 1740/1983d, §.668).

Proposing guidance rules for formulating intuitive judgments and notions *a posteriori*, Wolff devotes much time to specifying not only the different levels of our perceptions, on the basis of which we can have more or less distinct notions, but also the different features of those notes from which they can be derived, such as attributes or modes. Here observation plays a key role, because it can make us able or unable to distinguish a particular object and to formulate its definition (Wolff, 1740/1983d, §§.677–679). It is necessary to keep in mind the complexity of the phenomena and to articulate the analysis by paying attention to the elements at stake:

> The same method must also be applied to the soul. In fact, even if the different faculties are not determined by each other like the parts that contribute to form a material object, however, although considered individually, various elements can be distinguished that either exist together, or derive from each other. [...] Precisely because we are not warned of everything that happens in our mind, it happens very often that certain aspects that cannot be missing from a certain notion escape our attention. (Wolff, 1740/1983d, §.682, pp. 495–496)

The use of microscopes, as well as telescopes, has allowed us to expand our observations; in the same way, the search for the causes of the different phenomena has become decisive in the development of our knowledge. In this case, it is necessary to avoid mistaking the concomitance of phenomena with a relationship of mutual dependence: that is what happens, Wolff observes, in the hypothesis of the physical influence between soul and body, since perceptions in our minds are always connected with changes in the sensory organs. Even doctors sometimes pass off as experiences certain claims that are contrary to this rule, thus endangering the foundation of medicine (Wolff, 1740/1983d, §.702). Equally, it is necessary to carefully record our observations, in order to allow for comparison with previous experiences and to better define the universal notions from which they were derived. This is important not only in physics, but also in morality and psychology (Wolff, 1740/1983d, §.709).

In the next chapter, Wolff goes on to clarify the opposite path, which leads to discursive judgements and *a priori* definitions. The two paths had previously been indicated as moments of a single process, aimed at developing our knowledge either directly from experience, or through reason, building on existing knowledge (Wolff, 1740/1983d, §.663). The *a priori* is a second-level operation, which somehow presupposes a comparison with experience: this is confirmed by the indications suggested by Wolff. From the notions of species we can hark back to the notions of genus and vice versa, on the basis of the differences that arise. In this sense, universal notions can be obtained both by reflecting either on what has been perceived, or by abstracting what is common to several notions, or, finally, by exchanging with others the determinations that are variable (Wolff, 1740/1983d, §.716). Moreover, in the note, Wolff further defines this triplet of operations, now designated by the terms of "reflection", "abstraction" and "arbitrary determination", observing that the first two are now consolidated, while the last is still to be studied, although already known in the geometric field (Wolff, 1740/1983d, §.716). Descartes himself, reasoning about the idea of God, did not realise it, while he should have noticed that proceeding in this way can give rise to deceptive notions. Through the arbitrary determination of features that should belong to a notion, it is possible to compound something that involves a contradiction (Wolff, 1740/1983d, §.716). Once again, the criterion of possibility plays a fundamental role, ensuring the advancement of our knowledge, since building on notions that we obtain by means of reflection or abstraction, unlike those formed in an arbitrary way, are possible and true. In the same way, we must work on the search for the possible causes of a phenomenon (Wolff, 1740/1983d, §§.717–718).

We could summarise the link between *a priori* and *a posteriori* with the passage at the end of the *Latin Logic* concerning the method by which we can study this discipline. In astronomy and optics, observations and demonstrations are combined together, putting to work both kinds of knowledge, *a posteriori* and *a priori*. In all experimental philosophy, we thus find the example of the marriage of experience and reason which is precisely specified by logic, which clarifies the possibility of intuitive judgments (*a posteriori*) and the ways in which (*a priori*) we can develop our knowledge from those judgements (Wolff, 1740/1983d, §.1232).

4.5 Between Empirical Psychology and Rational Psychology

The two great Latin treatises, *Empirical Psychology* and *Rational Psychology*, are closely linked by virtue of the union of experience and reason, in the reciprocal reference between *a posteriori* and *a priori*, going beyond any dichotomy, typical of both rationalism and empiricism. Wolff programmatically reminds us of this in the note to the paragraph of *Empirical Psychology* dedicated to this topic:

> Of this marriage I hold the greatest account in the whole field of philosophy, precisely because it is not only decisive for the certainty of our knowledge, but also it is quite useful to the progress of the sciences. And this is the reason why even in this part of psychology, in which we focus on the *a posteriori* knowledge of the soul, we always seek reason's assistance, as if we were looking for *a priori* knowledge. And the same approach is maintained in experimental philosophy. (Wolff, 1738/1968, §.497, p. 379)

In the *Prolegomena* to *Empirical Psychology*, after having clarified the epistemological model to follow, he programmatically anticipates this intrinsic link between both kinds of research: "Thus the best thing is for one constantly to join the study of rational psychology with that of empirical psychology, even though we have considered it wise to treat them separately" (Wolff, 1738/1968, §.5, pp. 4–5).[3]

Indeed, Empirical Psychology does not just provide the mere account of experience, but follows a precise method in order to establish the basic propositions that will serve as a starting point for further research. And this is its very task. What has previously been said about the principle of reduction and the *vitium subreptionis* becomes fundamental here, since the scientific order of research is the necessary guarantee for achieving positive results. In the *Preface*, this is already emphasised as follows:

> Even if we do not propose in Empirical Psychology anything but what is evident on the basis of a sure experience and what everyone can experiment in himself, to the condition that one has developed his own skills so as to make them suitable for the exercise that it is required here, nevertheless, following the rules of our method, we have arranged each statement in a precise order, so as to demonstrate each other and what follows from what precedes. And we considered this aspect necessary in order to make evident that the individual propositions are duly determined and that the definitions are sufficient for what has to be demonstrated and therefore they are both useful principles for reasoning and can be applied fruitfully to the other disciplines. (Wolff, 1738/1968, *Preface*, p. 17*)

This task, as is clear, takes on a strictly philosophical value and is the indispensable premise for rational psychology itself, which is possible precisely because from these determinations one can begin an *a priori* in-depth analysis:

> The only way to form any essential concept and to derive *a priori* from it whatever we establish *a posteriori* in empirical psychology is to carefully examine those aspects of the soul observed in empirical psychology and, taking into account the distinct notions here developed, determine which of them can be demonstrated by others. (Wolff, 1740/1972b, §.5, p. 4)

In the *Prolegomena* of *Empirical Psychology*, Wolff highlights the need to pay attention to what occurs in our own soul: "We come to know the subjects dealt with in empirical psychology by attending to those occurrences in our souls of which we are conscious" (Wolff, 1738/1968, §.2, p. 2). The reference to experience appears to be closely connected with the possibility of reaching determined notions, since this is the only way to shift from the particular case to universal notions, drawing on evidence from ontology and its notions, to which Wolff, in the essay appearing in

[3] For the translations of the *Prolegomena*, both of *Rational* and *Empirical psychology*, see Richards (1980).

the *Horae subsecivae* (Hours of Leisure, 1729), refers to as "guiding notions", given that they are able to guide our research on the level of experience (Wolff, 1738/1968, §.3; see Wolff, 1729/1983e).

The close connection between *Empirical Psychology* and *Rational Psychology* thus appears clear. Indeed, the first can be called upon to provide principles to rational psychology, as in the field of experimental physics, which supplies principles for dogmatic physics (Wolff, 1738/1968, §.4; see Wolff, 1740/1972b, §.3). If the reference to experience can be the indispensable premise for any subsequent theoretical investigation; however, the role of control that this reference plays with respect to the theories that have been elaborated should not be forgotten:

> Empirical Psychology aims at controlling and confirming discoveries made *a priori* concerning the human soul. [...] Here again, Empirical Psychology is similar to experimental physics: for we use experiments—either directly or by deducing something from them—in order to examine the tenets of dogmatic physics. (Wolff, 1738/1968, §.5, pp. 3–4)

This controlling role had already been stressed in a 1725 addition to *German Metaphysics*, at the beginning of the fifth chapter dedicated to the essence of the soul:

> What has been previously stated about the soul starting from experience is a touchstone for what is taught here about its nature and essence, and on the acts that are founded in them, but in no way what is taught here is the touchstone for what experience teaches. (Wolff, 1751/1983a, § 727, pp. 453–454)

However, the epistemological model proved to be much more complex in 1732, since rational psychology not only derived the principles, found evidence and a touchstone in Empirical Psychology, but also played, in turn, an important role in the field of empirical investigation:

> Rational psychology contains those matters which we know *a priori* about the soul. Rational psychology obviously contributes to the progress of the empirical psychology, while borrowing principles from the latter: it returns with interest what it has borrowed. [...] Truths deduced *a priori* warn us about what we should observe and what otherwise would escape our notice. (Wolff, 1738/1968, §.5, p. 4)

The link between theory and empirical observation finds a fundamental paradigm in the model of astronomy, on the basis of the search for the possible which represents the characteristic feature of the entire Wolffian thought:

> Here the psychologist imitates the astronomer, who derives theory from observations and corroborates theory through observations, and who, by the aid of theory, is led to observations which he otherwise might not make. And thus the demonstrations of rational psychology suggest what ought to be considered in empirical psychology. And wherever empirical psychology is established and rational psychology cultivated, we are enriched by many principles which otherwise would never have been found. (Wolff, 1738/1968, §.5, pp. 4–5)[4]

What had been anticipated in *Empirical Psychology* was then confirmed in the *Prolegomena* of *Rational Psychology*, which underlines the importance of

---

[4] Referring to astronomy appears decisive for the definition of philosophy as a science of the possible as much as possible (see Wolff, 1735/1972a, pp. 107–108; 1729/1983f).

theoretical reflection in delving further into what experience offers us: "What is taught in Empirical Psychology is more completely and properly understood through Rational Psychology" (Wolff, 1740/1972b, §.7, p. 6). It is a matter of "increasing the acumen in observing what occurs in our soul" (§.8, p. 7), as we can all draw on our own personal experience, similarly to occurrences in the field of astronomy:

> If we experience within ourselves the content of a proposition, we will be convinced about its truth. Those who are experts in astronomy or who have investigated experimental philosophy through our method will apply what they have experienced to the field of psychology when they consider the universal aspect of a notion. (Wolff, 1740/1972b, §.8, p. 7)

Thus, there is a sort of virtuous cycle to Empirical Psychology and Rational Psychology, a relationship which obeys a precise scientific logic, the aforementioned art of discovery (*Ars inveniendi*) to which the entire path of Wolff's philosophy aims:

> Thus, since through the art of discovery unknown propositions are derived from already known propositions, the soul learns things which cannot be disclosed *a posteriori*. [...] Since our acumen for observing what occurs in our soul is increased through rational psychology, it can happen that those things that rational psychology found *a priori*, because they have become evident, are now easier to find through the observation; for we find easier to observe what has become evident than what is completely unknown. (Wolff, 1740/1972b, §.9, pp. 7–8)

In all this, the position of philosophical psychology remains central, beyond the distinction between Empirical Psychology and Rational Psychology: this distinction discerns two different moments of a single path, where the reference to experience is fundamental, without however limiting itself to a mere empiricism. Unsurprisingly, after the *Prolegomena*, Wolff takes up the starting point of *German Metaphysics* in *Empirical Psychology*, indicating the centrality of our own mind in the reconstruction of the whole reality (Wolff, 1738/1968, §.11, p. 9; see Wolff, 1751/1983a, §.1; Arnaud, 2004). Every other metaphysical truth finds its foundation and its reference point in the certainty of the awareness of ourselves and of our existence.

4.6 Conclusion

To sum up our inquiry, we can reaffirm some fundamental conclusions, at the same time leaving some useful questions open for further in-depth analysis. Taking into consideration the negative meaning of the term "empirical" at the beginning of the eighteenth century, the title of the two great Latin treatises written by Wolff on Empirical Psychology and Rational Psychology is a challenge in itself, since it stresses the importance of the marriage between experience and reason in each and every field of our knowledge. It is no coincidence that the reference to "experimental philosophy" seems key, it being part of an epistemological model in which the search of the possible materialises in the mutual exchange between *a posteriori* and *a priori*, following occurrences in the field of astronomy involving Galilei, one of the authors that Kant mentioned in the *Preface* to the second edition of the *Critique* (Kant, 1787/1911, p. 10). Differently to Kant, however, Wolff's epistemological model seems to favour a comparison with concrete experience, which serves as the cornerstone for all later developments. This does not, however, signifies that everything can limit itself to pure empiricism, due to the existing relationship between sensations and reflection, and between empirical content and rational elaboration.

Wolff's development of Empirical Psychology goes beyond the criticisms of his rationalisms, the controversies surrounding the pre-established harmony and the relationship between body and soul. Despite the debate Descartes left to us in his works, we must not ignore new notions that link the relationship between experience and reason and represent the strength of an updated epistemology in the field of human sciences. This epistemology relies on a quantitative contribution, without neglecting its qualitative counterpart and the importance of the great philosophical questions.

The issue concerning the bond between the great Latin treatises and literature in German language remains, since the former seems to linguistically favour the relationship with the great Aristotelic-scholastic tradition, while it is the German texts that have greater influence over later developments, above all in terms of philosophical lexicon.

References

Aristotle. (1997). Metaphysics. In W. D. Ross (Ed.), *Aristotle's metaphysics*. Clarendon Press.

Arnaud, T. (2002). Le critère du métaphysique chez Wolff. Pourquoi une Psychologie empirique au sein de la métaphysique? *Archives de Philosophie, 65*(1), 35–46.

Arnaud, T. (2004). Où commence la "Métaphysique allemande" de Christian Wolff? In O.-P. Rudolph & J.-F. Goubet (Eds.), *Die Psychologie Christian Wolffs. Systematische und historische Untersuchungen* (pp. 61–73). Niemeyer.

Blackwell, R. J. (1963). *Preliminary discourse on philosophy in general. Christian Wolff.* Bobbs-Merrill.

Feuerhahn, W. (2002). Comment la psychologie empirique est-elle née? *Archives de Philosophie, 65*(1), 47–64.

Feuerhahn, W. (2004). Die Wolffsche Psychometrie. In O.-P. Rudolph & J.-F. Goubet (Eds.), *Die Psychologie Christian Wolffs. Systematische und historische Untersuchungen* (pp. 227–236). Niemeyer.

Goubet, J.-F. (2018). Rationale Psychologie. In R. Theis & A. Aichele (Eds.), *Handbuch Christian Wolff* (pp. 153–174). Springer.

Kant, I. (1911). Kritik der reinen Vernunft. In Königlich Preussische Akademie der Wissenschaften (Ed.), *Kant's gesammelte Werke (Band III)* (2nd ed.). Reimer. (Original work published 1787).

Marcolungo, F. L. (2005). Tra fenomenologia e metafisica: Il compito della psicologia filosofica in Christian Wolff. In L. Cataldi Madonna (Ed.), *Macht und Bescheidenheit der Vernunft. Beiträge zur Philosophie Christian Wolffs. Gedenkband für Hans Werner Arndt* (pp. 151–169). Olms.

Marcolungo, F. L. (2007). Christian Wolff e il progetto di una psicologia filosofica. In F. L. Marcolungo (Ed.), *Christian Wolff tra psicologia empirica e psicologia razionale* (pp. 15–34). Olms.

Paccioni, J.-P. (2004). Wolff est-il "le vrai inventeur de la psychologie rationelle"? L'expérience, l'existence actuelle et la rationnalité dans le projet wolffien de la psychologie. In O.-P. Rudolph & J.-F. Goubet (Eds.), *Die Psychologie Christian Wolffs. Systematische und historische Untersuchungen* (pp. 75–97). Niemeyer.

Paccioni, J.-P. (2006). Psychologie empirique et psychologie rationelle: L'activité d'une raison non pure. In J.-P. Paccioni (Ed.), *Cet esprit de profondeur. Christian Wolff, l'ontologie et la métaphysique* (pp. 159–178). Vrin.

Richards, R. J. (1980). Christian Wolff's prolegomena to empirical and rational psychology: Translation and commentary. *Proceedings of the American Philosophical Society, 124*(3), 227–239.

Rumore, P. (2018). Empirical psychology. In R. Theis & A. Aichele (Eds.), *Handbuch Christian Wolff* (pp. 175–196). Springer.

Thümmig, L. P. (1982). Institutiones Philosophiae Wolfianae in usus academicos adornatae. In J. École (Ed.), *Christian Wolff: Gesammelte Werke* (Vol. 19.1, Pt. III). Olms. (Original work published 1725).

Wolff, C. (1965). Vernünftige Gedanken von den Kräften des menschlichen Verstandes und ihrem richtigen Gebrauche in Erkenntnis der Wahrheit. In H. W. Arndt (Ed.), *Christian Wolff: Gesammelte Werke* (Deutsche Logik) (Vol. 1, Pt. I, 14th ed.). Olms. (Original work published 1754).

Wolff, C. (1968). Psychologia Empirica, methodo scientifica pertractata, qua ea, quae de anima humana indubia experientiae fide constant, continentur et ad solidam universae philosophiae practicae ac theologiae naturalis tractationem via sternitur. In J. École (Ed.), *Christian Wolff: Gesammelte Werke* (Psychologia Empirica) (Vol. 5, Pt. II, 2nd ed.). Olms. (Original work published 1738).

Wolff, C. (1972a). Ratio praelectionum in mathesin et philosophiam universam. In J. École (Ed.), *Christian Wolff: Gesammelte Werke* (Ratio praelectionum) (Vol. 36, Pt. II, 2nd ed.). Olms. (Original work published 1735).

Wolff, C. (1972b). Psychologia Rationalis, methodo scientifica pertractata, qua ea, quae de anima humana indubia experientiae innotescunt, per essentiam et naturam animae explicantur et ad intimiorem naturae ejusque autoris cognitionem profutura proponuntur. In J. École (Ed.), *Christian Wolff: Gesammelte Werke* (Psychologia rationalis) (Vol. 6, Pt. II, 2nd ed.). Olms. (Original work published 1740).

Wolff, C. (1973). Ausführliche Nachricht von seinen eigenen Schrifften, die er in deutsche Sprache heraus gegeben. In H. W. Arndt (Ed.), *Christian Wolff: Gesammelte Werke* (Ausführliche Nachricht) (Vol. 9, Pt. I, 2nd ed.). Olms. (Original work published 1733).

Wolff, C. (1983a). Vernünfftige Gedancken von Gott, der Welt und der Seele des Menschen, auch allen Dingen überhaupt. In C. A. Corr (Ed.), *Christian Wolff: Gesammelte Werke* (Deutsche Metaphysik) (Vol. 2, Pt. I, 5th ed.). Olms. (Original work published 1751).

Wolff, C. (1983b). Philosophia rationalis sive Logica, methodo scientifica pertractata et ad usum scientiarum atque vitae aptata. Praemittitur Discursus praeliminaris de philosophia in genere. In J. École (Ed.), *Christian Wolff: Gesammelte Werke* (Discursus praemlinaris) (Vol. 1.1, Pt. II, 3rd ed.). Olms. (Original work published 1740).

Wolff, C. (1983c). Der Vernünfftigen Gedancken von Gott, der Welt und der Seele des Menschen, auch allen Dingen überhaupt, anderer Teil, bestehend in ausführlichen Anmerckungen. In C. A. Corr (Ed.), *Christian Wolff: Gesammelte Werke* (Anmerkungen zur deutschen Metaphysik) (Vol. 3, Pt. I, 4th ed.). Olms. (Original work published 1740).

Wolff, C. (1983d). Philosophia rationalis sive Logica, methodo scientifica pertractata et ad usum scientiarum atque vitae aptata. Praemittitur Discursus praeliminaris de philosophia in genere. In J. École (Ed.), *Christian Wolff: Gesammelte Werke* (Philosophia ratinalis sive Logica) (Vol. 1.2–3, Pt. II, 3rd ed.). Olms. (Original work published 1740).

Wolff, C. (1983e). De notionibus directricibus. In J. École (Ed.), *Christian Wolff: Gesammelte Werke* (Horae subsecivae marburgenses) (Vol. 34.1–3, Pt. II, pp. 310–350). Olms. (Original work published 1729).

Wolff, C. (1983f). De medico astronomos imitante. In J. École (Ed.), *Christian Wolff: Gesammelte Werke* (Horae subsecivae marburgenses) (Vol. 34.1–3, Pt. II, pp. 154–175). Olms. (Original work published 1729).

Zedler, J. H. (1731–1754). *Grosses vollständiges Universal-Lexicon aller Wissenschafften und Künste* (64 vols.). Zedler.

Chapter 5
Wolff and the Dogmas of Classical Rationalism

Corey W. Dyck

5.1 Introduction

Philosophical rationalism might be most generally characterized in terms of an intellectual attitude where the deliverances of common sense and ordinary experience are rigorously interrogated and simple or brute, inexplicable facts are rejected.[1] Historically speaking, in the seventeenth and eighteenth centuries this "rationalist impulse" was expressed in the form of endorsing any or all of the following claims: the doctrine of innate ideas (rather than the adventitious character of all ideas), the view that certain rather than merely probable knowledge is the aim of inquiry, the prioritizing of reason (or some form of intellectual intuition) over experience and sensation as a way of obtaining knowledge, and acceptance of the principle of sufficient reason (hereafter PSR).[2] So, we find Descartes (1596–1650) and Leibniz (1646–1716) defending innate ideas (the latter in explicitly dispositional terms, the former suggesting as much at one point); Spinoza (1632–1677) and Leibniz emphasizing certain rather than merely probable knowledge (the former in taking our knowledge to be of necessary connections within a single substance and the latter in taking even contingent truths to be analytic) and endorsing the PSR (with the latter explicitly elevating it to the status of the first metaphysical principle); and finally, with all three contending that foundational truths (in some cases disclosed by means of direct intellectual insight, such as the *cogito*) serve as the basis for a deductively connected system of knowledge.

---

[1] See Nelson (2005), and Della Rocca (2010).

[2] The first three claims are singled out explicitly in Dea et al. (2018). For a compatible presentation, see Huenemann (2008). Della Rocca (2003) emphasizes PSR as a central rationalist doctrine. Anderson (2015) provides a compatible characterization in the context of "rationalist ideal of knowledge from concepts alone" (p. 79).

C. W. Dyck (✉)
Western University, London, ON, Canada
e-mail: cdyck5@uwo.ca

S. d. F. Araujo et al. (eds.), *The Force of an Idea*, Studies in History and Philosophy of Science 50, https://doi.org/10.1007/978-3-030-74435-9_5

Some of course have taken issue with the accuracy and utility of this otherwise neat classification,[3] but whatever its applicability to the classical rationalist thinkers, it is nonetheless often thought that one thinker who unquestionably meets this description is the "*arch*-rationalistic" thinker Christian Wolff.[4] Indeed, Wolff appears to uphold *all* of these rationalist dogmas[5] and in some cases to an extent not met with in his esteemed predecessors. In §.306 of his *Anmerkungen zur Deutschen Metaphysik* (Remarks on the German Metaphysics), for instance, he pledges himself to the party of the Platonists and claims that "the soul already has its *ideas* or concepts in it" (Wolff, 1740/1983a, §.306, p. 511—my emphasis), which he claims he holds on the basis of the Leibnizian understanding of the soul as having a power through which it brings forth its own alterations. Likewise, Wolff notoriously champions the use of the mathematical method in all areas of philosophical investigation; thus, in domains such as metaphysics (including natural theology) but also ethics and other practical disciplines, Wolff contends that the conclusions reached through the use of demonstrations will be as certain as the axioms from which they proceed (Wolff, 1751/1983b, §.9). More generally, Wolff distinguishes between "two paths to truth" (§.372), namely reason and experience, and points out that the path of experience cannot yield properly philosophical cognition which consists in cognition of the ground or reason for why something is or takes place (Wolff, 1754/1965a, *Vorbericht*, §.6). Lastly, Wolff notoriously seems to outdo even Leibniz in his support for the principle of sufficient reason by attempting to prove it on the basis of the principle of contradiction (Wolff, 1751/1983b, §.30).

Despite his reputation for outdoing his predecessors with his rationalist commitments, however, Wolff's position on these classical rationalist dogmas is in fact more nuanced, and in what follows I will argue that Wolffian rationalism is distinguished by its *moderate* rather than its radical character. To this end, I will consider Wolff's views on each of these issues in turn and will endeavour to illustrate what I take to be the moderate character of Wolffian rationalism through examples drawn primarily from his rational psychology (which is likewise often characterized as a *purely* rationalistic enterprise). In the first section, I consider Wolff's treatment of the doctrine of innate ideas, which already amounts to a topic in rational psychology, and argue that Wolff's endorsement of this doctrine is qualified by the central role played by sensation in his discussion of the origin of the soul's thoughts, as well as by his hesitation in embracing the Leibnizian pre-established harmony. In the second section, I turn to Wolff's account of the utility of probable opinions as a supplement to the mathematical method in his treatment of "philosophical

[3] Anstey (2005), for instance, has argued for the rejection of the rationalism-empiricism distinction in favour of a distinction between speculative and experimental methodologies.

[4] See Gardner (1999) for such a characterization, and Vanzo helpfully catalogues a number of other instances of similarly immoderate characterizations of Wolffian rationalism (2015, note 8).

[5] I have adopted the term "dogmas" for these historical core contentions of philosophical rationalism (following the lead of Allison, 2005) for the sake of convenience and by way of reflecting their importance in the historical (and historiographical) debate, and so not in order to suggest that they were only uncritically adopted by these thinkers.

hypotheses." In the third section, I consider the role that experience plays within the mathematical method itself, a role that it likewise plays at key junctures in Wolff's *rational* psychology. Finally, I turn to Wolff's treatment of the PSR and show, first, that the PSR plays an important but limited role within the mathematical method itself, but also that Wolff's attempted derivation of the PSR from the principle of contradiction should not be taken as implying that all truths, particularly those known from experience, are ultimately conceptual.

5.2 Innate Ideas

Wolff was well aware of the controversy surrounding the doctrine of innate ideas in his own day. He was not only directly familiar with Locke's (1632–1704) *Essay* (and Locke's philosophical writings more generally) and its powerful attack on innate ideas and principles in its first book,[6] but was also acquainted with parts of Leibniz's defence of the doctrine against Locke's attacks. Of course, Wolff was not familiar with Leibniz's full-dress defence of (a dispositional account of) innate ideas presented in his *New Essays* which was only published in 1765. Yet he was aware of Leibniz's comparison of ideas in the mind to the shape of Hercules in the marble at the conclusion of the *Meditationes* (Meditations) of 1684 (see Leibniz, 1684/1989, p. 27), and more importantly he was familiar with Leibniz's initial draft of a response to Locke's *Essay* that was transmitted to Locke in 1697 and eventually printed in *Some Familiar Letters between Mr. Locke and Several of his Friends* in 1708, which Wolff reviewed in the following year for *Acta eruditorum*, in the issue of October 1711. In his response, Leibniz expresses sympathy with Locke's attack, given the evident abuse of the doctrine of innate ideas by the Cartesians, but contends that Locke goes too far in rejecting the innateness of some primitive principles, and even if he thinks that the problem of the origin of our ideas is not of first importance in philosophy, he confesses that he "finds something solid in what Plato calls *reminiscence*" (Locke, 1708, p. 157, original italics) as opposed to the Aristotelian *tabula rasa*. In his review, Wolff supplies a detailed summary (and Latin translation) of Leibniz's letter, which he thinks was unfairly dismissed by Locke and Molyneux (1656–1698) (Wolff, 1711/2001, pp. 592–594).

It seems likely then that Wolff himself endorses the doctrine of innate ideas, and his few published pronouncements on the topic seem to confirm this. So, in §.819 of the *German Metaphysics*, Wolff contends that the soul "unfolds [the images and concepts of corporeal things] from out of its essence" (Wolff, 1751/1983b, §.819), which position he proceeds to contrast with that of Locke and Aristotle (§.820), and this is followed with an express endorsement that "the soul has its ideas or concepts already within it" (§.306). However, Wolff's position on innate ideas is in fact more

[6] As Wolff writes in the first *Vorrede* (preface) to the *German Logic*: "In our land one takes Locke's work on human understanding for uncommonly sensible throughout, so that even those who despise all of his countrymen nonetheless hold him in high esteem" (Wolff, 1754/1965a, p. 107).

nuanced than initially thought, and I will show that this is in fact the case, though we will begin by considering a slightly different topic, namely Wolff's argument that the soul's nature consists in a power for representing the world as, rather surprisingly, the soul's capacity for *sensations* is the hinge on which this argument turns. Wolff's derivation of this claim in the context of his rational psychology proceeds from the claim that the soul has a power for the actualization of its accidents (§.744). In order to identify the power that constitutes the nature (see §.628) of the soul in accounting for the actuality of all its diverse representations, Wolff begins by considering its effects, first among which are sensations (see Wolff, 1740/1972, §.65). As he has previously noted, sensations represent those bodies to us which in some way induce changes in the organs of sense of our own body (Wolff, 1751/1983b, §.749); yet, the bodies that occasion these changes are in turn parts of the world (§.753), which is to say that by means of its capacity for sensation the soul represents a part (or parts) of the world in accordance with the position of its (own) body. From this Wolff concludes that the power of the soul, insofar as its activity accounts for the actuality of such representations, must be a power for representing the world in accordance with the position of the body (§.753; 1740/1972, §.62). However, given the immateriality and hence simplicity of the soul, it follows that it can have only a single power (Wolff, 1751/1983b, §.755) rather than possessing an individual power corresponding to each of its many faculties, such as the capacity for imagination. But, as Wolff proceeds to argue, the representations on the part of the soul's distinct faculties derive from sensations and, as a result, can be understood to amount to expressions of the same power. So, Wolff claims that imaginings agree (*kommen…überein*) with sensations given that they are representations of corporeal things that we have previously sensed (§.750). More generally, Wolff contends that the soul's thoughts can be traced back to sensations inasmuch as all of the soul's consciously-had representations amount to the application of memory and reflection to a sensation (§.752; 1740/1972, §.64). As a result, then, of the common origin of all of the soul's representations in sensations, Wolff concludes that the soul's single power must be that one which was already identified as involved in sensation.

The faculty of sensation thus features centrally in Wolff's account of the soul's nature, and his rational psychology more generally. Indeed, the importance of sensation for Wolff is also carried over into his epistemology, in connection with which Wolff contends that sensations are the foundation for all of the soul's thoughts. Wolff is quite explicit, for instance, in claiming that the entire series of thoughts in the soul, including its general concepts, begin with sensations and that the content of our concepts can be traced back to these (Wolff, 1740/1983a, §.271; 1751/1983b, §.831). Wolff later clarifies that it is our *intuitive* cognition, which is to say our cognition that is unmediated by signs (Wolff, 1751/1983b, §.316), that is ultimately traceable back to sensations (§.846), but with this restriction in place Wolff allows that the Aristotelian *nihil est in intellectu* dictum can be retained inasmuch as it is taken to amount to the claim that "all thoughts of the soul take their origin from sensations" (Wolff, 1740/1983a, §.299, p. 501). And while Wolff, like Leibniz in the *New Essays*, does not allow this dictum to hold in its unrestricted sense, Wolff would not exclude the concept of the intellect itself from its scope, since even if

Wolff does not identify a faculty of *inner* sensation,[7] the faculty of understanding is clearly not cognized figuratively (Wolff, 1751/1983b, §.191), but rather only mathematical cognition is excepted from an origin in sensation.[8]

Wolff's surprising endorsement of the Aristotelian dictum (see also Wolff, 1740/1972, §.429n), at least as far as thoughts are concerned, stands in an obvious tension with his documented support of innate ideas. This tension is resolved, however, insofar as Wolff holds that while all our thoughts are traceable back to sensations, our sensations are not necessarily *caused* by changes in external bodies but rather proceed from the soul's own nature. Wolff accounts for the generation of "material ideas" in mechanical terms, claiming that the external bodies affect the organs of sense, imparting motion to an intermediary, whether it is referred to as *spiritus animales* (animal spirits) or the oscillation of nerve fibres, that is thereby propagated to the brain, where a subtle motion is generated that corresponds to the sensation in the soul. While Wolff thinks this account is consistent with that offered by the *Physicorum* (physicians) but also with Melanchton's (1497–1560) account offered in his treatise *De anima* (Wolff, 1740/1983a, §.287; 1751/1983b, §.778), when it comes to accounting for how a representation is generated in the soul, he falls back on his own metaphysical conclusions and contends that a sensation's agreement with the associated states of the body is accounted for not through a causal relation between the two, nor immediately through God's activity, but rather through a ground in the soul itself, namely, its representative power which God has arranged such that its effects are generated at the same time as the corresponding changes in the body (Wolff, 1751/1983b, §§.766–767). This means that for Wolff, strictly speaking, sensations are functions of the *activity* of the soul rather than in some way passively received through the organs of sense:

> On account of the harmony with the body, sensations find their reason in the body (§.29, 765) and thus, it would appear [*dem Ansehen nach*], outside of the soul (§.45). On account of this they are numbered among the passions [of the soul]. However, since sensations are in fact produced by the soul (§.753), and are only set into harmony with the body, they are acts of souls (§.104), and reveal the soul to be an active thing when it senses. (Wolff, 1751/1983b, §.8418, pp. 507–508; cf. 1740/1983a, §.305)

As he here makes clear, it is Wolff's endorsement of the pre-established harmony in the context of rational psychology that allows him to hold both that the soul's thoughts stem from sensations and that its ideas, indeed all of its representations, are nonetheless innate. Given this, it is unsurprising to find that Wolff references his own defence of the harmony in §.765 of the *German Metaphysics* in the same passage when he endorses the doctrine of innate ideas: "Because the soul brings forth

[7] For a discussion of Wolff's treatment of inner sense, see Favaretti Camposampiero (2018).

[8] Wolff seems to exclude mathematical cognition (i.e. cognition in geometry and algebra) from that which is traceable back to sensation, since some of such cognition is figurative, but also because such cognition employs concepts obtained through arbitrary determination, which concepts are not formed by reflection on sensation or abstraction. So, concerning this cognition, he claims that "thus it is said that there is much in the intellect that was not in the senses" (Wolff, 1755/2003a, §.4, p. 14; cf. also Arndt, 1965, p. 24).

sensations by means of its own proper power (§.765), so the images and concepts of corporeal things do not come from without but rather the soul in fact already has them within it [...]" (Wolff, 1751/1983b, §.819, p. 508).

Wolff's position would thus seem to align closely with that of Leibniz (at least outside of the *New Essays*); yet, this would be to overlook Wolff's own reservations regarding the demonstrable truth of the pre-established harmony, and whether the systems of natural influence and occasionalism can be decisively set aside. In spite of his apparent refutations of the alternative systems in *German Metaphysics* (Wolff, 1751/1983b, §§.761–764), Wolff seems to think better of this in the *Anmerkungen* where he allows that each of these systems can be made consistent with the laws of physics, not to mention theology and morality (Wolff, 1740/1983a, §§.274–278).

Given, then, that all three systems are consistent with the observed agreement of our sensations with changes in the body, Wolff allows that it is possible that a sensation might have its ground in the changes in the body and not (merely) in the power of the soul (Wolff, 1740/1983a, §§.284–285). For his part, Wolff takes the pre-established harmony to be the most defensible of the three systems (§.277), but the fact that it is not demonstratively proven means that Wolff also cannot wholly endorse the doctrine of innate ideas which is founded on it. That this is so implies that the doctrine of innate ideas has no more central place in Wolff's system than the harmony, with respect to which Wolff emphasizes that nothing of philosophical significance follows from it as far as the rest of his system is concerned, and goes as far as to claim that it "is a matter of no importance to me whether one takes this system to be more probable than another," and for those who continue to have any apprehensions regarding the system of harmony he simply recommends endorsing one of the other two systems (Wolff, 1740/1983a, §.289, p. 487). Just as, then, Wolff objects to Leibniz's use of the merely problematically asserted system of pre-established harmony as a basis for the proof of God's existence, so he would likewise have to reject its use in any putative demonstration of the doctrine of innate ideas.[9] This is in any case a fairly considerable minimization of the importance of this doctrine within the context of philosophical rationalism, as its supposition is not required

[9] This suggests a resolution to the disagreement between Hans Werner Arndt and Gideon Stiening concerning Wolff's position on innate ideas. So, Arndt (1983), contends contrary to the passages cited that "there is hardly a plank of classical rationalism with respect to which Wolff expressed himself in a more hesitant way than the doctrine of the innateness of ideas" (p. 38) inasmuch as this issue has nothing to do with the formation of concepts which is Wolff's main interest (see also Beck, 1993, p. 10, for an even stronger formulation). Against this, Stiening (2004, p. 222) has emphasized Wolff's express endorsement of the innateness of the soul's ideas, though he limits these to certain *notiones communes* (or properly speaking a *capacity* to have these on the occasion of experience). However, *contra* Arndt, Wolff is clear in his endorsement of the doctrine of innate ideas (and this is *consistent with* his account of the origin of all the soul's thoughts in sensation), but *contra* Stiening, this amounts only to a *qualified* endorsement, and the fact that *all* of our thoughts (including of the intellect itself) have their source in sensation distinguishes Wolff's version of the Aristotelian dictum from the one later canvassed by Leibniz.

for the proof of any claim of significance in philosophy (and thus constitutes a departure as well from Leibniz's later emphasis on the utility of this doctrine in the *New Essays*).

5.3 Certain and Probable Cognition

This leads to the next dogma of classical rationalism, namely, the emphasis on certain as opposed to merely probable cognition, and here Wolff's championing of the mathematical method as the universal method for philosophy supplies a clear example. In general terms, Wolff understands the mathematical method as beginning by formulating (real) definitions, proceeding from these to principles, including axioms and postulates, and using these for deriving theorems and resolving problems. Taking these stages in turn, we arrive at definitions of terms by first acquiring and proving the possibility of the concepts that serve as these terms' referents (the methods for doing so will be considered below). Definitions, then, are formed through consideration of our distinct concepts, and indeed, the method for rendering our concepts distinct is also that for forming their definitions, which is why Wolff nearly identifies distinct concepts with definitions (as at Wolff, 1754/1965a, Chap. 1, §.36). Wolff further distinguishes between nominal and real definitions, or definitions of words (*Wort-Erklärungen*) and definitions of things (*Sach-Erklärungen*), where the former state sufficient properties by which the concept is distinguished from all other similar ones, and the latter show *how* the concept is possible by showing how an object corresponding to it can be brought about (Wolff, 1754/1965a, Chap. 1, §.41).

The discovery and formulation of definitions is crucial as (real) definitions serve as the grounds of demonstrations which are chains of valid arguments where the conclusions of the proofs earlier in the series serve as premises in a subsequent proof such that certainty is preserved throughout the inferences. Definitions serve as grounds, or first premises, in demonstrations in two different ways: either as first principles themselves (provided that the possibility of the concept has been proven) or insofar as further propositions are derived directly from one of them. Wolff refers to these derivative propositions generally as principles (*Grundsätze*) (Wolff, 1710/1999, §.29), and he divides them into theoretical and practical where the former assert that something is the case or pertains to some thing, and the latter that and how it could be done (Wolff, 1754/1965a, Chap. 3, §.12). Theoretical principles are called axioms (*axioma*), though Wolff sometimes simply refers to them as principles (*Grundsätze*) (Wolff, 1754/1965a, Chap. 3. §.13), whereas practical principles are called postulates (*postulata/Heische-Sätze*) (Wolff, 1754/1965a, Chap. 3, §.13; 1710/1999, §.30). Since both of these principles follow immediately from a definition, they are identified as "identical propositions" (Wolff, 1716/1965b, p. 224) and they do not require further proof as given the definition they are "clear in themselves", though Wolff comments that such a proof could readily be supplied (1754/1965a, Chap. 6, §.2).

Definitions thus provide starting principles for demonstrations (though we will see that this does not exhaust all such principles); yet it is theorems (*theorema/Lehrsätze*) and problems (*problema/Aufgabe*) that constitute the real core of Wolff's method and the focus of his scientific project. Theorems and problems are generally identified as propositions that make a claim (affirmative or negative) about what is or can be that are or can be derived from multiple definitions (Wolff, 1754/1965a, Chap. 3, §.14; 1716/1965b, pp. 1377–1378; 1710/1999, §.37). According to Wolff, a theorem is discovered through the comparison of a number of definitions with one another and through the method of repeated substitution; so, in comparing the definitions of *joy*—"an affect excited in us when we are convinced of a present good" (Wolff, 1754/1965a, Chap. 6, §.3, p. 191)—and *sadness*—"an effect expressed in us when we are convinced of a present evil" (p. 191)—Wolff contends that through a number of substitutions, the following theorem becomes evident: "*a joy will be changed into a sadness when one convinces another that something they took to be good is rather an evil*" (p. 191—my emphasis). Problems, by contrast, are practical propositions (Chap. 3, §.14) which simply state something that is to be done. While theorems clearly have priority in Wolff's demonstrative enterprise, we tend to find theorems by arriving first at problems and resolving them. So, we arrive at a problem through taking some effect as something to be accounted for, where this effect is something that we have inferred arbitrarily on the basis of other representations (without knowing whether it is in fact possible), that we have derived from known truths, or that we have a direct experience of (Chap. 6, §.5).

The elements of the mathematical method are rounded out by the discussion of corollaries (*Zusätze*), or an application of a theorem to a particular case, and scholia (*Anmerckungen*), or a further elucidation of something in the definitions or principles;[10] nonetheless, it will be largely on the basis of the previously enumerated elements that the method will attain its aim of discovering and justifying new truths. Wolff contends that the primary instrument for this will be "ordinary syllogisms," and as an example, Wolff considers a problem ("determine the sum of the angles in a triangle") where the answer is presumed as unknown and shows that, on the basis of a number of premises presumed from geometry, we can use syllogisms to derive the corresponding theorem (Wolff, 1754/1965a, Chap. 4, §.23). In any case, Wolff takes the successful application of the mathematical method to yield certain conclusions, and attaining this is the goal of science inasmuch as Wolff understands by it the skill (*Fertigkeit*) at making use of incontrovertible grounds to prove all of one's claims (Wolff, 1740/1983c, §.30, §.117). Consistent with this, Wolff emphasizes the need to strictly distinguish between the certain and the merely probable "because in philosophy one must strive for complete certainty" (§.125, p. 60).

In spite of this, Wolff does carve out an important, indeed, indispensable role for probable opinion in service of the overall aim of attaining certain cognition. Wolff does, for instance, allow for opinions that are merely probably asserted to be employed in philosophical contexts provided that their assumption is in some way

[10] On the former, see Wolff (1716/1965b, p. 433, p. 1231; 1710/1999, §§.48–50).

required for the course of ordinary life and affairs "the probable is permitted in philosophy principally on account of its use in life" (Wolff, 1740/1983c, §.125, p. 60).[11] Yet, Wolff also contends that there are opinions that can be admitted into philosophy the assumption of which is useful for attaining certain cognition. Wolff refers to these as "philosophical hypotheses," which he defines as "the assumption of something, which is not yet capable of demonstration, as if it were the case for the sake of providing the reason for something" (§.126, p. 60), and which he makes clear he considers to be a species of probable opinion, given that they cannot be sufficiently proven and so are capable of being true or false.[12] Such hypotheses are employed in philosophical contexts, including and particularly metaphysics, much as they are in natural scientific contexts. Thus, Wolff claims that through a philosophical hypothesis we presume some ground or reason for a given occurrence known from experience, then we proceed to derive what consequences would obtain were it the case that the assumed ground held, and on this basis, we turn to observation and experiment to determine whether they do in fact obtain—where the consequence is contradicted by our experience, the hypothesis is rejected as false, but if it agrees with it, then, according to Wolff, "the way to manifest truth [*veritatem liquidam*] is paved" (1740/1983c, §.127, p. 61), and we can proceed to determine whether the hypothesis might itself be demonstrated as a theorem (Wolff, 1729/1983d, §.6, p. 188). Moreover, in recognition of the potential for the abuse of these propositions, Wolff subjects their use to a number of restrictions, including that the state of affairs posited through the hypothesis must be possible in itself, that it is free of contradiction, but also cannot involve the assumption of anything inconsistent with what is known to exist in the world (Wolff, 1729/1983d, §§.8–9, pp. 196–205). Most importantly, as merely probable opinions, philosophical hypotheses cannot be employed as principles in demonstrations (Wolff, 1740/1983c, §.128).

Among examples cited by Wolff of philosophical hypotheses are Descartes' conjecture that there are three elements from which all natural occurrences can be derived (Wolff, 1729/1983d, §.1, p. 179), as well as all three systems purporting to account for the ground of the agreement between the states of the body and the soul, including Leibniz's pre-established harmony. Concerning the latter, Wolff notes that the fact that changes in the soul agree with changes in the body is confirmed through experience (and is appropriately catalogued in the context of empirical psychology) (see Wolff, 1751/1983b, §§.527–528). However, Wolff notes that the activity of the soul on the body, or vice versa (and presumably the soul's own activity through its own power), is not similarly available to observation such that the actual ground of this agreement can only be the subject of a philosophical hypothesis. Moreover, each of the three systems (natural influence, occasionalism, and the pre-established harmony) are possible in themselves (though Wolff admits he has difficulty

[11] This likely constitutes a reference to the notion of *moral* certainty, which however Wolff does not devote much discussion to; on this consult Fonnesu (2011).

[12] For a detailed discussion of Wolff's notion of a philosophical hypothesis and the development in his views concerning it, see Leduc (2017).

conceiving of the possibility of natural influence but concedes that this might be his own limitation), and each can be made consistent with the laws of nature and with God's wisdom and power (Wolff, 1740/1983a, §§.273–277), which is to say that each remains (only) a plausible hypothesis. It is this status of the system of pre-established harmony as an hypothesis that ultimately accounts for the restrictions Wolff applies to it, as we noted above; namely, that it cannot be used as a basis for a demonstration (of God's existence, for instance) (cf. Wolff, 1740/1983c, §.128n), and indeed for the diminished role that it plays in his rational psychology and ethics.

As such, when employed correctly, philosophical hypotheses are an indispensable and powerful supplement to the mathematical method (which method Wolff conceives of as a method of discovery *and* justification). Of course, this is not to say that Wolff in any way challenges the priority of certain to probable knowledge since the former remains the undisputed goal of science. Even so, the addition of hypotheses to the philosopher's quiver, as a way to attain certainty, does mitigate what might otherwise be taken as an unadulterated apriorism in Wolff's use of the mathematical method. As Wolff makes clear, hypotheses are only useful when they are properly formulated, that is, on the basis of observed phenomena rather than for the sake of other speculative ends:

> Who, then, is so bold as to doubt that certain cognition is to be preferred to uncertain? Now in physics it is possible to have cognition of many certain things if that is derived from certain observations [*observationibus certis*] which permits of being known through a correct inference; but that way of proceeding on the part of those who feign general hypotheses from which they may derive all cognition of natural things a priori, or in order to uncover the reason of observed phenomena from them, finds little approval with me (1729/1983d, §.3, p. 182).

Properly employed, philosophical hypotheses serve to ensure that the overall aim of attaining certainty is not achieved at the expense of the phenomena themselves. The role of hypotheses also serves to illustrate that Wolff does not conceive it to be possible, or desirable, for us to frame a philosophical system from the comfort of the armchair.[13]

5.4 Reason and Experience

The foregoing might suggest that experience figures prominently in Wolff's philosophical method. And yet Wolff seems to deny this when, for instance, he contrasts experience and reason as two "paths for the cognition of truth" (Wolff, 1751/1983b, §.372, p. 228), where experience supplies us with cognition of particulars by means of the senses (either through simple observation or as the result of staged experiments—§.325), and where reason involves an insight into the connection of truths, particularly insofar as they follow from principles through inference. As Wolff writes:

[13] See Vanzo (2015), for a detailed treatment of Wolff's experimentalism in various parts of his philosophical system.

> we have two ways in which we can attain to cognition of the truth. The first is grounded in the senses (§.220, 325), the second however in the understanding (§.277, 368). For example, most people cognize that the sun rises repeatedly in the morning from experience (§.338) and they cannot say why it happens; by contrast, an astronomer who has insight into the cause of heavenly motions and into the connection of the earth with the heavens cognizes this through reason and is able to demonstrate that, why, and at what time it must take place. (Wolff, 1751/1983b, §.372, p. 228)

It is in particular through reason that we are able to cognize the reason why something is or takes place, whereas experience only discloses the fact of something occurring. In the terms that Wolff will later introduce, this is to say that experience yields *historical* cognition whereas reason provides us with properly *philosophical* cognition (Wolff, 1740/1983c, §.3, §.6). Given this, and given the connection Wolff draws between the grounded inferences produced through the mathematical method and the use of reason (Wolff, 1751/1983b, §.383), it would seem that the importance assigned to experience in testing philosophical hypotheses does not extend to a significant role within the mathematical method itself.

Even so, and as has been recognized by commentators, Wolff "hardly keeps to this distinction" (Kuehn, 1987, p. 255) between the two paths, and the "distinction between rational [philosophical] knowledge and historical [...] is not a sharp one" (Beck, 1969, p. 268). Rather, and quite generally, Wolff denies that the use of reason is "not always pure particularly in the cognition of nature and of ourselves" (Wolff, 1751/1983b, §.382, p. 235); more narrowly, Wolff claims that philosophical cognition nonetheless depends on historical cognition, since it is often through observation that we discover the ground for something else that is or occurs. Wolff is quite explicit in thus identifying historical cognition as the *fundamentum* of philosophical cognition (Wolff, 1740/1983c, §.10), and claiming that "historical cognition should precede philosophical cognition and be constantly conjoined with it so that it does not lack a firm foundation" (§.11, p. 5).[14] In fact, far from minimizing the importance of experience for philosophy, Wolff contends that the philosopher should aim at the combination of observation and demonstration, setting out from what is known a posteriori and seeking its grounds in accordance with a priori rules, as is the case in astronomy and optics. The result, then, is a "marriage of reason and experience [*connubium rationis et experientiae*],"[15] in which the philosopher makes use of both ways of cognizing truth.

Wolff goes further, however, than simply claiming that the philosopher ought to keep an eye on experience when conducting his demonstrations. Rather, and significantly, at a number of junctures within the mathematical method Wolff carves out a crucial role for experience. First, as has been recently and effectively explored by Katherine Dunlop (2019, pp. 163–169), Wolff takes reference to experience to be important for certifying the possibility of our distinct concepts and for discovering their real definitions for use in demonstrations. For Wolff, a concept is distinct when

---

[14] For more on the significance of historical cognition for Wolff, see Kreimendahl (2007, p. 97).

[15] See Cataldi Madonna (2007), for discussion and an account of the historical context of Wolff's notion of such a *connubium*.

it is clear, that is, when it suffices for recognizing the things that it is a concept of, and when we are aware of the marks by which its objects are so distinguished (Wolff, 1754/1965a, Chap. 1, §.9, §.13). In terms of how we go about acquiring distinct concepts, Wolff considers three different ways: first, through careful attention to the parts of our representation of some thing and to how they are connected and ordered (Chap. 1, §.19); second, by attending to commonalities between this representation and various others (Chap. 1, §.26); and third, by arbitrarily removing and adding determinations (Chap. 1, §.30). Concerning these concepts, however, there is still a question as to whether they are possible, that is, whether they contain marks that do not conflict with one another (and so that they may refer to things—Chap. 1, §.31; 1751/1983b, §.16), and Wolff further considers three ways in which this possibility might be demonstrated. In the case of concepts that are abstracted directly from sensations, the concept's possibility is certain since the fact that they are taken from actual objects of experience ensures their possibility (Wolff, 1754/1965a, Chap. 1, §.32); however, in the case of arbitrarily determined concepts, we must either make use of a proof or find an instance within experience that, as in the previous case, certifies the concept's possibility (Chap. 1, §.34). These distinct (and indeed *complete* such that they apply only to a single sort of things—Chap. 1, §.36) concepts will in any case serve as the basis for definitions, and insofar as it will be admissible as a ground in a demonstration, the definition will need to be *real*, which is to say, they must suffice to disclose the possibility of the concept whose term is defined (Chap. 1, §.41). Unsurprisingly, as before we might have recourse to a careful analysis of the marks of the concept to determine whether there are any inconsistencies (Chap. 1, §.54), but we can also make use of experience to determine whether there are such things that correspond to the concept, making use of technical means as necessary, or by carefully attending to the generation of a given thing (Chap. 1, §.51, §§.56–57).

In this way, recourse to experience is involved in the formation of the definitions that will serve as the first premises in any demonstration. Yet Wolff carves out another and arguably more important role for experience, as he allows that some experiences can serve as principles in demonstrations alongside definitions and axioms; thus, he writes, "one calls it a *demonstration* when one can conduct one's proof so far until, in the last inference, one has nothing else than definitions, *clear experiences*, and other identical propositions as premises" (Wolff, 1754/1965a, Chap. 4, §.21, p. 172—my emphasis). That Wolff refers to *clear* experiences (*klare Erfahrungen*) here suggests that the experience he has in mind as serving as the ground of a demonstration is something distinct from mere or "common experience [*gemeine Erfahrung*]" (Wolff, 1740/1983a, §.99, p. 170), which presumably consists in the immediate deliverances of observation. A clear experience, by contrast, requires careful vetting if it is to be allowed to serve as a principle in a demonstration; specifically, since experience is a source of singular propositions or judgments (Wolff, 1754/1965a, Chap. 5, §.2), such an experience will need to be put in a suitably universal form in order to be employed in a demonstration.

Appropriately, Wolff devotes considerable attention (that is, all of the fifth chapter of the *German Logic*) to the strictures that must be observed if a given proposition from experience is to thus serve as a principle. So, Wolff identifies three different cases of experience that are suitable: the experience of a thing along with its properties; the experience of the changes that something undergoes; and the experience of a thing's effects upon another thing (Wolff, 1754/1965a, Chap. 5, §.5). What is then to be determined is whether our experience in each of these cases is, for instance, in fact of an attribute (*Eigenschaft*) that must belong to that (sort of) thing as long as it exists inasmuch as it finds its sufficient reason solely in that thing's essence (Wolff, 1751/1983b, §.44). The point then is to outline a procedure such that we can be confident that our experience discloses that some feature we have observed some thing to possess is the experience of some attribute of that thing and so can be suitably universalized to hold of all such things in general. In order to do this, Wolff recommends that we consider the object in different circumstances to rule out some ground external to it (Wolff, 1754/1965a, Chap. 5, §.6). Alternatively, in cases where the thing under consideration cannot be easily moved, we might observe it over an extended period of time (Chap. 5, §.7), or we might opt to investigate the matter a priori and, after acquiring a distinct concept of the thing, elaborate the marks of the concept to determine whether the connection between it and the alleged property is found (Chap. 5, §.8).[16] When employed successfully, these methods certify that the property which we observe some thing to have is in fact grounded in its essence and so belongs to it under any circumstances; in this way, Wolff claims that these singular experiences can be changed into universal propositions (Chap. 5, §.15).

Again, this distinctively Wolffian emphasis on experience is borne out even and especially in his *rational* psychology. Wolff sets out from the foundational experience that "we are conscious of ourselves and of things outside of us" (Wolff, 1738/1968, §.11, p. 9; 1751/1983b, §.1, p. 1), which experience is already framed in universal form given that Wolff takes it as obvious that it is such that anyone could institute it for themselves regardless of external circumstances (Wolff, 1754/1965a, Chap. 5, §.2; 1751/1983b, §.7). This experience in turn serves as a principle in Wolff's demonstration of the truth that "we exist" at the outset of the *German Metaphysics* (Wolff, 1751/1983b, §.6, §.8), but also informs Wolff's later (nominal) definition of the soul at the outset of the empirical psychology proper, as "that thing [...] which is conscious of itself and other things outside of it" (§.192, p. 107). This experience is likewise put to work in discovering the essence and nature of the soul (and so in discovering its *real* definition) (Wolff, 1754/1965a, Chap. 1, §.48). So proceeding from the experience that we are conscious, Wolff considers what faculties on the part of the soul are required in order to account for this, and he determines that reflection and memory are required inasmuch as we need to be able to recall and compare given representations in order to differentiate them which is just what consciousness consists in (Wolff, 1751/1983b, §§.733–735; cf. §.201). Wolff then argues that a body is incapable of consciousness given that the reflection that is required for it cannot be accounted for in terms of mechanical properties (§.738), which is to say that the soul, inasmuch as

[16] For instructive discussion of this, see Dunlop (2019).

we experience that it is so conscious, cannot be a body, or even a composite, but must be simple (§.742), and as such a substance with a single power that is the ground of the actuality of its representations (§§.743–745). As we have already seen, this power is further identified as a power of representing the world in accordance with the position of the body, in virtue of the fact that all thoughts on the part of the soul can be traced back to sensations (see Sect. 5.2 above), and it is this power that is ultimately identified as the essence (and nature) of the soul (§§.755–756).

It might be thought odd, or even inconsistent that Wolff should assign such an important role to experience in the context of *rational* psychology. Yet, Wolff is clear that rational psychology should make use of what is disclosed in empirical psychology to expand what can be known of the soul:

> Since insofar as something more than we are conscious of is to be encountered in us, we have to bring it out through inferences and, indeed, inferences from that of which we are conscious for otherwise we would not have any basis for them. (Wolff, 1751/1983b, §.193, p. 108)

Rational psychology is thus a "rational" discipline in a distinctly Wolffian sense, namely, that it sets out from experience but is not strictly limited to it as it can employ inference (and hypothesis) to advance beyond what is immediately disclosed to observation. Moreover rational psychology continues to remain subject to the touchstone (*Probier-Stein*) of experience as its speculative results must still be confirmed, or at least not refuted, by what is observed of the soul (§.727). Wolff thus distinguishes rational psychology, which is rational in a mixed sense (§.382), from disciplines like arithmetic and algebra which are rational sciences in a pure sense such that nothing is admitted in the course of reasoning "except definitions and propositions cognized *a priori*" (Wolff, 1738/1968, §.495, p. 378—my emphasis). Despite its mixed status, however, rational psychology does not sacrifice anything regarding the certainty of its results since the experiences it relies upon for its demonstrations are themselves of undoubted certainty.

5.5 Principle of Sufficient Reason

The final dogma of classical rationalism to be considered is the principle of sufficient reason. Wolff follows Leibniz in endorsing the PSR and assigning it the status of one of the "first principles" of human cognition. However, where Leibniz was content to point to examples of the principle holding, and noting that we have yet to observe any counterexample,[17] by way of its justification, Wolff goes further and attempts to prove the principle itself, claiming that the PSR must hold inasmuch as denying it to hold would require one to likewise deny the truism that nothing comes from nothing (Wolff, 1751/1983b, §.30). Few have been impressed by Wolff's proofs which are beset by numerous ambiguities, but *that* Wolff should even attempt

[17] Leibniz (1710/1952, Pt. I, §.44, pp. 147–148); and the fifth letter to Clarke, §.129 (Ariew, 2000, p. 65).

a proof of the principle, and indeed on the basis of the meagre resources of the principle of contradiction (since it is from this that we know that nothing comes from nothing—see Wolff, 1751/1983b, §.28), might nonetheless be taken as indicative of his commitment to a particularly radical philosophical rationalism.

Unfortunately, a detailed treatment of Wolff's views concerning the PSR cannot be entered into here. However, we might briefly note that Wolff tasks the principle with a fairly narrow responsibility within the framework of his mathematical method. Indeed, it is in virtue of this role that Wolff first introduces the principle of sufficient reason, though without naming it as such and without an attempt at proof, at the outset of his *German Logic*: "Everything has a reason [*raison*] why it is" (Wolff, 1754/1965a, *Vorbericht*, §.4, p. 115). Specifically, the PSR is introduced in connection with *theorems*, which Wolff claims stipulate the "ground [*Grund*] why something can pertain to a thing or not" (1754/1965a, Chap. 3, §.5, p. 158);[18] similarly, problems clearly presuppose the availability of a cause for a given effect, where the resolution of the problem enumerates everything that is sufficient to bring the desired effect about (Wolff, 1710/1999, §.47). Indeed, this ground-consequent relation is reflected in the very syntax of theorems, as Wolff divides them into a condition (*hypothesis*/*Bedingung*) and an assertion (*thesis*/*Aussage*), where the former states the grounds under which some feature or property can pertain or not pertain to a thing, and the latter asserts that that same feature or property pertains to a thing (or not) (Wolff, 1754/1965a, Chap. 3, §.6).[19] Wolff apparently takes the PSR to be a necessary metaphysical presupposition for the truth of theorems inasmuch as we would not be warranted in asserting that some occurrence has a reason for why it is so and not otherwise unless it was the case that *all* occurrences are so grounded. Thus, in the context of a discussion of theorems, Wolff cites the claim that "Nothing is absolutely possible except the self-sufficient being; otherwise, everything has its causes why it exists" (Wolff, 1710/1999, §.39, p. X), and in the *German Logic,* this same claim is framed in terms of all beings apart from the self-sufficient being having a "*reason* [*raison*] why it rather is than not" (Wolff, 1754/1965a, Chap. 1, §.47, p. 146). In any case, given that the PSR is explicitly invoked only in connection with the truth of theorems, it follows that the truth of other principles that serve as the grounds of demonstrations, including definitions, axioms, and clear experiences, is not similarly dependant on the PSR. It is not clear that Wolff would accord these experiences the status of "brute facts" since, considered as an occurrence, my consciousness of myself would still find a reason in some antecedent ground. Nonetheless, such experiences (and definitions and axioms) do not depend on the PSR by way of validating their truth, and in fact any attempt to justify them in

[18] Cf. also Goubet (2011, especially, pp. 81–82).

[19] While Wolff first introduced this analysis of theorems in 1707 (cf. Wolff, 1755/2003b, Section I, Num. II, p. 7), in the *German Logic*, Wolff presents this structure as a feature of propositions (*Sätze*) generally, but elsewhere indicates that it pertains to theorems specifically (as in Wolff, 1710/1999, §.39 and 1716/1965b, pp. 1377–1378). Note as well that even in the *German Logic*, Wolff later identifies the propositions under discussion in Chap. 3, §.6 as *theorems* (cf. Wolff, 1754/1965a, Chap. 3, §.14, where he refers back to Chap. 3, §.5 to which §.6 in turn refers).

accordance with that principle could only serve to mislead us as to the nature and source of the certainty with which they are held.

That Wolff countenances limits to the use of the PSR, in the sense of admitting brute facts is clear in the context of his metaphysics proper. So, in his *Ontologia* (Ontology), Wolff introduces the notion of *essentialia* (by which he understands the essential determinations that constitute the essence of a thing) (Wolff, 1736/1962, §.143). These essential determinations contain the sufficient reason for the attributes of a thing (§.157); however, Wolff is clear that there is no sufficient reason for why these *essentialia* belong to a given thing (§.156). Interestingly, later in his *Theologia naturalis* (Natural Theology), Wolff urges this point against Spinoza who is accused of a promiscuous misemployment of the PSR precisely because he erroneously holds that there must be a sufficient reason for all the determinations of a given finite thing, including essential determinations. In virtue of this, and because Spinoza could not find an *internal* reason for these determinations, he took there to be an external reason in the form of some other finite thing serving as its cause:

> No doubt the notion of a sufficient reason had occurred to Spinoza, just like it had to Descartes (*Ont.* §.321n), whose principles he was educated in (§.667n), and through it he confusedly cognized the principle of sufficient reason as well, but a distinct notion of the same eluded him; from where, having erred in the principle's application, he sought a sufficient reason for the original limitations which are connected with the essences of finite things. Since no internal reason is given (*Ont.* §.156), he concluded that there must be an external one, which he wrongly confuses with a cause (*Ont.* §.321), as it were, with a source of its determination, which contradicts this principle because it does not permit such a cause (*Ont.* §.883). (Wolff, 1741/1981, §.686, p. X)

Whatever the effectiveness of Wolff's criticism, in thus exempting *essentialia* from the scope of the PSR, Wolff is clearly admitting what amounts to a "brute fact" at the very core of his metaphysics, an admission that at least sets him apart from Spinoza, if not necessarily from Leibniz.

These limits that Wolff sets to the PSR might be taken to be besides the point, as far as determining the character of his rationalism is concerned, given his efforts to derive this principle from the principle of contradiction. Indeed, that Wolff would assign the principle of contradiction the status of *the* fundamental principle in his system threatens to undermine any contribution on the part of experience as Wolff goes as far as to contend that, just as the principle of contradiction is the ground of all certainty in syllogisms, so it is the ground of all certainty in cognition from *experience* (Wolff, 1751/1983b, §.391),[20] as Wolff takes himself to have shown when he claimed that the truth of our experience that we are conscious of ourselves and other things cannot be denied on pain of contradiction (§.10). Against this, however, we might note the following points (though this falls well short of an adequate discussion of the issue). First, any role that the principle of contradiction might play with respect to our experience that we are conscious of ourselves and other things should not be extended to *every* truth gained by experience. Wolff's point is that any

[20]This passage is particularly emphasized by Anderson (2015, pp. 80–81) as evidencing Wolff's commitment to the claim that all truths are conceptual.

experience that is certain "has [the principle of contradiction] to thank for it" (§.391, p. 239), which presumably holds for any experience that can serve as a principle in a demonstration, but of course does not imply that our "common experience," for which Wolff does not claim certainty, are in some way dependent on that principle (as ultimately conceptual truths). Second, Wolff contends that the principle of contradiction *itself* is grounded on experience, identifying our experience that when we judge that something is we cannot at the same time judge that that same thing is not as the "foundation of the principle of contradiction [*fundamentum principii contradictionis*]" (Wolff, 1736/1962, §.27, p. 15). This would suggest that the principle of contradiction itself constitutes, or is grounded on, a kind of clear experience, a universalization from our individual experience of the impossibility of such a contradiction, where any other experience would have to likewise involve such an express contradiction in thought in order to be held as certain (and in this way can be said to "owe" its certainty to the principle of contradiction without in some way amounting to a conceptual truth).[21] This is all to say that the role Wolff assigns to the principle of contradiction in certifying the certainty of (some) propositions known by experience is entirely consistent with the admission that the way of experience is an indispensable source of cognition.

5.6 Conclusion

With respect to these four dogmas of rationalism, then, we can see that Wolff stakes out a much more nuanced, and in most cases more moderate, positions than those typically ascribed to his fellow rationalist thinkers. By way of a general characterization of Wolffian rationalism, we might conclude that it is distinguished not by an unbending faith in the powers of unaided human reason, but by a tempered optimism about our capacity to attain to cognition of a rationally-ordered world, but only insofar as we make use of all of the resources available to us, including and especially what can be known of ourselves and of the world through our experience, and where Wolff recognizes that, in some cases, inquiry has to terminate with the recognition either of a brute fact or at least the inability of our finite intellects to penetrate any further. In light of this, the caricature of Wolff as an arch-rationalist thinker must be rejected as not only widely inaccurate but also profoundly unfair given that, in many ways, Wolff intends his distinctive rationalism as a corrective to what he regarded as the excessive and highly unscientific apriorism of his predecessors.

[21] For further, and what I take to be complementary discussions of this issue, see Cataldi Madonna (2019, pp. 209–212) and Dunlop (2019, pp. 173–175).

References

Allison, H. E. (2005). Kant and the two dogmas of rationalism. In A. Nelson (Ed.), *A companion to rationalism* (pp. 343–359). Blackwell.

Anderson, R. L. (2015). *The poverty of conceptual truth*. Oxford University Press.

Anstey, P. (2005). Experimental versus speculative natural philosophy. In P. Anstey & J. A. Schuster (Eds.), *The science of nature in the seventeenth century: Patterns of change in early modern natural philosophy* (pp. 215–242). Springer.

Ariew, R. (Ed.). (2000). *G. W. Leibniz and Samuel Clarke correspondence*. Hackett.

Arndt, H. W. (1965). Einführung. In H. W. Arndt (Ed.), *Christian Wolff: Gesammelte Werke* (Deutsche Logik) (Vol. 1, Pt. I). Olms.

Arndt, H. W. (1983). Rationalismus und Empirismus in der Erkenntnislehre Christian Wolffs. In W. Schneider (Ed.), *Christian Wolff, 1679–1754: Interpretationen zur seiner Philosophie und deren Wirkung* (pp. 31–47). Meiner.

Beck, L. W. (1969). *Early German philosophy: Kant and his predecessors*. Harvard University Press.

Beck, L. W. (1993). From Leibniz to Kant. In R. C. Solomon & K. M. Higgins (Eds.), *Routledge history of philosophy* (Vol. 6, pp. 5–39). Routledge.

Cataldi Madonna, L. (2007). *Christian Wolff und das System des klassischen Rationalismus*. Olms.

Cataldi Madonna, L. (2019). *Erfahrung und Wissenschaftstheorie bei Christian Wolff*. Olms.

Dea, S., Walsh, J., & Lennon, T. M. (2018). Continental rationalism. In E. N. Zalta (Ed.), *The Stanford encyclopedia of philosophy*. (Winter 2018 Edition). Retrieved form https://plato.stanford.edu/archives/win2018/entries/continental-rationalism/

Della Rocca, M. (2003). A rationalist manifesto: Spinoza and the principle of sufficient reason. *Philosophical Topics, 31*, 75–93.

Della Rocca, M. (2010). PSR. *Philosophers' Imprint, 10*(7), 1–13.

Dunlop, K. (2019). Definitions and empirical justifications in Christian Wolff's theory of science. *Logical Analysis and History of Philosophy, 21*, 149–176.

Favaretti Camposampiero, M. (2018). Anthropology from a logical point of view: The role of inner sense from Jungius to Kant. In G. Lorini & R. Louden (Eds.), *Knowledge, morals and practice in Kant's anthropology* (pp. 43–61). Palgrave Macmillan.

Fonnesu, L. (2011). Kant on "moral certainty". In L. C. Madonna & P. Rumore (Eds.), *Kant und die Aufklärung*. Olms.

Gardner, S. (1999). *Routledge philosophy guidebook to Kant and the critique of pure reason*. Routledge.

Goubet, J.-F. (2011). Qu'y a-t-il d'ontologique dans la logique wolffienne? Sur la *veriori logica*. In F. Fabbianelli, J.-F. Goubet, & O.-P. Rudolph (Eds.), *Zwischen Grundsätzen und Gegenständen. Untersuchungen zur Ontologie Christian Wolffs*. Olms.

Huenemann, C. (2008). *Understanding rationalism*. Acumen.

Kreimendahl, L. (2007). Empiristische Elemente im Denken Wolffs. In J. Stolzenberg & O.-P. Rudolph (Eds.), *Christian Wolff und die europäische Aufklärung: Akten des 1. Internationalen Christian-Wolff-Kongresses, Halle (Saale), 4.-8.April 2004* (Vol. 1, pp. 95–112). Olms.

Kuehn, M. (1987). *Scottish common sense in Germany, 1768–1800: A contribution to the history of critical philosophy*. McGill-Queen's University Press.

Leduc, C. (2017). Wolff on hypothesis. In A. Pelletier (Ed.), *300 years Christian Wolff's German logic* (pp. 77–97). Olms.

Leibniz, G. W. (1952). *Theodicy* (A. Farrer, Ed. & E. M. Huggard, Trans.). Yale University Press. (Original work published 1710).

Leibniz, G. W. (1989). Meditations on truth, knowledge and ideas. In R. Ariew & D. Garber (Eds.), *Philosophical essays* (pp. 23–27). Hackett. (Original work published 1684).

Locke, J. (1708). *Some familiar letters between Mr. Locke and several of his friends*. A. and J. Churchill.

Nelson, A. (2005). The rationalist impulse. In A. Nelson (Ed.), *A companion to rationalism* (pp. 3–11). Blackwell.

Stiening, G. (2004). Psychologie als metaphysica specialis. In O.-P. Rudolph & J.-F. Goubet (Eds.), *Die Psychologie Christian Wolffs. Systematische und historische Untersuchungen* (pp. 207–226). Max Niemeyer Verlag.

Vanzo, A. (2015). Christian Wolff and experimental philosophy. In D. Garber & D. Rutherford (Eds.), *Oxford studies in early modern philosophy* (Vol. 7, pp. 225–255). Oxford University Press.

Wolff, C. (1962). Philosophia prima, sive Ontologia, methodo scientifica pertractata, qua omnis cognitiones humanae principia continentur. In J. École (Ed.), *Christian Wolff: Gesammelte Werke* (Prima philosophia sive Ontologia) (Vol. 3, Pt. II, 2nd ed.). Olms. (Original work published 1736).

Wolff, C. (1965a). Vernünfftige Gedancken von den Kräfften des menschlichen Verstandes und ihrem richtigen Gebrauche in Erkänntnis der Wahrheit. In H. W. Arndt (Ed.), *Christian Wolff: Gesammelte Werke* (Deutsche Logik) (Vol. 1, Pt. I, 14th ed.). Olms. (Original work published 1754).

Wolff, C. (1965b). Mathematisches Lexicon, darinnen die in allen Theilen der Mathematick üblichen Kunst-Wörter erkläret, und zur Historie der Mathematischen Wissenschaften dienliche Nachrichten ertheilet, auch die Schrifften, wo iede Materie ausgeführet zu finden, angeführet werden. In J. École (Ed.), *Christian Wolff: Gesammelte Werke* (Mathematisches Lexicon) (Vol. 11, Pt. I). Olms. (Original work published 1716).

Wolff, C. (1968). Psychologia Empirica, methodo scientifica pertractata, qua ea, quae de anima humana indubia experientiae fide constant, continentur et ad solidam universae philosophiae practicae ac theologiae naturalis tractationem via sternitur. In J. École (Ed.), *Christian Wolff: Gesammelte Werke* (Psychologia Empirica) (Vol. 5, Pt. II, 2nd ed.). Olms. (Original work published 1738).

Wolff, C. (1972). Psychologia rationalis, methodo scientifica pertractata, qua ea, quae de anima humana indubia experientiae fide innotescunt, per essentiam et naturam animae explicantur, et ad intimiorem naturae ejusque autoris cognitionem profutura proponuntur. In J. École (Ed.), *Christian Wolff: Gesammelte Werke* (Psychologia rationalis) (Vol. 6, Pt. II, 2nd ed.). Olms. (Original work published 1740).

Wolff, C. (1981). Theologia naturalis, methodo scientifica pertractata. Pars Posterior qua existentia et attributa Dei ex notione entis perfectissimi et natura animae demonstrantur, et Atheismi, Deismi, Fatalismi, Naturalismi, Spinosismi, aliorumque de Deo errorum fundamenta subvertuntur. In J. École (Ed.), *Christian Wolff: Gesammelte Werke* (Theologia naturals, pars posterior) (Vol. 8, Pt. II, 2nd ed.). Olms. (Original work published 1741).

Wolff, C. (1983a). Anmerckungen über die vernünfftige Gedancken von Gott, der Welt und der Seele des Menschen, auch allen Dingen überhaupt, zu besserem Verstande und bequemeren Gebrauche derselben heraus gegeben. In C. Corr (Ed.), *Christian Wolff: Gesammelte Werke* (Anmerkungen zur Deutschen Metaphysik) (Vol. 3, Pt. I, 4th ed.). Olms. (Original work published 1740).

Wolff, C. (1983b). Vernünfftige Gedancken von Gott, der Welt und der Seele des Menschen, auch allen Dingen überhaupt. In C. Corr (Ed.), *Christian Wolff: Gesammelte Werke* (Deutsche Metaphysik) (Vol. 2, Pt. I, 11th ed.). Olms. (Original work published 1751).

Wolff, C. (1983c). Philosophia rationalis sive Logica, methodo scientifica pertractata et ad usum scientiarum atque vitae aptata. Praemittitur Discursus praeliminaris de philosophia in genere. In J. École (Ed.), *Christian Wolff: Gesammelte Werke* (Discursus praemlinaris) (Vol. 1.1, Pt. II, 3rd ed.). Olms. (Original work published 1740).

Wolff, C. (1983d). De Hypothesibus philosophicius. In J. École (Ed.), *Christian Wolff: Gesammelte Werke* (Horae subsecivae Marburgenses anni MDCCXXIX) (Vol. 34.1, Pt. II). Olms. (Original work published 1729).

Wolff, C. (1999). Die Anfangs-Gründe aller mathematischen Wissenschaften: Welcher Einen Unterricht von der Mathematischen Lehr-Art, die Rechen-Kunst, Geometrie, Trigonometrie, und Bau-Kunst in sich enthält. In J. École (Ed.), *Christian Wolff: Gesammelte Werke* (Anfangs-Gründe aller mathematischen Wissenschaften) (Vol. 12, Pt. I). Olms. (Original work published 1710).

Wolff, C. (2001). Some familiar letters between Mr. Locke and several of his friends. In H. A. Laeven & L. J. Laeven-Aretz (Eds.), *Christian Wolff: Gesammelte Werke* (Sämtliche Rezensionen in den Acta Eruditorum (1705–1731) (Teil II, 1711–1718)) (Vol. 38.2, Pt. II, pp. 591–597). Olms. (Original work published 1711).

Wolff, C. (2003a). Solutio nonnullarum difficultatum circa mentem humanam obviaruum, ubi simul agitur de origine notionum & facultate ratiocinandi (1707). In J. École (Ed.), *Christian Wolff: Gesammelte Werke* (Meletemata mathematico-philosophica) (Vol. 35, Pt. II, pp. 11–17). Olms. (Original work published 1755).

Wolff, C. (2003b). Methodus demonstrandi veritatem religionis christianae (1707). In J. École (Ed.), *Christian Wolff: Ge sammelte Werke* (Meletemata mathematico-philosophica) (Vol. 35, Pt. II, pp. 5–7). Olms. (Original work published 1755).

Chapter 6
Wolff's Idea of *Psychometria*

Manuela Mei

6.1 Introduction

The history of psychology is very difficult to approach. That's why it would be better to speak about traditions, trends, and theories without using the terms "paradigm" or "scientific revolution" (Mecacci, 2008). Certainly, an important "trend" in the history of psychology has concerned the distinction between rational and empirical psychology. One attempt to subdivide psychological knowledge was made by Johann Heinrich Alsted (1588–1638), who in his 1630 *Encyclopedia* appears to separate psychology into *pneumatological* and *physical* forms. *Pneumatological* psychology involves the essence of the soul, whereas *physical* psychology concerns the relationship between body and soul (Alsted, 1630/1989, vol. 2, p. 730). Although Alsted postulates a double system of psychological knowledge, some difficulty in considering psychology as a science can be detected in his work.

According to the *Geschichte der Psychologie* (History of psychology) of Friedrich August Carus (1770–1807), the effective subdivision of psychology into empirical and rational forms probably dated back to Christian Wolff (Carus, 1808/1990, p. 545).[1] In 1732, this *praeceptor Germaniae* (Educator of German Nation) wrote the *Psychologia empirica, methodo scientifica pertractata* (Empirical psychology, treated according to the scientific method) and in 1734 his *Psychologia rationalis, methodo scientifica pertractata* (Rational psychology, treated according to the scientific method). As may be assumed from the titles, these works pursued different objectives even though they featured the same basic methodological

[1] Jean École does not share the same view. According to him the terminological distinction between *empirical* and *rational* psychology appears for the first time in Thümmig's *Institutiones* in 1725 (École, 1990). Even so, "One must assume … that much of what is new in the 'Institutiones' goes back to Wolff's oral explanations, not to Thümmig's own elaborations" (Albrecht, 2010, p. 1178).

M. Mei (✉)
Liceo Scientifico Statale "Marco Vitruvio Pollione", Avezzano, Italia

S. d. F. Araujo et al. (eds.), *The Force of an Idea*, Studies in History and Philosophy of Science 50, https://doi.org/10.1007/978-3-030-74435-9_6

structure applied through a scientific method. The *Psychologia empirica* deals with what is acquired by our soul through experience and derives its notions in the soul from what we observe in ourselves. By contrast, the *Psychologia rationalis* leads us to the most intimate knowledge of our soul and reveals the *ratio* (reason) of what we comprehend through experience in empirical psychology.

If Wolff regards "psychology" as the science of what is possible through the human soul (*per animam humanam*) (Wolff, 1740/1972, §.1, p. 1)—in its turn, "science" is defined as the ability to obtain what remains on indubitable principles by means of correct arguments (Wolff, 1751/1983a, §.361), as can be found in the first edition of his *Vernünfftige Gedancken von Gott, der Welt und der Seele des Menschen auch allen Dingen überhaupt* (Rational Thoughts on God, the world and the soul of men, also on all things in general) of 1720. The concept of psychology and, more generally, of philosophy, intended as a "science"—and in reference to the mathematical method—is always developed in his works. However, the use of the mathematical method "is in no way incompatible with the philosopher's indispensable reliance on experience" (Dyck, 2014, p. 23) and it is in no way incompatible—that is what we would like to focus on—with the possibility of measuring some effects of our psychological life through a *psychometria* (psychometrics). Thomas Hobbes (1588–1679), René Descartes (1596–1650), Gottfried Wilhelm Leibniz (1646–1716), and Nicolas Malebranche (1683–1715)—to mention just some of the most important philosophers—have been regarded as having a "pessimistic" attitude toward *psychometria*. By contrast, Wolff has an "optimistic" approach to it (Sturm, 2006, pp. 357–362). Indeed, from the beginning of 1730, Wolff considers the possibility of *measuring* psychic phenomena by means of a newly born *psychometria*.

The goal of the first part of this chapter is to demonstrate that, according to Wolff, it is possible to have the idea of a *psychometria* through the measurement of the soul's effects. To reach this goal, we will analyze the following issues: (1) the possibility of measuring the duration and clearness of psychic phenomena (disposition and habit); (2) the possibility of measuring the intensity of psychic phenomena (imagination, memory, psychology, and physiology); and (3) the possibility of measuring degrees of attention (using the concrete example of the quantity of qualities). The goal of the second part of this chapter is to highlight the similarities between Wolff and Robert Greene (1678–1730) with regard to their ideas of a *psychometria*.

6.2 The Measurement of the Soul's Effects

In 1738, Wolff maintains that actions may not only be divided into *free* actions and *non-free* actions, but also that they feature different *degrees* of freedom corresponding to different degrees of moral responsibility. The *ratio* is proportional: the freer an action is, the greater the responsibility of the acting subject (Wolff, 1738/1971, §.607). This implies that if the degree of freedom can be measured, then *psychometria* may also be applied to measure the attribution of responsibility to the subject (Métraux, 1983). Regardless, it is quite clear that the most difficult part in such a program would be to identify the correct way of measuring the above-mentioned qualities (Wolff, 1738/1971, §.608).

In his *Discursus praeliminaris de philosophia in genere* (Preliminary discourse on philosophy in general), Wolff had already made clear that mathematical knowledge—that is, knowledge about the quantity of things (Wolff, 1740/1983b, §.14)—must be combined with philosophical knowledge in order to achieve a higher degree of certainty (Wolff, 1740/1983b, §.28). However, there seems to be a further step needed. *Psychometria* seems to provide the methodological tool that can lead to the establishment of a full-fledged measure for the qualities of our soul.

Let us proceed step by step. As Wolff theorizes the measurement of perfection and imperfection, he also tries to determine the measurement of degree of responsibility. In this regard, in paragraph 522 of the *Psychologia empirica,* he maintains some peculiar and relevant postulates (Wolff, 1732/1968, §.522): (a) *Psychometria* teaches us how we should go about measuring perfection and imperfection; (b) in order to apply *psychometria,* a psychologist needs to identify the appropriate measure for the object of his/her search; (c) the soul follows mathematical laws; and (d) mathematical truths merge with contingent truths in the human mind and in the material world.

In contrast to previous philosophers, Wolff assumes that the application of mathematical truths to the mind involves a further step: physical and psychic matters focus on *psychometria*, a form of methodological mediation that implies the ability to measure the *effects* of the soul rather than its *substance*. In other words, *psychometria* allows us to take into scientific consideration the possibility of a first form—albeit theoretical—of the *naturalization* or *mathematization* of the mind.

It should be pointed out that Wolff's resort to mathematics to deal with psychological matters must be regarded with some caution. It is our understanding that Wolff seems to highlight the importance of mathematical measuring not so much with regard to psychology *tout court* but rather through specific themes of a psychological nature (indeed, if the term *psychometria* had been a pivotal element in his psychology, Wolff would at the very least have included it in the table of contents of his *Psychologia empirica*).

How is the idea of *psychometria* possible? How is it possible to have a measure for the qualities of our soul? In this regard, as early as in 1730—in his *Philosophia prima sive Ontologia* (First philosophy or Ontology) (Wolff, 1736/1962)—Wolff maintained that (a) qualities are measurable (Wolff, 1736/1962, §.757); (b) it is possible to achieve a mathematical knowledge of quality (Wolff, 1736/1962, §.755); and (c) the way to find the measure for a quality is particularly challenging (Wolff, 1736/1962, §.752).

Concerning point (a), Wolff states that each quality is measurable (Wolff, 1736/1962, §.757). For example, the density of fluids is a quality and can be measured by means of aerometers, heat and cold can be measured by means of a thermometer, the density of air can be measured by means of a manometer, and the gravity of air can be measured with a barometer (Wolff, 1736/1962, §.757). It thus most probably does not imply a contradiction to think of the possibility of measuring qualities by means of a *psychometria*.

Concerning point (b), Wolff explains that the mathematical knowledge of things consists in the knowledge of the quantity of things; the qualities have a quantity or a degree, and therefore, it is possible to have a mathematical knowledge of degrees or qualities (Wolff, 1736/1962, §.755). He also states that there is nothing in any "thing" for which mathematical knowledge is not possible (Wolff, 1736/1962, §.756) and that the size of its degree corresponds to *"notiones imaginariae"* (imaginary notions) (Wolff, 1736/1962, §.753, p. 558).[2] However, such notions should not lead us to suppose that this is a kind of knowledge without relevance. For example, without imaginary notions there would be no increase in knowledge that could lead to a *psychometria*.

As a matter of fact, in point (c), Wolff opens the way to the development of the knowledge of natural philosophy and psychology (Wolff, 1736/1962, §.755), while stressing that there is a "common prejudice" that not all qualities are measurable (Wolff, 1736/1962, §.757). If the idea of a *psychometria* is possible, however, then it is necessary to specify how this can happen. For this purpose, Wolff first explains the possibility of measuring the duration of psychic phenomena and their degree of clearness.

6.2.1 The Possibility of Measuring the Duration and Clearness of Psychic Phenomena: Disposition and Habit

According to the possibility of measuring the *duration* of psychic phenomena, Wolff clearly states that thoughts are not immediate and that "some" time is required to allow human thought to proceed (Wolff, 1740/1972, §.27). According to Wolff, "time" is given by existing things that occur in continuous series (Wolff, 1736/1962, §.574), while "duration" is the simultaneous existence of several successive things (Wolff, 1736/1962, §.578).

Time can be represented through the imaginary notion of a straight line consisting of a continuous series of points (Wolff, 1736/1962, §.582). In this regard, Wolff points out that the size of time refers to the concept of measurement in mathematics (Wolff, 1736/1962, §.582) and that there is an analogy between time and number (Wolff, 1736/1962, §.587). Now,

[2] For example, the degree of speed cannot be divided into parts of which one exists outside the other; for this reason, the notion that we have of it is imaginary (Wolff, 1736/1962, §.752). Moreover, the notion of substance is also imaginary. Wolff also cites Robert Greene, who, in his work *The Principles of Philosophy of Expansive and Contractive Forces or an Inquiry into the Principles of Modern Philosophy,* defines the notion of substance as the result of imagination (Wolff, 1736/1962, §.773). Among imaginary notions, Wolff also mentions, for example, the notion of substance (Wolff, 1736/1962, §.773).

- Perceptions occur over time and can be partial or composite (Wolff, 1738/1968, §.40),
- Each composite entity can be measured (Wolff, 1736/1962, §.628), because it has several parts and a specific size too (Wolff, 1736/1962, §.628),
- A size can be attributed to all the things if it is possible to recognize any of those parts (Wolff, 1740/1972, §.442), and,
- If each composite entity can be measured as consisting of parts, then, in some way, the composite perceptions could also be measured, as parts of partial perceptions.

How is it possible to measure composite perceptions? First, it is possible to measure the time required for them. To clarify this, Wolff distinguishes between disposition (*dispositio*) and habit (*habitus*).[3] If the ideas corresponding to a visible object and its corresponding word are clearer over time, it is because the movement of the corresponding material ideas is faster (Wolff, 1740/1972, §.291). Not surprisingly, illness or old age can determine irregular motions in the brain; it alters knowledge and some cognitive processes connected to it (Wolff, 1740/1968, §.299). Otherwise, there is a greater or lesser speed of the material ideas present in our brain that underlies the distinction between disposition and habit so that the same subject can take a shorter time to carry out the same activity—in this case, it is a matter of an intellectual activity, but this also applies to manual activities—if it passes from condition of disposition to one of habit. By contrast, it can take a longer time if it regresses from a condition of habit to one of disposition. This is precisely the difference in terms of time that can be measured.

Moreover, as we have observed, a size can be attributed to every different part of all things. A composite perception consists of several partial perceptions (Wolff, 1738/1968, §.40), and if the partial perceptions are clear, then the corresponding composite ones are perceived distinctly (Wolff, 1738/1968, §.41). For example: I have a distinct perception of a tree if I clearly perceive the trunk, and then the branches and then the other parts that are in a tree. In fact, a perception is not detected in itself, but it becomes "a perception" in a subsequent way—"successively" (Wolff, 1740/1972, §.287, p. 229). The cause of the growing perception is given precisely by the subsequent partial perceptions (Wolff, 1740/1972, §.287). But what

[3] Wolff uses the following arguments. We compare two composite entities: a visible object—consisting of several parts, to which several partial perceptions correspond, such as a tree (*arborem*)—and the corresponding word—the word *arborem,* consisting of several syllables arranged in a specific sequence. We can more easily imagine a tree and, in general, objects that are visible, than we can smells, tastes, and tactile qualities, much as we can more easily imagine words than we can confused qualities. The reason for this is that we can have confused perceptions of qualities but distinct perceptions of visible objects and words (Wolff, 1740/1972, §.289). The example of words used by Wolff is even more significant because words are given to us by dual material ideas that are perceived through sight in a successive (*successive*) manner and through hearing in a simultaneous (*simul*) manner. However, we have to learn how to read written words, and we must develop a habit that reduces the amount of time it takes to perceive the written word, in the same way as happens with a musical instrument or with the educational process, in contrast to the conditions in which savages live, for example (Wolff, 1740/1972, §.290).

is the degree of distinct perceptions? The greater the number of the particular, clear perceptions, the greater the degree of the distinctiveness of the subsequent composite perception (Wolff, 1738/1968, §.42); the *ratio* is also inversely proportional.

Therefore, the degree of distinct perceptions changes not only from subject to subject but also in the same subject if the corresponding conditions of the cognitive process vary.[4] In any case, it is the time—by means of which you have a perception—or the number of required perceptions that allows some kind of measurement.

6.2.2 The Possibility of Measuring the Intensity of Psychic Phenomena: Imagination, Memory, Psychology, and Physiology

In the previous section, we saw the following:

- If the ideas corresponding to a visible object are clearer over time, it is because the movement of the corresponding material ideas is faster (Wolff, 1740/1972, §.291).
- There is a greater or lesser speed of the material ideas present in our brain that underlies the distinction between disposition and habit. This depends on the fact that if the motion of the sensory nerves is impressed more quickly, then the corresponding sensory idea is clearer and vice versa (Wolff, 1740/1972, §.125).

Each of the above steps helps Wolff to clarify that if something is distinctly perceived, it is easier for the memory to retain (Wolff, 1740/1972, §.293). According to Wolff, a good memory admits of different degrees that vary from subject to subject (Wolff, 1738/1968, §.190) but can be identified by a series of parameters (Wolff, 1738/1968, §.191):

- The time spent holding onto an idea in the mind.
- The number of acts by which the reproduced ideas are delivered to memory.
- The number of acts with which they are held in the memory.

For this reason, it can be affirmed that those people with a "great" memory can reproduce the ideas of many things and recognize the ideas reproduced, like those who can remember the whole Bible and can quote each part of it in the right order or those with a "long" memory who can remember a long series of things (Wolff, 1738/1968, §.192). This is why Wolff states that it is possible to have a dual measure of the size of the memory that depends on (1) the number of things previously perceived and (2) the number of things perceived in a continuous series at the same

[4] Wolff also points out that partial perceptions are not only perceptions of the parts of which the perceived thing is composed but also of the determinations or quantity and quality of the same thing (Wolff, 1738/1968, §.40).

time (Wolff, 1738/1968, §.194). Once again, time is the central pivot of Wolff's thought.

However, memory is not the only faculty to have different degrees: imagination also has degrees, to the extent that it reproduces the ideas of many things; memory, on the other hand, recognizes the ideas reproduced (Wolff, 1738/1968, §.193). Wolff tries again to offer criteria to split memory into different subjects:

- The first difference concerns the number of requested acts.
- The second difference concerns the number of reproduced ideas.
- The third difference concerns a greater or lesser repetition of ideas.
- The fourth difference is the ability to reproduce ideas that have not been reproduced for a period of time (Wolff, 1738/1968, §§.182–185).

The goodness of an individual's memory depends more generally on a disposition of the soul and in particular on the qualities of the soul (Wolff, 1738/1968, §.190), which can be subject to gradual diversification. This is due to the fact that in different subjects, there is a diversity of nerve fibrils (Wolff, 1740/1972, §.307), such that when the quality of this gradual diversification in the soul varies, the goodness of memory will also vary (Wolff, 1740/1972, §.310). Body and soul are once again closely connected and interdependent, and it is precisely because of this interconnection that the physical and the psychic can converge in a *psychometria.* That means that through a *psychometria,* it is possible to offer a measurement of the *effects* of the soul, such as the duration and clearness of psychic phenomena, dispositions, and habits, and the degrees of imagination and memory. It is like a form of mediation that seems to imply the ability to measure not the soul as *substance*, but rather its *effects.*

6.2.3 The Possibility of Measuring Degrees of Attention: A Concrete Example of the Quantity of Qualities

In the *Philosophia prima sive Ontologia,* Wolff states that degrees are the "quantity of qualities" (Wolff, 1736/1962, §.747, p. 555), and every time we talk about degrees we do not refer to objects, subjects, and activities—manual or intellectual—but to relations between objects, subjects, and activities. For example, we say that the thickness of a line is twice, three times, or four times as thick if compared to another one; in the same way, we can say that the degree of temperature in a subject is twice, three times, or four times that of another subject (Wolff, 1736/1962, §.749). The same is true for intellectual activities: we say that Tito's intellect is greater than Mevio's and thus that Mevio's intellect is smaller than Tito's. Alternatively, we say that Tito's brain is larger than Mevio's, or that Mevio's is smaller (Wolff, 1736/1962, §.751). Furthermore, the degrees do not have parts, properly called, but are imaginary; for this reason, when speaking of degrees, the notions we refer to must be imaginary, as we have already seen (Wolff, 1738/1968, §.752).

Now, let us look at degrees of attention. Wolff talks about five differences, as follows:

1. In some cases, attention is greater or smaller, depending on how much the sense organs are involved in perception (Wolff, 1738/1968, §.243).
2. In some cases, a mental content is "preserved" for a long time, while in others it is easily extinguished (Wolff, 1738/1968, §.243).
3. In some cases, people can pay attention simultaneously to several things, whereas in other cases people can pay attention to only one (Wolff, 1738/1968, §.245).
4. In certain cases, some people may pay attention to any object, while other people, such as poets, may pay attention only to poetic meditations and hardly or not at all to daily activities—toward which a certain "indifference" is manifested (Wolff, 1738/1968, §.246). In this case, the determination of the attention is strictly tied to the appetite (Wolff, 1738/1968, §.246).
5. In certain cases, some people may pay attention to a present object, whereas other people, such as poets, may pay attention to the objects typical of poetic meditation (Wolff, 1738/1968, §.247).

With regard to point 1, degree of attention can be measured by the force with which the sense organs are aroused; in point 2, by the time required for "conservation" in the mind of a perception; in point 3, the number of things upon which the attention is focused; and in points 4 and 5, the degree of attention depends on the subject's appetite. After discussing these five cases, Wolff states that if the attention varies between the different subjects—this can also happen in the same subject through exercise (Wolff, 1738/1968, §.248)—this means that the levels of attention are not equal and therefore that they are different in quantity. Thus, attention has quantities; indeed, it is a specific example of the "quantity of qualities" (Wolff, 1738/1968, §.249), which can be diversified by starting from the five criteria listed above.

This thesis is based on two statements. One is in the *Ontologia*, where Wolff states that forces can be measured through actions and actions through forces (Wolff, 1736/1962, §.743). The other is in the *Psychologia empirica*, where he states that the reflection of the soul on its acts is the ideal tool for reaching knowledge of the soul itself (Wolff, 1738/1968, §.261) and therefore of its representative force. In fact, according to Wolff, the soul is provided with a *vis repraesentativa* (representative force) through which we represent the world materially (*materialiter*), according to the position of the body in the world, and formally (*formaliter*), according to the constitution of our sense organs (Wolff, 1740/1972, §.63) and of our nerve fibrils, as we have seen.

We can try to understand this *vis repraesentativa* of the soul through its effects, that is, the perceptive acts of our cognitive faculties, which continually follow upon each other in our mind. What emerges from these considerations is a range of forces, thrusts, and tensions that come into play on a psychological level that lead ideas to overtake one another (Wolff, 1740/1972, §.138). Moreover, there is an undeniable analogy between the motive power of bodies and the perceptive force of the soul (Wolff, 1740/1972, §.77), such that both these forces are subject to certain laws. Concerning the perceptive force of the soul, the laws in question are the laws of perceptions and the operations of the mind and appetites; as for the strength of bodies, the laws in question are those of movement (Wolff, 1740/1972, §.76). This analogy leads to a consideration of the hypothesis of a *psychometria* because the mind does not proceed randomly; on the contrary, its way of functioning can be expressed through some form of measurement of its effects. The faculty of attention is the most illustrative way of proceeding, potentially offering some form of measurement of mental acts and of a *psychometria*.

On the other hand, it is true that from Wolff's perspective, resorting to mathematics does not seem to fully explain the field of psychology, the explanation of which sets out questions and issues of such complexity that—according to Wolff—mathematical measurement is not suitable. Reading the *Psychologia empirica* and *Philosophia practica universalis* (Universal practical philosophy), works that contain explicit references to the term *psychometria*, we are under the impression that not every aspect of psychology may be accessed with the certainty of truth and mathematical laws. It is sufficient, for instance, to think about the emotional components of our psychic life.[5] In other words, we can understand that it would not be correct to speak of an identity between *psychometria* and psychology. In addition, as pertinently stated by Wolff himself, the hard part of such a thesis lies in the fact that a psychologist who means to measure psychological events with a mathematical paradigm does not have a user's manual to resort to but must find the appropriate measure for the object of his/her analysis every time. The psychologists' task of finding criteria time after time that are as objective as possible for achieving scientificity in their research is thus a long one.

6.3 The Possibility of a *Psychometria*: Greene and Wolff

In light of the references to the *Psychologia empirica* and *Philosophia practica universalis*, Wolff would rightfully be seen as the *creator*, so to speak, of the particular methodological structure according to which it is possible to investigate psychic phenomena using the concepts of measurement and quantity, concepts that had been considered exclusively pertinent to the physical world up to that point. This view is supported by the *Grosses Universal-Lexicon* (Great Universal Lexicon) of Johann Heinrich Zedler (1706–1751), in which the term "Psychometria" is directly related to Wolff's philosophy and more specifically to his *Philosophia practica universalis* (Zedler, 1731–1754, vol. 29, column 1090). It should be borne in mind that the credit for the *psychometric* approach in psychology should not be attributed to the *praeceptor Germaniae*, however, because the dating is wrong (Métraux, 1983; Feuerhahn, 2003; Gundlach, 2006). This may be confirmed if we analyze one of Wolff's reviews published in 1729 in the *Acta Eruditorum* (Wolff, 1705–1731/2001,

[5]Above all, it would be difficult to establish an exact relationship between the perfection or willingness on one side and the imperfection or tedium on the other (Wolff, 1738/1968, §.522).

pp. 1839–1853) and, more precisely, the review of Robert Greene's most important work.

Greene was an eccentric philosopher, physicist, and mathematician who lived from the end of the seventeenth century and the first half of the eighteenth; his book *The Principles of Philosophy of the Expansive and Contractive Forces or an Inquiry into the Principles of the Modern Philosophy* was published in Cambridge in 1727. The secondary bibliography has analyzed attentively only a few issues with regard to Greene's natural philosophy work, such as those relating to contractive and expansive forces in the anti-Newtonian movement of the early eighteenth century (Schofield, 1970; Thackray, 1970; Heimann & McGuire, 1971; Heimann, 1978). In all those cases, the intuition of a *psychometria* is entirely neglected; by contrast, in our view, it is the distinctive trait of Greene's thought.

In Greene's most successful publication, the British author outlines a very interesting theory with regard to the leveling of psychic and physical phenomena: "[…] the Powers and Faculties of it [men's mind] are as much Reducible to Mathematicks and Proportions, or a *Psycheometria*, as any other Quantities" (Greene, 1727, p. 127).

Behind this thesis concerning a *psychometria,* there lies Greene's theory of expansive and contractive forces. According to Greene, these expansive and contractive forces cannot be self-existent but "they must be deriv'd from the Will and Pleasure of God" (Greene, 1727, p. 109). Through these two forces, we can explain every aspect of reality.[6] However, what we are most interested in is how Greene applies this theory of contractive and expansive forces to the human mind and, in particular, to perception.[7] It is precisely the different proportions between the expansive and contractive forces in human minds that provide us with "a New Set of Mathematick Propositions and Give a farther Advance to the Elements of a Psycheometria" (Greene, 1727, p. 640).

According to Greene, the foundation of a *psychometria* is possible if five benchmarks are respected: the first, "THAT the Mind does always Think"; the second, "THAT all our Ideas are not Originally from Sensation" […] because "the Mind was first Endued with an Active Power of it's [sic] own"; the third, "THAT it would be no Objection against this Original Action of the Mind"; the fourth, "THAT, therefore, if there is an Innate Intelligence and Action belonging to the Mind, it is not

[6]For example, the relationship between the Sun and the Moon: "[T]he Sun has an Expansive and Dissipating Force, as is Evident in Fact, the Earth has a Contractive, as is Evident from Gravitation, and the Moon has a Contractive, but which acts in a quite contrary Direction to that of the Earth, as is manifest from the Tides produc'd by their Contranitent Forces" (Greene, 1727, p. 106). Another example is the animal system: "[T]he Animal System has a Contractive, as well as an Expansive Force belonging to it, as is Evident from the Systole and Diastole of the Earth, from the Pulse or Contraction and Dilatation of the Arteries, and from the Expansion and Contraction of the Muscles" (Greene, 1727, p. 108).

[7]In this regard he states, "As to the Doctrine of the Expansive and Contractive Forces, Perception seems to Participate of both; It has the Greatest Expansive Force, when it Exerts it's [sic] self into Business and Action, and the Greatest Contractive, when it Retires from it, and Resigns it's [sic] self to Study and Contemplation" (Greene, 1727, p. 638).

certain but it might Pre-exist to its Union with the Material Systeme"; and the fifth, "THAT from this Innate Action and Intelligence of the Mind are in a great Measure Derived the Various Properties, which belong to Animate Beings" (Greene, 1727, pp. 621–622).

Another important element for the constitution of a *psychometria,* "or a Reduction of the Forces of the Mind to a Mathematick Reasoning," deals with "the Intension and Force of Perception" that make it possible to distinguish men from beasts and individual men from one another (Greene, 1727, p. 629). In this case, eight points are analyzed by Greene:

> 1st. THE Greater the Quickness of Apprehension is, the Greater will be the Capacity of Retaining the Idea in our Minds, and of Reviving it upon Occasion.
>
> 2nd. THE Less the Quickness of Apprehension is, the Less will be the Capacities of Retaining or Reviving our Ideas.
>
> 3rd. THE Apprehension may be Quicker, when the Faculty of Retaining an Idea may be Weaker, or of Recollecting it Slower.
>
> 4th. OR, The Apprehension may be Slower, when the Faculty of Retaining is Stronger, and that of Recollecting is Quicker.
>
> 5th. OR, The Faculty of Retaining may be Stronger, when the Apprehension may be Slower, and the Faculty of Recollecting Slower.
>
> 6th. OR, The Faculty of Retaining may be Weaker, when the Apprehension may be Quicker, and the Faculty of Recollecting Quicker.
>
> 7th. OR, The Faculty of Recollecting may be Quicker, when the Apprehension may be Slower, and the Faculty of Retaining Weaker.
>
> 8th. OR, The Faculty of Recollecting may be Slower, when the Apprehension may be Quicker, and the Faculty of Retaining Stronger (Greene, 1727, p. 629).

According to Greene, these seem to be the cases that can occur when we speak about Apprehension, Retention, and Recollection, and each of them can be considered according to different proportions. Retracing the scheme examined by Greene we may, in the first case, for example, find ourselves faced with a type of balanced and rational man; in the second, with a stupid one; in the third, with a brilliant and broad-minded man struggling to remain intent in his contemplation; in the fourth case, a scholar but "Dull in Apprehending Things" (Greene, 1727, p. 630); in the fifth, a man who has to "Retain" the ideas in his mind because he has some contemplations and apprehensions "Languid and Weak" (Greene, 1727, p. 630); in the sixth, a man who hardly manages to devote himself to the study or intense contemplation of his ideas and therefore who has "a Quick and Lively Apprehension" (Greene, 1727, p. 630); in the seventh, a man who must have a good memory because his "Apprehensions are not Extraordinary" (Greene, 1727, p. 630); and in the eighth, a man with a fallible memory and who, precisely for this reason, must be "Tenacious of the Ideas" (Greene, 1727, p. 630).

According to Greene, the differences between men are based on the different proportions and relationships between their cognitive faculties. For example, with regard to our Understanding, Greene speaks of different typologies:

> The First might be called, The Apprehensive-Retentive-Recollective.
>
> The Second, The Non Apprehensive-Non Retentive-Non Recollective.
>
> The Third, The Apprehensive-Non Retentive-Non Recollective.

> The Fourth, The Retentive-Recollective-Non Apprehensive.
> The Fifth, The Retentive, Non Apprehensive-Non Recollective.
> The Sixth, The Apprehensive-Recollective-Non Retentive.
> The Seventh, The Recollective-Non Apprehensive-Non Retentive.
> The Eighth, The Apprehensive-Retentive-Non Recollective (Greene, 1727, p. 630).

For all these cases, Greene says, there can be thousands of different proportions and relationships that—if analyzed with a new cognitive approach—would open the way to a new mathematical knowledge of the human mind.[8] The intent—undoubtedly complex but in any case ambitious—is to reach a level of truth in the future with regard to what Wolff called the "quantity of qualities." The *praeceptor Germaniae* insists on different concepts—such as strength, speed, time, intensity, number, and duration—in order to study the functioning of the human mind with a scientific approach. Through his investigations, Greene is trying to find the criteria to bring the mind back to proportions, patterns, and mathematical rigor. He analyzes the mind starting from the concepts of strength, intensity, and speed, even going so far as to distinguish different types of understanding that correspond to different types of men.

The starting points of Greene and Wolff are certainly different, but the common denominator seems to be always the same: to identify one or more ways of studying the functioning of the human mind from a mathematical point of view. This is possible for both of them, at least on a theoretical basis, precisely by means of a *psychometria.*

6.4 Conclusion

In Wolff's writings of the 1730s, the development of the conception of psychology toward *psychometria* is important because it gives some idea of the way in which certain fortunate intuitions were often the result of relentless collaboration among the scientists of the time. This is important because it allows us to understand why some of Wolff's ideas on psychology—already found in outline in his philosophy and starting from a particular historical and cultural time—took a specific direction rather than others.

We may say that Wolff's *Psychologia* may be considered scientific knowledge not because it turns out to be measurable through *psychometria* but rather, as we have seen, because it is part of a *system* in which mathematical knowledge—used to attain a higher level of scientificity—was matched with philosophical knowledge (Wolff, 1740/1983b, §.28). However, it is also true that by considering the possibility of measuring psychic phenomena, Wolff enriched the conception of psychology

[8] This aspect is strongly affirmed by Greene when he states that all the cases analyzed above "Furnish abundant Matter for Mathematicians to Employ their Speculations about; and which would as much Answer the Purposes of a True Knowledge and a much more Important one, as the Proceeding (which is at Present done) in the Inquiry only into the Relations and Proportions of Abstracted Quantities in Extension" (Greene, 1727, p. 629).

as *scientia,* using the mathematical calculation in the psychological field.[9] The elaboration and development process of a psychometric theory—first formulated by Greene and which, in eighteenth-century Germany, found a new theorization in Wolff's psychology—came to a halt mostly because of Kant's hegemony.

According to the philosopher from Königsberg, empirical psychology may have a marginal place in metaphysics. The role that Wolff gave to psychological knowledge has a completely different relevance; the knowledge of the soul may be perfected and enriched by resorting to mathematical laws and truths that allow us to measure psychic phenomena from a quantitative perspective, such as from the *number* of psychic actions required for one or more series of perceptions, and from a qualitative perspective—with regard to the higher or lower *intensity* of our perceptions.

Thus, if it is true that psychology was not considered as a *discipline* any earlier than the nineteenth century (Gundlach, 2006), it is also true that the so-called *historical antecedents* of psychology as *scientific* knowledge are undoubtedly to be found in Wolff's *Psychologia empirica*, that is, the work in which the author detects the possibility of analyzing psychic phenomena through mathematical, albeit still theoretical, study (Sinatra, 2005).

In this regard, Schwaiger asserts, "Finally he sketches a theory of psychometry, which gave birth to the highly influential idea of the quantification of psychic phaenomena" (Schwaiger, 2010, p. 1295). This optimism about a *psychometria* comes from the fact that it is possible to analyze the mind through its effects, treating psychological studies as a beautiful *theatrum anatomicum* (anatomical theater) of the human mind (Kertscher, 2018, p. 175). As in a *theatrum anatomicum,* in which corpses are dissected to analyze the anatomy of the human body and to help improve medicine, Wolff tries to analyze in greater depth the human mind through the observation of its effects—intensity, degrees, and time—with the purpose of helping to advance the field of psychology. It was not a "Copernican revolution." It was a new "trend" in psychological knowledge but nonetheless significant in the history and development of psychology and philosophy. This importance comes from the fact that Wolff tries to find criteria that are as objective as possible to study the soul through its effects in the increasingly close relationship between physical and psychic phenomena.

References

Aalto, S. K. (1981). The rise and fall of moral algebra: Francis Hutcheson and the mathematisation of psychology. *Journal of the History of the Behavioural Sciences, 17*, 343–356.

Albrecht, M. (2010). *Thümmig, Ludwig Philipp* 1697–1728. In H. F. Klemme & M. Kuehn (Eds.), *The dictionary of eighteenth-century German philosophers* (Vol. III, pp. 1177–1181). Continuum.

[9] In fact, there had been attempts at mathematical calculation in the moral field by, for example, Christian Thomasius (1655–1728) and Francis Hutcheson (1694–1746) (see Aalto, 1981, pp. 343–56; Wood, 2003, p. 815). However, these attempts are not regarded as an integral part of the *scientific* conception of psychology, in contrast to Wolff's.

Alsted, J. H. (1989). *Encyclopaedia* (4 Vols.). Frommann-Holzboog. (Original work published 1630).

Carus, F. A. (1990). *Geschichte der Psychologie*. Springer. (Original work published 1808).

Dyck, C. W. (2014). *Kant and rational psychology*. Oxford University Press.

École, J. (1990). *La métaphysique de Christian Wolff* (2 Vols.). Olms.

Feuerhahn, W. (2003). Entre métaphysique, mathématique, optique et physiologie: la psychométrie au XVIIIe siècle. *Revue Philosophique de la France et de L'étranger, 128*, 279–292.

Greene, R. (1727). *The principles of philosophy of the expansive and contractive forces or an inquiry into the principles of modern philosophy*. Cambridge University Press.

Gundlach, H. (2006). Psychology as science and as discipline: The case of Germany. *Physis, 43*, 61–89.

Heimann, P. M. (1978). Voluntarism and immanence: Conceptions of nature in eighteenth-century thought. *Journal of the History of Ideas, XXXIX*, 271–283.

Heimann, P. M., & McGuire, J. E. (1971). Newtonian forces and Lockean powers: Concepts of matter in eighteenth-century thought. In R. McCormmach (Ed.), *Historical studies in the physical sciences* (pp. 233–306). University of Pennsylvania.

Kertscher, H. J. (2018). *"Er brachte Licht und Ordnung in die Welt". Christian Wolff – Eine Biographie*. Mitteldeutscher Verlag.

Mecacci, L. (2008). *Storia della psicologia del Novecento*. Laterza.

Métraux, A. (1983). An essay on the early beginnings of psychometrics. In G. Eckardt & L. Sprung (Eds.), *Advances in historiography of psychology* (pp. 241–247). Deutscher Verlag der Wissenschaften.

Schofield, R. E. (1970). *Mechanism and materialism. British natural philosophy in an age of reason*. Princeton University Press.

Schwaiger, C. (2010). Wolff, Christian (von) (1679–1754). In H. Klemme & M. Kuehn (Eds.), *The dictionary of eighteenth-century German philosophers* (Vol. 3, pp. 1290–1297). Continuum.

Sinatra, M. (2005). *Storia della psicologia prescientifica. Lo sperimentale prima di Wundt*. Progedit.

Sturm, T. (2006). Is there a problem with mathematical psychology in the eighteenth century? A fresh look at Kant's old argument. *Journal of the History of the Behavioral Sciences, 42*(4), 353–377.

Thackray, A. (1970). *Atoms and powers: An essay on Newtonian matter-theory and the development of chemistry*. Harvard University Press.

Wolff, C. (1962). Philosophia prima, sive Ontologia, methodo scientifica pertractata, qua omnis cognitionis humanae principia continentur (New edition emended from the prior). In J. École (Ed.), *Christian Wolff: Gesammelte Werke* (Ontologia) (Vol. 3, Pt. II). Olms. (Original work published 1736).

Wolff, C. (1968). Psychologia empirica, methodo scientifica pertractata, qua ea, quae de anima humana indubia experientiae fide constant, continentur et ad solidam universae philosophiae practicae ac theologiae naturalis tractationem via sternitur (New edition emended from the prior, with privileges). In J. École (Ed.), *Christian Wolff: Gesammelte Werke* (Psychologia empirica) (Vol. 5, Pt. II). Olms. (Original work published 1738).

Wolff, C. (1971). Philosophia practica universalis, methodo scientifica pertractata, pars prior (New edition emended from the prior, with privileges). In J. École (Ed.), *Christian Wolff: Gesammelte Werke* (Philosophia practica universalis) (Vol. 10, Pt. II). Olms. (Original work published 1738).

Wolff, C. (1972). Psychologia rationalis, methodo scientifica pertractata, qua ea, quae de anima humana indubia experientiae fide innotescunt, per essentiam et naturam animae explicantur, et ad intimiorem naturae eiusque autoris cognitionem profutura proponuntur (New edition emended from the prior, with privileges). In J. École (Ed.), *Christian Wolff: Gesammelte Werke* (Psychologia rationalis) (Vol. 6, Pt. II). Olms. (Original work published 1740).

Wolff, C. (1983a). Vernünfftige Gedancken von Gott, der Welt und der Seele des Menschen, auch allen Dingen überhaupt. In J. École (Ed.), *Christian Wolff: Gesammelte Werke* (Deutsche Metaphysik) (Vol. 2, Pt. I, 11th ed.). Olms. (Original work published 1751).

Wolff, C. (1983b). Philosophia rationalis sive logica, methodo scientifica pertractata et ad usum scientiarum atque vitae aptata. Praemittitur discursus praeliminaris de philosophia in genere. In J. École (Ed.), *Christian Wolff: Gesammelte Werke* (Discursus praemlinaris) (Vol. 1.1, Pt. II). Olms. (Original work published 1740).

Wolff, C. (2001). The principles of the philosophy of the expansive and contractive forces &c. In H. A. Laeven & L. J. M. Laeven-Aretz (Eds.), *Christian Wolff: Gesammelte Werke*. Sämtliche Rezensionen in den Acta Eruditorum (Vol. 4, pp. 1839–1853) (Vol. 38, P. II, 1–5). Olms. (Original work published 1705–1731).

Wood, P. (2003). Science, philosophy, and the mind. In R. Porter (Ed.), *The Cambridge history of science. Eighteenth-century science* (Vol. 4, pp. 800–824). Cambridge University Press.

Zedler, J. H. (1731–1754). *Grosses vollständiges Universal Lexicon aller Wissenschaften und Künste* (64 Vols.). Zedler.

Chapter 7
Wolff on Monadology and "Materialisterey"

Falk Wunderlich

7.1 Introduction

This chapter deals with a few core metaphysical aspects of Wolff's psychology, specifically with his understanding of monadology and the doctrine of pre-established harmony.[1] I will try to specify to what extent Wolff repudiates Leibniz's understanding of monads as the basic elements the world is composed of, and what the alternative he puts forward consists in (Sects. 7.2 and 7.3). I will also reconstruct Wolff's motivation behind this repudiation that I believe is based on his opposition to materialism and his supposition that Leibniz's doctrine of monads might not be able to avoid it (Sect. 7.4). My topic is thus not restricted to rational psychology but touches on general ontology and cosmology as well. For, although Wolff restricts pre-established harmony to the interaction between human mind and body, his repudiation of Leibniz's version of it is based on broader metaphysical considerations. Leibniz's monadology is a general ontological doctrine, and what Wolff finds dubious about it pertains to his alternative general ontology (which he develops mostly within his cosmology). I will thus look more closely into Wolff's theory of simple substances.

In Wolff's own interpretation, Leibniz is straightforward as to how simple things act: Leibniz argues, according to Wolff, that each simple thing *represents* the entire world from its own perspective. Wolff considers this to be a consistent explanation of how each simple thing differs from all other simple things, and how each simple

[1] A preliminary version of this chapter was presented at the conference "Christian Wolffs Deutsche Metaphysik/Christan Wolff's German Metaphysics" at Martin-Luther-Universität Halle-Wittenberg. I would like to thank the participants of this conference for their helpful comments, as well as Eric Watkins for his detailed and thorough comments on a later draft.

F. Wunderlich (✉)
Martin-Luther-Universität Halle-Wittenberg, Halle, Germany
e-mail: falk.wunderlich@phil.uni-halle.de

S. d. F. Araujo et al. (eds.), *The Force of an Idea*, Studies in History and Philosophy of Science 50, https://doi.org/10.1007/978-3-030-74435-9_7

thing is related to the entire world. However, he remains hesitant: "But I still have reservations about adopting this" (Wolff, 1751/1983, §.598, p. 368), and he becomes more outspoken in the *Anmerckungen*, where he remarks that "he cannot endorse what Herr von Leibniz has taught about the *monadibus*" (Wolff, 1724, p. 462). So while Wolff appreciates Leibniz's monadology and deems it a consistent philosophical theory, he ultimately remains unconvinced. Nonetheless, he states that it is not only internally consistent, but also compatible with his own doctrine: it is "not contrary to what we have established with regard to the elements of the world" (Wolff, 1751/1983, §.599, p. 369). But what are the reasons for these "reservations"? And since Wolff does not reject monadology in general, how does he modify and appropriate it? The concept of "element" is Wolff's replacement for that of monad, as I will discuss in more detail below. The question thus is to what extent Leibniz's doctrine of monads goes together with Wolff's doctrine of elements, and where their differences lie. Wolff's own take on Leibniz's monadology has been discussed in the literature on various occasions. Brandon Look, for instance, argues that Wolff advances a "non-Leibnizian theory of simple substances" by rejecting their representational nature (Look, 2013, p. 199). According to Look, this explains why Wolff is a weak advocate of the doctrine of pre-established harmony: Wolff claims that it is a likely hypothesis, but remains mostly *agnostic* about further details (p. 201). In a similar fashion, Eric Watkins argues that Wolff is agnostic about whether *all* simple substances must have a representative power (see Watkins, 2006, pp. 275–90, in particular p. 281).[2] So according to this view, Wolff is mainly less straightforward than Leibniz and rather remains agnostic about some of its more ambitious aspects.

In this chapter, I want to defend a slightly different though not entirely incompatible view on this score: Wolff's "doctrine of elements" is a *minimal theory* of simple things and their operation. This doctrine of elements is minimal, compared to Leibniz's monadology, in that it does *not necessarily include* that simple things as such have mental attributes. However, it does *not exclude* mental features either. According to the minimal theory, all simple things have a certain set of attributes, but only some of them have mental attributes *in addition*.[3]

My suggestion is thus more in line with the (albeit brief) accounts of Christian Leduc and Gualtiero Lorini (Leduc, 2018; Lorini, 2016). Leduc also argues that Wolff with the "elements" introduces a basic concept that is quite different from Leibnizian monads but not incompatible with them, and not just a less ambitious or more cautious version of Leibniz's conception.[4] I thus call Wolff's theory of simple

[2] See also Fabian (1925, pp. 34–40) who notes Wolff's hesitation concerning the powers of simple substances and holds that Wolff eventually advances a form of dualism. On Wolff's doctrine of soul and his relation to Leibniz compare also Blackwell (1961), Corr (1975) and École (1986).

[3] Watkins would agree thus far, I take it that the main point of disagreement here is whether this is primarily an agnostic thesis about the ultimate constituents of the world, or an actual, positive claim about them.

[4] Leduc (2018) argues that Wolff introduces the concept of element to designate the first constituents of bodies in the first place (i.e. without mental attributes at all) and mentions Leibniz just as one example of such a theory among others.

substances a minimal theory because it is partly, but not entirely agnostic about their essence. Hence, although we do not know much about them, we do know something and can establish this as a minimal metaphysical theory.

7.2 Wolff's Minimal Theory of Elements

The passages where Wolff notes his hesitations about Leibniz's monadology that I have discussed so far remain rather vague as to what exactly bothered Wolff. There is one passage in a recently edited letter, though, that Leduc (2018) mentions and that is quite informative in this regard:

> I have known the abbot Mr. Conti for many years, and I have received letters from him for almost 30 years, or even longer, dating back to when Herr von Leibnitz' [sic!] monads still used to be a puzzle; although even now only a few know them and have an appropriate notion of his system which only begins where mine ends. The confusion, however, is due to Herr Bülffinger who first came up with the concept of a Philosophia Leibnitio-Wolfiana. And thus one could still say that the Leibnizian monads on which his actual system is built are a puzzle that has not been fully resolved, and that I do not like to resolve although I could, because I do not need it in my endeavours and therefore I leave it at that, in its value and its disvalue. (Middell & Neumann, 2019, vol. 2, pp. 287–292).

This letter provides some interesting information. First, Wolff obviously takes issue with being too closely aligned to Leibniz. Second, Leibniz's monadology is a puzzle that Wolff has no intention to solve (though he claims that he could), because it is unnecessary for his purposes. So neither his doctrine of elements nor the purported pre-established harmony between human mind and body require a full understanding and acceptance of Leibniz's doctrine. This suggests that Wolff intends his own theory to be more independent and different from Leibniz's one than one might expect.

According to Wolff, simple things must exist because the world is a composite thing, and composite things must ultimately consist of simple things. There must be simple things ultimately because otherwise, the parts of the composed things would have to consist in yet smaller parts, according to Wolff. If that were so, however, "we could give no reason where the composed parts ultimately come from", in a similar way as "we could not comprehend how a composite number originates if it did not contain any units in it" (Wolff, 1751/1983, §.76, p. 36). The principle of sufficient reason demands that such a reason exists. Thus, there must be simple things, according to Wolff, because only simple things can serve as the sufficient reason of composed things whose parts they are. Watkins (2006, p. 276f.) has provided a concise formal reconstruction of this argument, the essence of which is, in abbreviation:

1. If there were no simples, everything would be composite.
2. If everything were composite, all parts of composites would be further divisible (in infinitum).
3. If all parts were divisible into parts, there would be no ultimate reason for the existence of the parts.

The key to Wolff's argument seems to consist in the third premise: infinitely divisible parts lack a proper reason of their existence. But why is that, i.e. what exactly would qualify as a sufficient reason of parts? Watkins (2006) discusses a few clues from Wolff's *Ontologia* that are helpful here although they ultimately rest on a premise that is itself problematic (Wolff, 1736/1977a, §§.686, 533, 789–791). In brief, in the *Ontologia* Wolff argues that composition as a mere relation is *fundamentally accidental*, i.e. the essence of a composite being must be something else and thus requires a substance, according to Wolff. If composition is only accidental, only the entities that are combined can be substances, and thus composition ultimately depends on the things that are combined.

This line of reasoning illustrates why Wolff thought that combination requires non-composite ultimate parts, although it seems problematic in itself. One might object that physically indivisible atoms could serve as a foundation as well, though Wolff would reply that physical atoms are still divisible geometrically. Another potential objection is that Wolff still does not explain how exactly unextended parts can be combined in such a way as to yield extended composites, i.e. that the most pressing issue is not solved in the way suggested. But these issues are not important for the purposes of this chapter.

Having established the necessity of simple things, Wolff calls those simple things the world is composed of "elements" in the cosmology section of *Deutsche Metaphysik* (Wolff, 1751/1983, §.582). These elements exist by themselves (compare Wolff, 1751/1983, §.127) and can only cease to exist by annihilation (Wolff, 1751/1983, §.102), i.e. they are substances. Wolff infers from the simplicity of simple substances that (1) they have no size and no parts, (2) they are not composed of other things, (3) they do not occupy space, and (4) they do not have inner physical motion (Wolff, 1751/1983, §.73).[5] These are the attributes of extended, composite things; thus, simple things are heterogeneous with them; i.e., "we cannot attribute anything we perceive in them [composite things] to simple ones" (Wolff, 1751/1983, §.82, pp. 40). As the elements are simple substances, the same applies to them (compare Wolff, 1751/1983, §§.583ff.), and as simple things and thus the elements are indivisible, they can only be limited by degrees (Wolff, 1751/1983, §.125); i.e., they only have different intensive magnitudes. Things that are both indivisible and limited by degrees have a power, according to Wolff:

1. Indivisible things contain a manifold because otherwise there would be no ground for the different degrees that are the basis of their limitations.
2. Because this manifold is grounded in a simple thing but is also mutable and thus not absolutely necessary, it can become real only through the action of the simple thing.
3. This action arises from a continuous effort, and thus a simple thing must have a power.

[5] Wolff (1751/1983, §.583) makes it clear that the internal power of simple things is not to be confused with an inner *motion*, i.e. a kind of physical motion, as that would require real parts, which a simple thing obviously does not have.

Wolff seems to argue here that due to their simplicity, simple things cannot be acted upon "from outside" like material objects that can be pushed or otherwise mechanically impacted by other material objects. But simple things such as the human soul or the elements of material objects do undergo modifications. If there is no intelligible way of there being an external influence on simple things, though they do change, this change must have an internal source. Thus, simple things can only have an *internal* source of modifications, while at the same time, they must have a source of modifications since they are not absolutely necessary and thus mutable (Wolff, 1751/1983, §.127).

In §.597 of *Deutsche Metaphysik*, Wolff discusses these internal sources in more detail: the inner state of a simple thing is nothing but the kind of "limitation" of the grounds of its existence, and its modifications are changes in its limitations—in the same way as, for instance, concepts are limitations of the soul (Wolff, 1751/1983, §.121). The modifications of a simple substance are the changes of its limitations, and at the same time, they are nothing but the alterations of the degree of its powers, according to Wolff (compare Wolff, 1751/1983, §.115). Simple things continuously act in this way, and thus, their actions must have an effect not only on all other simple things, but also on all composite things in the entire world since they are all connected. But Wolff deliberately leaves it open at this stage what *exactly* the effects of the actions of the simple things are, thus he is agnostic in this quite specific regard.

In the *Anmerckungen*, Wolff states that there is a general harmony between the states of simple things, in the sense that their states correspond to each other (Wolff, 1724, p. 337). But it remains an open question what exactly this harmony consists in. Wolff thus acknowledges that he has not determined yet what the inner state of elements and their powers consists in. So, again, Wolff's ontology is clear about some things: that there are simple elements composite things are composed of and that they have a power, but agnostic about others, e.g. what that power is, or whether it is one and the same for all elements. The latter point—that there may be different kinds of elements with different kinds of powers—is an idea Wolff discusses in another passage in the *Anmerckungen*, in fact in one of the few passages where he reveals a little more of his metaphysics of elements (Wolff, 1724, §.215). Wolff here again protests against the "allegation" that he identified Leibnizian monads with the basic elements of the world. While he acknowledges that the elements must have a power that causes the modifications of their states, he finds fault with the Leibnizian idea that all elements have *one and the same kind of power*. He here ponders whether the elements of *corporeal things* may have a specific power (distinct from the power of mind-like things) "wherefrom the power of the bodies that reveals itself in addition to their modification in motion can be deduced in an intelligible way" (Wolff, 1724, §.215, p. 335). So, the specific power of the elements of corporeal things would be one to which modifications of material objects can be *reduced*, and it would allow one to explain the modifications of material objects *other than* motion.

Moreover, Wolff maintains that he also has a proof of the mental nature of many other simple things that represent the world in a less perfect way than the human soul. Wolff hence does allow for simple things that only have dark, i.e. unconscious representations, and are thus in a persistent sleep-like state (Wolff, 1751/1983,

§.900). This description matches the one Leibniz has given for the lowest kind of monads, "simple" monads. But even though Wolff believes in the existence of permanently "sleeping" simple things, he explicitly rejects Leibniz's view here that *all* elements of the world are of this kind: "I have mentioned above already … that I will leave it undecided for the time being whether the elements are the kind of things that represent the world in an obscure way, that is without being conscious of it" (Wolff, 1751/1983, §.900, p. 560). So according to Wolff, there may be simple things that only have unconscious mental states, but there may also be simple things that have no mental states at all.

Wolff concludes his discussion of simple things or elements on the cosmological level by announcing that he will prove in his rational psychology (Wolff, 1751/1983, §.742) that the human soul belongs to the class of simple things. Significantly, this proof consists in an anti-materialist argument; i.e., Wolff seeks to prove the simplicity of the soul *by way of* proving that matter cannot think and that the soul has a power to represent the world (Wolff, 1751/1983, §.753).[6] Wolff thus claims that he has a proof for the mental nature of *some* of the simple substances, namely human souls, whereas for Leibniz, *each* simple thing represents the entire world (Wolff, 1751/1983, §.598). Hence, it seems obvious that Wolff mainly takes issue with Leibniz's view that all basic elements in the world have mental features, a view that has often been called Leibniz's "panpsychism" or "idealism". I will use the former term here because Wolff sets aside idealism for George Berkeley (1685–1753) alone (Wolff, 1740/1994, §.36).[7]

For Wolff, Leibnizian monads are merely a *possibility*: "such things are possible, like the *Leibnizian unities of nature*" (Wolff, 1751/1983, §.599), because monads have all the features required for simple substances (according to Wolff, 1751/1983, §.597). These common features are that (1) the inner state of a simple thing is its kind of limitation; (2) its modifications are the changes of its limitations; and (3) its modifications are nothing but the modifications of the degrees of its powers. This much holds true of all simple things per se, regardless of their further metaphysical nature. So according to Wolff, we know quite a few things about the basic elements of the world, significantly more than a thoroughly agnostic theory would maintain.

Leibniz's monadology thus turns out to be a *more specific version* of Wolff's minimal theory of elements: according to the minimal theory, the inner state of all elements refers to all other things in the world, and Leibniz *explains this relation* more specifically as one of *representation*, the representation of the entire world in every simple thing according to its location: "and *Herr von Leibniz* had explained this [relation] in such a way that in each simple thing the entire world is represented according to the point where it is" (Wolff, 1751/1983, §.600, p. 370). The minimal theory of elements, on the contrary, merely states that the simple elements are related to all other things while leaving it undetermined or more abstract in what

[6] I will analyze this peculiar argument in Sect. 7.3 in more detail.

[7] For an overview of the discussions of Leibniz's "idealism" or "panpsychism", see Smith (2011, pp. 101–105).

that relation consists. Remarkably, though, Wolff does not only cite Leibniz's monadology as a kind of theory that shares a common ground with his theory of elements, but also that of Henry More (1614–1687) (Wolff, 1737/1977b, §.182), albeit in a rather superficial way.[8] More (1671, pp. 75–87) does speak of monads, mostly "monades physicae", but merely as a synonym for atoms.[9]

Wolff could conceive of Leibniz's doctrine of pre-established harmony as one *potential application* of his own theory of simple things: it consists in the representations of the elements matching with the states of the world, whereas Wolff allows for the possibility that this relation may be explained differently, i.e. not as one of representation. One might call this indeterminacy "agnostic", but it is a rather local agnosticism that does not extend any further. Wolff also readily acknowledges that *some* of the elements of the world, namely souls, do relate to other things by way of representation; he only questions whether all elements of the world operate in this way. This explains why there is not a lot of disagreement between Wolff and Leibniz when it comes to rational psychology as such, as I will discuss in what follows.

7.3 Wolff on Simple Souls

I will address two points here: First, briefly what Wolff positively has to say about simple souls; and second, how he conceives of pre-established harmony in relation to mind–body interaction.

Wolff concludes that because material bodies and matter in general cannot think, the entity underlying thought must be simple. Since all simple things are substances and thus exist per se, also the soul is a substance (Wolff, 1751/1983, §.743). Because every simple thing has a power, it can be the source of its own modifications (Wolff, 1751/1983, §.127). Thus, the status of the soul as a substance comes with a power that is the source of its modifications (Wolff, 1751/1983, §.744), and it can only have one single power since the soul has no real parts.

In the *Psychologia rationalis* Wolff states: "The elements of material things are not spirits" (Wolff, 1740/1994, §.644, p. 588). It is important to note that in Leibniz's terminology, a spirit is a specific kind of monad; i.e., not all monads are spirits in spite of their mental features.[10] Only monads that have apperception, beyond perception and appetition, qualify as spirits, for Leibniz. Wolff explicitly acknowledges this and states: "Leibniz admittedly attributed perception and appetition to his monads, from which he thought that they are the elements of material things, but without apperception" (Wolff, 1740/1994, §.644n, p. 589). Thus, Wolff does not reject the Leibnizian concept of monad in this very passage but rather acknowledges that Leibniz does not hold the problematic view that brute material things have fully

[8] I take this reference to More from Leduc (2018, p. 46).

[9] On this, see Reid (2012, pp. 44–51), in particular p. 51.

[10] Compare Leibniz' *Monadology* (1714/1885, p. 609).

developed mental states—notwithstanding Wolff's general reservation about Leibnizian pre-established harmony.

So, what is Wolff's view of pre-established harmony when applied to the mind–body relation alone? Notably, experience plays a crucial role in Wolff's argument. He maintains that it is a basic fact of experience that there is often a "harmony" between the thoughts (or mental states) of our soul and bodily events. That is, we often experience accurate mental representations of the actual state of our body, for instance when a part of our body is modified in a way that results in an injury. Our representation of our bodily states is correct more often than not.[11] That is what he has in mind when he speaks of the experiential fact of harmony. But experience alone just reveals that there is *some* harmony, i.e. that mental states match with bodily ones, but it does not include any hint as to the metaphysical grounds of it, according to Wolff.

On the basis of this experiential advantage for pre-established harmony, Wolff criticizes the two other "systems" of mind–body interaction, i.e. physical influx and occasionalism. Against physical influx, or *natürlicher Einfluß*, Wolff objects that it is a naive common sense doctrine whose followers *believe* that it is based on experience while it is really not. He sees important theoretical arguments against physical influx, primarily the familiar one that minds acting on bodies would increase the motive force in the world, which however should remain constant (Wolff, 1740/1994, §.762). According to occasionalism (in Wolff's reconstruction under the name of "Cartesianism"), God causes thoughts in the mind and motions in the body, and the soul is only the occasional cause of certain motions in the body such as voluntary ones. Wolff objects that the actions of body and soul are not sufficiently distinct from those of God, according to this system, and since they are not primarily grounded in the nature of body and soul, they are miracles.[12]

Wolff insists that it is not enough to maintain that God has set up the harmony between mental and bodily states, which would make pre-established harmony collapse into occasionalism. To distinguish pre-established harmony from occasionalism, it is crucial for Wolff that the chains of modifications both in bodies and in minds follow their immutable orders respectively, so that it suffices that they have been put into harmony *only once*, i.e. at the very creation. After this, they have to be able to proceed based on their own laws paralleling each other, without further divine intervention. But these demands are fulfilled by Leibnizian pre-established harmony, and Wolff thus states that "in the system of pre-established harmony the interaction [commercium] between soul and body … is explained in an intelligible way" (Wolff, 1740/1994, §.620, p. 552) and that as a consequence, the "system of pre-established harmony … is to be preferred in rational psychology over the other systems explaining the interaction between soul and body" (Wolff, 1740/1994,

[11] Wolff discusses various empirical cases of harmony between mind and body in the section on empirical psychology in *Deutsche Metaphysik* (Wolff, 1751/1983, §§.527–539).

[12] This criticism of Malebranche is neither original (i.e. adopted from Leibniz) nor fair, because in Malebranche, God acts through natural laws and not by miraculous intervention (see Perler & Rudolph, 2000, pp. 229–234).

§.639, p. 581). It thus turns out that Wolff can easily accommodate Leibnizian pre-established harmony to his own views of mind–body interaction and thus prefer it *within rational psychology*, and why he can do so.

7.4 Against "Materialisterey": Why Did Wolff Find Fault with Leibnizian Panpsychism?

After having discussed the internal mechanics of Wolff's theory of simple elements and the arguments he makes explicit against Leibnizian panpsychism, it remains to be seen what the philosophical motivation of his rejection of the latter could be. Watkins has pointed out that Wolff asked Leibniz for a proof of the mental nature of the basic elements but never received an answer. Thus, Wolff's motivation could be that while he grants that some simple things, namely souls, have the power to represent the world, "he notes that it has not yet been proved that all simples must have such a power" (Watkins, 2006, p. 282). Leibnizian panpsychism is thus problematic plainly because there is no appropriate proof available. In what follows, I would like to discuss another reason why Wolff would consider Leibniz's proposal problematic (not in contrast but in addition to Watkins).

It is first striking that Wolff's rejection of materialism plays a central role within the very arguments of *Deutsche Metaphysik* and *Psychologia rationalis*. In the rational psychology part of *Deutsche Metaphysik*, the rejection of materialism figures at the beginning right after the introductory definitions of consciousness and thought (Wolff, 1751/1983, §.727–737). Similarly, the claim that material bodies cannot think is the first actual claim Wolff makes in *Psychologia rationalis*: "The body cannot think" (Wolff, 1740/1994, §.44, p. 29). Moreover, Wolff explicitly seeks to establish the simplicity of the soul *by way of rejecting materialism*, i.e. in an indirect way and distinctly not by a positive argument like, for instance Descartes does. As Wolff puts it:

> Because a body can, according to its essence and nature, neither think (§.738.739) nor can a power to think be communicated to it or to matter (§.741); the soul cannot be anything corporeal nor consist of matter (§.192). And since it becomes evident from the proofs of the reasons discussed that thoughts cannot inhere in a composite thing; the soul must be a simple thing. (Wolff, 1751/1983, §.742, p. 463)

And since the soul is a simple substance, Wolff concludes: "Materialism is a false hypothesis" (Wolff, 1740/1994, §.50, p. 33). There are other indications that Wolff did see materialism as the foremost opponent of his theory, for instance when he complains in the *Anmerckungen* that "materialism [die Materialisterey] gets out of control these days—unfortunately!—, and thereby people drawn to salaciousness are pulled away from religion and call the immortality of the soul into question" (Wolff, 1724, p. 413).

The main point here is that Wolff's hesitation about panpsychism and his rejection of materialism are in fact two sides of the same coin. Even though there is no doubt that Leibniz opposed materialism, one could still see materialist potential in the claim that all ontologically basic things in the world, monads, have mental states because it includes that also the elements material objects are composed of have them. It is thus possible that Wolff thought that Leibniz's doctrine is not strong enough against materialism, or that it even unintentionally provides a foundation materialists could make use of. First, since monads are the only basic elements of the world, all elements of the world have mental features, including the monads material objects are composed of. But this comes dangerously close to claiming that matter can think in an at least rudimentary way, which Wolff avoids by claiming that the elements of material objects do not have this feature. Second, monads are inherently active, since all monads have appetitions. As matter too is based on monads, materialists could exploit this feature for their purposes, since many of them (with the potential exception of Hobbes) consider matter to be fundamentally active.[13] This is not to say that Wolff or anyone else constructs Leibniz as a secret materialist, but rather that Wolff saw that materialists could disingenuously use Leibnizian panpsychism as a source of inspiration.[14]

There are also more general reasons for Wolff to put particular emphasis on the rejection of materialism during the first decades of the eighteenth century. As he himself notes in the passage quoted above, *die Materialisterey* was on the rise. Although early modern materialism is often associated with Thomas Hobbes (1588–1679) and Baruch Spinoza (1632–1677), John Locke's (1632–1704) proposal that God could have endowed matter with the capacity of thought was the most important origin of materialist theorizing. John Yolton, for instance, has thus argued that the

> story of the thinking matter controversy in eighteenth-century Britain is largely the story of reactions to Locke's suggestion. While Hobbes and Spinoza are routinely cited as the arch materialists, it is to Locke's suggestion that most of the reactions were directed. (Yolton, 1983, p. xi).

It is thus apt that Wolff pays particular attention to Locke-inspired materialism, as I will discuss in what follows.

The turmoil caused by "die Materialisterey" occurred not only in Britain, however, but also in Wolff's immediate academic context. The best-known case is that of the clandestine pamphlet *Zweyer guten Freunde vertrauter Brieff-Wechsel vom Wesen der Seele* (The intimate correspondence of two good friends on the nature of soul) that was written by Urban Gottfried Bucher (1679–?), a medical student at Wittenberg and Halle between 1704 and 1707, and published anonymously in 1713,

[13] On this aspect, compare Wunderlich (2016) and Wunderlich (forthcoming). On Leibniz as a potential resource for materialism see also Wolfe (2014, pp. 96–99, in particular p. 97f.) on Leibnizian influences on Diderot and Montpellier vitalism.

[14] A case in point would be John Toland (1670–1722) who appropriated and modified Leibnizian theorems into a partly Spinozistic, materialistic doctrine; compare, for instance, Leask (2012) and the literature quoted there.

probably without Bucher's knowledge.[15] Wolff had been a university professor at Halle since 1706. Bucher's text was widely known and attacked immediately. It was mentioned by the theologian Johann Franz Budde (1667–1729) as early as 5 February 1713, a lecture was devoted to it in the summer semester of 1713 at Jena (by Johann Jacob Syrbius, 1674–1738), and it was criticized in Valentin Ernst Löscher's (1673–1749) periodical *Unschuldige Nachrichten von alten und neuen theologischen Sachen*. These were only the beginnings of a then widely known controversy.[16] Bucher's pamphlet is straightforwardly materialistic in claiming, for instance, that there is no human soul (not even a material one, Bucher, 1713, p. 17) and that instead all mental content derives from sensation, which in turn derives from the sense organs and the nerves (Bucher, 1713, pp. 19–21).

Hence, there was ample reason for Wolff to address materialism in particular, which would explain why its discussion figures so prominently in *Deutsche Metaphysik* and *Psychologia rationalis*. Turning towards Wolff's actual arguments against materialism, it is striking that he presents three different ones: the first against materialism in general and two further ones against specific forms of materialism, both in *Deutsche Metaphysik* and in *Psychologia rationalis*. The first, general argument is presented in a rather confusing way in §.738 of *Deutsche Metaphysik* and in almost identical fashion in §.44 of *Psychologia rationalis*. Essentially, Wolff seeks to make the following points here:

1. According to Wolff, all modifications of a material body occur through motion and are based on the size, shape and position of the parts of the body. If a body could think, its thoughts would have to be modifications based on the position of some of its parts, and the modifications would have to be caused by a determinate motion.
2. If the body were to become conscious of these modification and thus think—thoughts are, by Wolff's definition, modifications of the soul it is conscious of (Wolff, 1751/1983, §.194)—the different states of the body would have to be *compared* and their differences noticed.
3. This cannot be accomplished by the motion of parts, "because they cannot do anything other than represent something composite by means of their size, figure and position" (Wolff, 1751/1983, §.738, p. 461).

So Wolff seems to allow for the possibility that material objects represent something, but he denies that they can become aware of it, however without providing a more detailed argument than the one outlined. The argument is reminiscent to some extent of what has come to be called the "Achilles argument" according to which the unitary character of thought requires a unitary substratum, but it would be too

[15] For a detailed reconstruction of the contents of this pamphlet and its publication history, see Mulsow (2002) and Mulsow (2018, vol. 2, pp. 11–96). Bucher apparently discussed the manuscript with his teacher Johann Baptist Roeschel (1652–1712). It was found in Roeschel's estate and then later published by an unknown editor (see Mulsow, 2018, vol. 2, pp. 27, 37f).

[16] For details of the reception of Bucher's pamphlet, see Mulsow (2018, vol. 2, pp. 73–77).

speculative to treat it as an actual instance of this argument.[17] Another possibility is that Wolff simply wants to make the point that consciousness and physical motion are too heterogeneous to render one of them the cause of the other, similar to Leibniz's mill example in the *Monadology* (Leibniz, 1714/1885, p. 609).

Wolff's arguments against specific forms of materialism are in fact more interesting. First, Wolff addresses the widespread idea that human thought is based on a *specific kind of matter* that differs from ordinary, tangible matter. He thus acknowledges: "I know well that those who attribute thoughts to the body fancy that thoughts consist in the motion of a subtle matter in the brain" (Wolff, 1751/1983, §.739, p. 461). Against this proposal, Wolff indicates that his refutation of materialism is not restricted to "gross" matter and insists that also subtle matter can only yield representations of something composite: "For we cannot make it any further than that a representation of a composite is best obtained in this way; but consciousness, which is still required for thought (§.194), is missing as it is earlier [i.e. in the first argument]" (Wolff, 1751/1983, §.739, p. 461). Historically, a materialist recourse to subtle matter was often combined with the view that there is a soul distinct from the body that is composed of subtle matter.[18] This seems to be Wolff's target in §.47 of the *Psychologia rationalis* where he makes a separate point that the "soul cannot be material or a body", although he has previously established that bodies cannot think (Wolff, 1740/1994, §.44).

Wolff's second, more specific argument is intended to reject the possibility that God could have endowed a material body or matter in general with the capacity to think, i.e. Locke's proposal. He thus acknowledges that "I know well that some believe that God could communicate the power of thinking to a body, or, as they argue even more awkwardly, to matter" (Wolff, 1751/1983, §.741, p. 462). What Wolff has in mind here is Locke's proposal that the latter conceived rather as a theoretical possibility than actually advocating it: Locke argues that our cognitive means do not allow us to exclude the possibility that God has endowed matter with the capacity of thought by way of "superaddition". He argues that it "being, in respect of our Notions, not much more remote from our Comprehension to conceive, that GOD can, if he pleases, superadd to Matter a Faculty of Thinking, than that he should superadd it to another Substance, with a Faculty of Thinking" (Locke, 1690/1975, p. 541).[19] Although this proposal was often seen as a contribution to materialist theory, if not a concealed endorsement, Locke himself does *not* intend to establish materialism but rather argues that our cognitive means are insufficient to decide between the options of dualism and materialism, in part because we do not even understand the idea of thinking. His argument is thus both directed against dogmatic materialists and dualists

[17] Compare, for instance, Lennon and Stainton (2008).

[18] Compare Wolfe and van Esveld (2014).

[19] Recent contributions to this debate are Stuart (2013) and Jolley (2015).

> who … finding not *Cogitation* within the natural Powers of Matter, examined over and over again, by the utmost Intention of Mind, have the confidence to conclude, that Omnipotency it self, cannot give Perception and Thought to a Substance, which has the Modification of Solidity. (Locke, 1690/1975, p. 542).

In detail, Wolff argues against Locke:

1. By claiming that God should endow matter with the capacity of thought, (crypto-) materialists of this kind acknowledge that material things as such cannot think, since they need an external power to enable them to do so. Similarly, Wolff states against this theory: "The faculty of thinking cannot be communicated to the body or to matter, which they do not have by themselves" (Wolff, 1740/1994, §.46, p. 31).
2. If the (crypto-) materialists of this camp were right, God would have to "occasion that from the essence of a body something follows that cannot follow from it and thus either change the essence of that thing, or to at the same time communicate the essence of another thing that has the capacity of thought to it" (Wolff, 1751/1983, §.741, p. 463).
3. But it is known both that the essence of a thing is invariant and that it cannot be communicated to another thing.
4. Therefore, arguing that God could communicate the power of thought to matter is tantamount to arguing that God could turn one thing into another one in such a way that it would have the essential features of both things at the same time.

In sum, Wolff argues that it is impossible for God to attribute faculties to material bodies that they do not already have by their own nature, which is exactly what Lockean superaddition would amount to. Providing a material body with the capacity of thought would be tantamount to either changing the essence of that body or to bestow the essence of another thing on it at the same time. But essences are immutable, according to Wolff, and also one thing cannot have two different essences concurrently. Accordingly, Wolff argues that maintaining that God communicates the power of thought to matter implies "demanding that GOD should turn iron into gold at the same time, so that is would be iron and gold at the same time" (Wolff, 1751/1983, §.741, p. 463). Apparently, essence means something like substance concept here, in the sense of a concept that provides the most fundamental answer to the question: what is that thing? There can be only one answer to this question for a given thing, although it may be controversial what this answer is. For instance, one could argue that human beings are most fundamentally organisms, or one could argue that they are persons most fundamentally. Either way, they cannot be *both* most fundamentally.[20]

[20] For this use of substance concepts, see Olson (1997, pp. 27–31). Thanks to Eric Watkins for pointing me to this problem.

7.5 Conclusion

Although Wolff deems Leibnizian panpsychist monadology a consistent philosophical theory, he refuses to adopt it. But instead of merely remaining agnostic about whether the most basic things in the world have mental features, he sets up a more abstract theory of "elements". Whereas monads have a power to represent the world, elements have an indeterminate power, or different kinds of them could have different kinds of powers, including but not restricted to representative ones. Thus, all monads (including basic ones that just have unconscious representations) are elements, in Wolff's ontology, but not all elements are monads. Monads or "mental elements" are just ordinary elements with some additional, mental features, whereas human souls, or more precisely: spirits, are elements with additional, more advanced features, and thus Wolff is content to hold both monadology and pre-established harmony when taken as an explanation of the human mind and its interaction with the body. To conceive of all basic elements of the world as having mental features, on the other hand, would open a door to materialism, which makes it plausible why Wolff was at great pains to avoid the panpsychist aspects of Leibnizian monadology.

References

Blackwell, R. (1961). Christian Wolff's doctrine of the soul. *Journal of the History of Ideas, 22*(3), 339–354.

Bucher, U. G. (1713). *Zweyer Guten Freunde vertrauter Brief-Wechsel vom Wesen der Seelen*. Peter von der Aa (forged publisher).

Corr, C. (1975). Christian Wolff and Leibniz. *Journal of the History of Ideas, 36*(2), 241–262.

École, J. (1986). Des différentes parties de la métaphysique selon Wolff. In W. Schneiders (Ed.), *Christian Wolff 1679–1754* (pp. 121–128). Meiner.

Fabian, G. (1925). *Beitrag zur Geschichte des Leib-Seele-Problems*. Hermann Beyer und Söhne.

Jolley, N. (2015). *Locke's touchy subjects: Materialism and immortality*. Oxford University Press.

Leask, I. (2012). Unholy force: Toland's Leibnizian 'consummation' of Spinozism. *British Journal for the History of Philosophy, 20*(3), 499–537.

Leduc, C. (2018). Sources of Wolff's philosophy: Scholastics/Leibniz. In R. Theis & A. Aichele (Eds.), *Handbuch Christian Wolff* (pp. 35–53). Springer.

Leibniz, G. W. (1885). Principia philosophiae, Seu Theses in gratiam Principis Eugenii &c. [Monadology]. In C. Gerhardt (Ed.), *Die philosophischen Schriften von Gottfried Wilhelm Leibniz* (Vol. 6, pp. 607–623). Weidmann. (Original work published 1714).

Lennon, T., & Stainton, R. (2008). *The Achilles of rationalist psychology*. Springer.

Locke, J. (1975). *An essay concerning human understanding*. Clarendon Press. (Original work published 1690).

Look, B. (2013). Simplicity of substance in Leibniz, Wolff and Baumgarten. *Studia Leibnitiana, 45*(2), 191–208.

Lorini, G. (2016). Receptions of Leibniz's pre-established harmony. Wolff and Baumgarten. In M. Camposampiero, M. Geretto, & L. Perissinotto (Eds.), *Theodicy and reason: Logic, metaphysics, and theology in Leibniz's Essais de Théodicée (1710)* (pp. 163–179). Edizioni Ca' Foscari.

Middell, K., & Neumann, H.-P. (Eds.) (2019). *Briefwechsel zwischen Christian Wolff und Ernst Christoph von Manteuffel. 1738–1748* (3 Vols.). Olms.

More, H. (1671). *Enchiridion metaphysicum*. Flesher.

Mulsow, M. (2002). Säkularisierung der Seelenlehre? Biblizismus und Materialismus in Urban Gottfried Buchers Brief-Wechsel vom Wesen der Seelen (1713). In L. Danneberg, S. Pott, J. Schönert, & F. Vollhardt (Eds.), *Säkularisierung in den Wissenschaften seit der frühen Neuzeit* (Vol. 2, pp. 145–173). De Gruyter.

Mulsow, M. (2018). *Radikale Frühaufklärung in Deutschland 1680–1720* (2 Vols.). Wallstein.

Olson, E. T. (1997). *The human animal. Personal identity without psychology*. Oxford University Press.

Perler, D., & Rudolph, U. (2000). *Occasionalismus. Theorien der Kausalität im arabisch-islamischen und im europäischen Denken*. Vandenhoeck und Ruprecht.

Reid, J. (2012). *The metaphysics of Henry More*. Springer.

Smith, J. E. H. (2011). *Divine machines. Leibniz and the sciences of life*. Princeton University Press.

Stuart, M. (2013). *Locke's metaphysics*. Oxford University Press.

Watkins, E. (2006). On the necessity and nature of simples: Leibniz, Wolff, Baumgarten, and the pre-critical Kant. *Oxford Studies in Early Modern Philosophy, 3*, 261–314.

Wolfe, C. (2014). Materialism. In A. Garrett (Ed.), *The Routledge companion to 18th century philosophy* (pp. 91–118). Routledge.

Wolfe, C., & van Esveld, M. (2014). The material soul: Strategies for naturalising the soul in an early modern Epicurean context. In D. Kambaskovic (Ed.), *Conjunctions of mind, soul and body from Plato to the enlightenment* (pp. 371–421). Springer.

Wolff, C. (1724). *Der vernünfftigen Gedancken von Gott, der Welt und der Seele des Menschen, auch allen Dingen überhaupt, anderer Theil, bestehend in ausführlichen Anmerkungen*. Andreae.

Wolff, C. (1977a). Philosophia prima, sive Ontologia, methodo scientifica pertractata, qua omnis cognitiones humanae principia continentur. In J. École (Ed.), *Christian Wolff: Gesammelte Werke* (Prima philosophia sive Ontologia) (Vol. 3, Pt. II, 2nd ed.). Olms. (Original work published 1736).

Wolff, C. (1977b). Cosmologia generalis, methodo scientifica pertractata, qua ad solidam, inprimis die atque naturae, cognitionem via sternitur. In J. École (Ed.), *Christian Wolff: Gesammelte Werke* (Cosmologia generalis) (Vol. 4, Pt. II). Olms. (Original work published 1737).

Wolff, C. (1983). Vernünfftige Gedancken von Gott, der Welt und der Seele des Menschen, auch allen Dingen überhaupt. In C. Corr (Ed.), *Christian Wolff: Gesammelte Werke* (Deutsche Metaphysik) (Vol. 2, Pt. I, 11th ed.). Olms. (Original work published 1751).

Wolff, C. (1994). Psychologia rationalis, methodo scientifica pertractata, qua ea, quae de anima humana indubia experientiae fide innotescunt, per essentiam et naturam animae explicantur, et ad intimiorem naturae ejusque autoris cognitionem profutura proponuntur. In J. École (Ed.), *Christian Wolff: Gesammelte Werke* (Psychologia rationalis) (Vol. 6, Pt. II, 2nd ed.). Olms. (Original work published 1740).

Wunderlich, F. (2016). Varieties of early modern materialism. *British Journal for the History of Philosophy, 24*(5), 797–813.

Wunderlich, F. (forthcoming). Materialism. In D. Jalobeanu & C. Wolfe (Eds.), *Encyclopedia of early modern philosophy and sciences*. Springer.

Yolton, J. (1983). *Thinking matter: Materialism in eighteenth-century Britain*. University of Minnesota Press.

Chapter 8
Wolff and the Logic of the Human Mind

Matteo Favaretti Camposampiero

8.1 Introduction

Christian Wolff saw a close connection between his logical and psychological doctrines. His characteristic claim that logic must borrow its principles from psychology (Wolff, 1738/1968, §.9; 1740/1983a, vol. 1, §.89) earned him a charge of psychologism (Engfer, 1992; but cf. Goubet, 2004). Although there may be good reasons to describe Wolff's treatment of logic as psychologistic, this label simplifies what is in fact a complex issue. In particular, it obscures the other side of the coin, namely Wolff's deeply rooted idea that our cognitive and especially inferential operations are governed by the laws of logic, so that (genuine) logic provides the tools for analysing mental activity. In this perspective, it is logic that seems to play a foundational role with respect to psychology.

This chapter argues that Wolff's final views on the relation between logic and psychology are to be found in his doctrine of natural logic. Although he began to use this traditional term in 1724, the underlying theory had been developing for nearly 20 years. This long gestation began in 1705 with Wolff's re-evaluation of syllogism, which led him to discover the hidden syllogistic structure of geometrical demonstrations (see Sect. 8.3 below). Some 10 years later, he announced his further discovery that syllogism also informs all of our inferential processes; the laws of syllogistics are indeed the very laws of reasoning (see Sects. 8.4 and 8.5). Finally, he drew on the scholastic distinction between natural and artificial logic, as well as on the division of the former into innate and acquired, to explain how the principles of logic can govern our higher cognitive activity even without our knowing them (see Sects. 8.6–8.8). There is a logic embedded in the human mind which determines the laws of thought. My suggestion is that, instead of wondering whether it is

M. Favaretti Camposampiero (✉)
Ca' Foscari University of Venice, Venice, Italy
e-mail: matteo.favaretti@unive.it

S. d. F. Araujo et al. (eds.), *The Force of an Idea*, Studies in History and Philosophy of Science 50, https://doi.org/10.1007/978-3-030-74435-9_8

psychology that grounds logic or vice versa, we should rather consider the possibility that natural logic grounds both.

After the serious criticisms raised in recent decades (notably by Harman, 1986), the identification of the principles of demonstration or argument with the principles of reasoning or inference can no longer be taken for granted. In this respect, what makes Wolff's position interesting is the fact that it offers one of the most radical versions of the identity thesis: his psychological analysis of reasoning purports to show that every inferential chain of thought is actually a chain of syllogisms. This may sound far too naive, but Wolff had strong metaphysical reasons to ascribe a syllogistic structure to human reasoning. From his early endorsement of modern mechanism, he thought that even the mind—on a par with the body—can be described as a machine whose present state is determined by a series of previous states and a set of fixed laws (see Sect. 8.9). The rules of syllogistics must have appeared to Wolff as the only plausible candidate for the role of genuine laws of thought. If my reconstruction is sound, a fundamental assumption of modern psychology—the assumption that there are psychological laws—owes something to both early modern mechanism and the doctrine of natural logic.

8.2 The Dual Character of Wolff's Logic

Wolff's logic is connected with psychology by the very role it plays in the system of science. The Wolffian soul has two main faculties, the cognitive and the appetitive, both of which can fail with respect to their goals. Just as the appetitive faculty can stray from the good and choose evil, the cognitive faculty can deviate from truth and embrace error instead. Practical philosophy and logic are precisely what the soul needs in order to avoid moral and cognitive errors, respectively (Wolff, 1740/1983a, vol. 1, §.60). Both have therefore a normative function with respect to the exercising of mental faculties. Logic is "the part of philosophy that teaches the use of the cognitive faculty in knowing the truth and avoiding error"; it can thus be characterized as "the science of directing the cognitive faculty in knowing the truth" (Wolff, 1740/1983a, vol. 1, §.61, p. 30). Other characterizations of logic are less straightforwardly normative. *Deutsche Logik* (German Logic), for instance, characterizes logic as the part of philosophy that shows "the forces of human understanding and their correct use in the knowledge of truth" (Wolff, 1754/1965a, *Vorbericht von der Welt-Weisheit*, §.10, p. 118). Here, logic appears to have a descriptive and explanatory function too, insofar as it provides knowledge of our cognitive faculties. On the other hand, Wolff's 1726 *Ausführliche Nachricht von seinen eigenen Schrifften* (Detailed Report of His Own Writings) maintains that "logic should show how we use the intellect in the knowledge of truth" (1733/1973a, §.56, p. 182). Is logic concerned with the correct or the actual use of understanding?

Consider, first, that Wolff's concept of use is intrinsically normative. By "use" of an instrument (be it an organ of the body, a technical device, or a mental faculty) Wolff means its genuine destination, the function for which it is designed (Favaretti

Camposampiero, 2016, p. 89). In this respect, "use" and "correct use" are nearly equivalent expressions. Consider, second, that several passages concerning the laws of thought allow for both a prescriptive and a descriptive reading. For instance, logic is the discipline that teaches and proves the rules "to which the mind is bound in the cognition of things" (Wolff, 1740/1983a, vol. 2, §.11n, p. 112). Does this mental bond express a physical or a deontic necessity? Is Wolff talking about the (psychological) laws that the mind inevitably follows in its cognitive operations or the (logical) laws that the mind should respect in order to achieve its cognitive goals? Both, I think.

Ausführliche Nachricht suggests that the intertwining of the descriptive and normative sides of logic is rooted in Wolff's concept of natural logic: "And since in the *Logic* I show what natural logic is, that is, to what rules the intellect ordinarily conforms, or rather must conform, in its operations, one evidently recognizes what I have mentioned concerning natural and artificial logic" (Wolff, 1733/1973a, §.93, p. 258). Artificial logic is logic in the ordinary sense; it is the discipline taught in logic classes and expounded in logic handbooks, whereas natural logic is an inborn disposition of the mind. This idea that logic is a natural endowment of the human mind is paramount not only to Wolff's foundation of the laws of logic but also to his psychological account of cognitive processes and consequently to his foundation of psychology as a science.

Although the distinction between natural and artificial logic was a core topic in the late-medieval and early modern philosophy of logic, its first explicit mention in Wolff's works dates from 1724. But far from revealing a sudden theoretical turn, this relatively belated appearance represents the final outcome of the long reflection on the inner logic of the human mind that Wolff began in 1705. The next sections reconstruct this gradual development.

8.3 Mental Operations and Syllogisms

Natural logic is the privileged theoretical locus of Wolff's connection of logic and psychology. This connection, however, has both a broader and a narrower scope. The latter concerns reasoning in particular; the former embraces all three operations of the mind: (1) simple apprehension or the formation of concepts, (2) judgement, and (3) inference or reasoning. All mental operations have their rules (Wolff, 1740/1983a, vol. 2, §.3). Natural logic is precisely the natural disposition of the mind to direct these cognitive operations towards the acquisition of truth (§.6). Historically, the link between the doctrine of natural logic and the doctrine of the three mental operations appears to date from the early sixteenth century, when logical treatises gradually abandoned the structure of Peter of Spain's (first half of the thirteenth century to second half of the thirteenth century) *Tractatus* (Treatise) and adopted the Ockhamist division modelled on the mind's activity (Hoenen, 2010, p. 112). Wolff himself credits the Aristotelians with assuming the three mental operations as the basis of their logic (Wolff, 1733/1973a, §.56). He agrees with them that logic only needs to explain these operations (§.58), for all cognitive processes can ultimately be reduced to them and the task of logic is to clarify "what happens in the soul when we know something" (§.56, p. 182). If artificial logic did not stick to the three operations, it would depart from natural logic, but artificial logic "is no different from *natural logic*. It does not give other rules from those that nature prescribes to us, but only clarifies them[1]" (§.56, p. 182). The rules of artificial logic are the rules of natural logic, which are the rules of mental operations.

However, it is under the narrower scope of the third operation that Wolff's connection of logic and psychology takes its first steps and gradually acquires full significance. At the very beginning of his career, Wolff shares the mainstream post-Cartesian contempt for scholastic logic and enthusiasm for mathematical reasoning—most notably illustrated by E. W. von Tschirnhaus (1651–1708) (see Wolff, 1735/1972a, section II, ch. 2, §§.6–8, §§.11–16). In December 1704, he starts his correspondence with Leibniz by sending him, along with a letter and another dissertation, a dedicated copy of his fresh from the press dissertation *De algorithmo infinitesimali differentiali* (On the Infinitesimal Differential Algorithm), whose final corollaries includes the claim that "syllogism is not a means to discover truth" (Wolff, 1755/1974, section II, p. 289). In his reply, Leibniz politely points out that he does not approve of this dismissive attitude: "As concerns your corollaries, I would not dare say that, absolutely speaking, syllogism is not a means to discover truth" (Gerhardt, 1860, p. 18). The episode is well known, but its long-term consequences have not yet been fully appreciated. Wolff's initial puzzlement soon evolves into two distinct trains of thought. On the one hand, he rethinks his teenage use of syllogistic chains to defeat his opponents in disputations. On the other, he reconsiders geometrical demonstrations and realizes that they, too, consist of interconnected syllogisms (Wolff, 1735/1972a, section II, ch. 2, §.26). Through this insight into the formal structure of demonstrations and its identity with the ordinary syllogistic form, Wolff begins to understand the role of syllogism in the logic of discovery.

By 1710, Wolff has come to the conclusion that "the only way to produce a convincing proof is when our thoughts follow one another according to the syllogistic rules" (Wolff, 1750/1973b, §.46, p. 28). In 1713, *Deutsche Logik* spells out the epistemic role of syllogism in four points:

> 1) in geometrical demonstrations, we really conceive syllogisms composed in the correct form; 2) in mathematics itself, nothing is found except by means of such syllogisms; 3) in other disciplines, if we wish to demonstrate and expound something in mathematical fashion, we can only do so through syllogisms composed in the correct form; 4) with the aid of these syllogisms, we can resist the most subtle errors. (Wolff, 1754/1965a, ch. 4, §.22, p. 173)

Syllogism is not merely one means among others—it is the only means available to human understanding for both proof and discovery. Every truth we demonstrate, every new truth we discover, is but the conclusion of a syllogistic chain, "even though neither in discovering nor in proving do we always have the form of syllogisms distinct before our eyes" (Wolff, 1754/1965a, ch. 4, §.20, p. 171). For in

[1] I read "dieselben" instead of "dieselbe", which would otherwise refer to "nature".

the meantime, informed by Leibniz in 1705 about the theory of small perceptions (Gerhardt, 1860, p. 32), Wolff has also departed from the Cartesian view that all mental contents and processes are necessarily conscious (Favaretti Camposampiero, 2009, pp. 623–625).

8.4 A Logic for Human Reasoning

The further, decisive step towards the constitution of a doctrine of natural logic is documented by the 1716 *Mathematisches Lexikon* (Mathematical Lexicon). Here, in the entry on demonstration, Wolff not only reaffirms the syllogistic character of geometrical proofs, but also asserts for the first time the syllogistic character of every inferential process. He now views syllogistic logic as the key to unlock the inner workings of the human mind. Since the structure of syllogism is intrinsic to the third intellectual operation as such, geometrical demonstrations offer simply an instance of the general procedure that our mind follows in deriving conclusions from premises. First, the direct perception or intuition (*Anschauen*) of a geometrical figure (e.g. a triangle) elicits an empirical judgement, which is assumed as the minor premise. This proposition calls to our mind another, more general proposition (the major premise) because of a term (e.g. "triangle") that occurs in both, and which thus becomes the middle term of the syllogism. By joining the two extreme terms of the premises, we form a conclusion which in turn calls to mind another proposition, and so forth until this chain of syllogisms attains the conclusion that we wanted to prove (Wolff, 1716/1965b, columns 502–503).

It should be clear that Wolff's syllogistic analysis of demonstrations is actually a psychological analysis of inferences as chains of thought. Consequently, Wolff points out that his analysis applies not only to demonstrations but to the very process of everyday human reasoning: "However, not only do geometers take this into consideration in demonstrating, but all humans every day, nay, every hour and every moment, as often as they come from one present thought to another one" (column 503). To illustrate this idea, Wolff proposes the first version of a striking example that later (see Wolff, 1738/1968, §.393) becomes a strongpoint of his treatises on psychology:

> For instance, Tom [*Titius*] hears the bells ringing, puts small change into his pocket, and goes to church. Here is the question: How can the ringing of bells induce Tom to put small change into his pocket and go to church? For, it does not have this effect on other people. I say that this happens because Tom always infers [*schliesset*] one thing from the other in precisely the same way as geometers do in demonstrating. (Wolff, 1716/1965b, column 503)

The bells inform Tom that it is Sunday, 7 am. If Tom pays attention to his sensory perceptions, he is led to form the following factual judgement: "It is Sunday, 7am". This proposition reminds him of a conditional judgement stored in his memory: "If it is Sunday, 7am, then it is time to go to church". From both premises, Tom easily concludes by *modus ponens* that it is time to go to church (and so on for the decision of taking money with him).

First, the example aims to show that the inferential behaviour of Tom's mind is no different from the deductive procedures of geometers (column 504). Both everyday and mathematical reasoning ultimately conform to "the nature of human reason" (column 505) in that both follow the natural order of reasoning, "the natural order [...] by which thoughts follow one another in the mind" (Wolff, 1735/1972a, section II, ch. 3, §.34, p. 154).

Second, Wolff's reconstruction of Tom's cognitive behaviour aims to show that the psychological structure of natural reasoning is identical to the logical structure of syllogism. The decisive step Wolff makes in the 1716 work is precisely to argue that the syllogistic rules are explanatory of thought processes. Some 4 years after, the 1720 *Deutsche Metaphysik* (German Metaphysics) emphasizes the groundbreaking character of this idea in the following terms:

> Although I have talked there [in *Deutsche Logik*] about the manifold usefulness of *Schlüsse* [syllogisms] in the sciences [...], here I still have to discuss a particular usefulness they have, of which nobody has thought yet. I first talked about it in the entry *demonstration* of my *Mathematisches Lexicon*, and then mentioned it in the *Ratio praelectionum*, sect. II, ch. 3, §.20, p. 147. Indeed, syllogisms help us understand how one thought always follows from another in an immovable series, and in this way we can indicate the reason for all the thoughts that have originated from another thought and represent something to us that is not present to our senses. (Wolff, 1751/1983b, §.341, pp. 194–195)[2]

The paragraph from the 1718 *Ratio praelectionum* (Plan of Lectures) that Wolff evokes here is also worth quoting, for it reveals that the laws of syllogism—which we would regard as pertaining to pure logic—cooperate with eminently psychological laws to explain what goes on in the mind:

> Inquiring into the rules from which one could always explain the reason for the sequence of perceptions in the mind, I eventually observed that three things are sufficient for this purpose: 1) the harmony of perceptions with the changes that happen in the sense organs; 2) the fortuitous connection of ideas in the imagination; 3) the syllogism. (Wolff, 1735/1972a, section II, ch. 3, §.20, p. 149)

For every thought that occurs to my mind, it must be possible to give a sufficient reason why it occurs at that precise point in the series of my thoughts. This sufficient reason lies in the functioning of either the sense faculty or the imagination or reason. As *Deutsche Metaphysik* makes clear in its chapter on rational psychology, the three factors mentioned in the passage are actually psychological laws, "rules of thought":

> Thus, if we wish to indicate each time a sufficient reason for the thoughts that occur in our soul, then concerning the cognition of things[3] we have to look at three types of reasons. The first reason is the harmony of the soul's sensations with the changes in the sense organs. The

[2] See also Wolff (1738/1968, §.392n, §.395n).

[3] This qualification means that the three reasons given here are sufficient to explain cognitive processes but not other psychological phenomena like volitional processes and emotional states (which require the representation of good and evil) or pleasure and sorrow (which arise from the feeling of perfection and imperfection, respectively). See Wolff (1735/1972a, section II, ch. 3, §.20).

> second reason is the rule of the imagination. The third reason is syllogisms, which are based on the two previous reasons […]. (Wolff, 1751/1983b, §.847, pp. 525–526)

To explain why I am having a certain thought, I may refer either (1) to the law of sensation, stating that our perceptions always harmonise with (i.e. correspond to) what happens in our sensory organs, or (2) to the law of the imagination, which governs all the processes of mental association, or (3) to the laws of syllogism, which explain how a third proposition results from two previously assumed propositions that share one term. As is clear from Tom's example, the first proposition is provided by sensation while the second proposition if provided by associative imagination. Human reason produces syllogisms by drawing on both these lower faculties, so that in the final analysis even our inferential activity can be explained as a product of the representational force that constitutes the essence of our soul (Wolff, 1751/1983b, §.841). Along with the psychophysical law of sensation and the psychological law of the imagination, the laws of syllogistics contribute to determining (and thus serve to explain) the stream of our thoughts (see also Wolff, 1740/1983c, §.105). The latter laws are embedded in our mind just like the former.

Wolff never abandoned this sort of logicism inherent to his psychology, just as he never rejected his psychologistic assumptions about logic. Many years later, in the 1741 *Programma de necessitate methodi scientificae* (Programme Concerning the Necessity of the Scientific Method), right after repeating that his logical theory can be demonstrated from "the very nature of the human mind", Wolff claimed once again to have been the first to "publicly point out that the series of perceptions in the soul is explicable by the rules of logic" (Wolff, 1755/1974, section III, p. 185).

8.5 Implicit Cognition and Psychological Plausibility

Wolff invokes the explanatory character of the syllogistic structure also to forestall the objection that his reconstruction of Tom's chain of thought lacks psychological evidence and is therefore arbitrary. How can Wolff be certain that "all these thoughts must come to Tom's mind in the order" in which Wolff relates them? His argument is that "if one of the preceding thoughts is omitted, we can no longer give the reason why the following thought comes to Tom's mind" (Wolff, 1716/1965b, column 504). Tom's chain of thought must follow the logical order of syllogism because there is no way to explain how the mind comes to a certain conclusion without assuming that it infers that conclusion from its logical premises, which must therefore be themselves thought in the right order. Thus, Wolff appears to base his syllogistic account of inferential processes directly on the principle of sufficient reason.[4] If our mental activity did not incorporate the rules of logic, its actual course would be inexplicable.

[4] "Every thought must have its sufficient reason. But if we omit one of the previous propositions, we will no longer be able to intelligibly show how we have come to this or that thought" (Wolff,

A different version of the objection of psychological implausibility runs as follows: "Some might come to think that the soul does not really make all these *Schlüsse* (syllogisms), since it does not conceive all the propositions with the words in which we have formulated them" (Wolff, 1751/1983b, §.343, p. 197). If Wolff's theory postulated or entailed full awareness by the mind of all its inferential steps, it would be highly implausible. This leads Wolff to specify the required degree of awareness by suggesting a distinction between implicit (i.e. obscure, non-verbalized) and explicit cognition:

> It is not necessary that, for instance, when I hear [the clock] strike five, I conceive these words, 'it is striking five', or 'it strikes five', but it is sufficient that I hear the striking, am conscious of it, and remember the meaning of the bell ringing only, as it were, from a distance in my memory. [...] Likewise, it is not required that [...] I conceive the words, 'when it strikes five, I want to get up', but again it is enough that I only represent to myself the previous situation in which I made the resolution. For, what the words mean is contained in this representation, even though this is quite obscure then. (1751/1983b, §.343, pp. 197–198)

Implicit premises are also postulated by Wolff's account of so-called immediate (i.e. non-syllogistic) consequences, which apparently draw the conclusion from just one premise. According to Wolff, an immediate consequence is in fact a *förmlicher Schluß* (formal syllogism), one of whose premises is left implicit (1751/1983b, §.354).[5] Consider, for instance, an inference by conversion of terms like the following: "All learned people are mortal. Thus, some mortal people are learned". Since Wolff maintains that the validity of the conversion depends on "universal *Gründe* [principles] of conversion, which are known through ordinary syllogisms" (§.359, p. 218), his idea seems to be that, in this case, the suppressed premise is the rule of inference that allows the conversion by subalternation.

Also relevant to our topic is the reason why Wolff treats immediate consequences as mutilated syllogisms. He argues that our thinking of the explicit premise alone is not sufficient to explain why we come to think of the conclusion; for, if that premise was a sufficient reason for inferring the conclusion, then we would think of the conclusion whenever we think of the premise, which is obviously not the case (1738/1968, §.410; 1751/1983b, §.354, §.359). This argument raises the question of whether Wolff conflates the relation of logical consequence and the relation of psychological consequence. The conclusion follows from the premise(s) if and only if thinking of the premise(s) determines the mind to think of the conclusion. Since it is never the case that one premise is sufficient to perform such a determination, Wolff maintains that the third operation of the mind, inference, always has the threefold structure of syllogism. However, this is precisely what distinguishes inference as the work of reason from mental associations brought about by the imagination. As we have seen, when the mind forms the minor premise, the imagination retrieves

1751/1983b, §.343, p. 198).

[5]As far as I can see, the analysis Wolff advances here of the oppositional inference "All envious people are miserable; thus, it is false that some envious people are not miserable" (Wolff, 1751/1983b, §.354, p. 209) contains a serious fallacy.

the major premise from memory. But the minor premise simply calls to mind one proposition or the other according to the circumstances, the personal history of the reasoner, etc. It does not by itself invariably determine the mind to think of the major premise, as is the case with both premises and their conclusion. For, the third operation of the mind is always determined by the logical laws of syllogism, even though the mind may be unaware of these laws as well as of part of the premises. Thus, although the relation of logical consequence is implemented by psychological operations of reasoning, its nature and cogency are entirely different from merely psychological phenomena like mental associations.

Nevertheless, the complex relationship that these early works establish between logic and psychology raises further difficulties. First, there is the problem of circularity. By praising the use of syllogistics to explain mental inferential processes, Wolff seems to imply that logic plays a foundational role with respect to psychology, in that it helps the doctrine of the soul to fulfil the task of "discovering the reasons for any and all modifications that can happen to the mind" (Wolff, 1735/1972a, section II, ch. 3, §.33, p. 154). The laws of logic are explanatory of the functioning of the human mind. On the other hand, just three lines earlier Wolff unequivocally affirms that the doctrine of the soul also allows us "to give a priori reasons for logical rules" (Wolff, 1735/1972a, section II, ch. 3, §.33, p. 154). The doctrine of the soul indicates "the *Grund* (fundament) of what is taught in logic" (Wolff, 1740/1983c, §.55, p. 125). As noted in the 1725 third edition of *Deutsche Metaphysik*, the truths that Wolff demonstrates about the soul "are the foundation of the rules according to which the powers of the soul are directed both in cognition and in willing or not willing, hence they are the foundation of logic, ethics, and politics" (Wolff, 1751/1983b, §.191, p. 107). If psychology is expected to ground logic, the use of logic to ground psychology appears at first sight to be circular.

Second, the question is still open of whether the rules of logic *do* or rather *should* direct our mental activity. On the one hand, Wolff is clearly committed to the view that it is the task of logic to exert a normative pressure on our inferential processes. On the other hand, the psychological foundation of logical rules has the effect of naturalizing logic, so to speak, thereby making the laws of logic similar in character to the laws of nature, which are obviously not normative. Briefly, Wolff appears to maintain both that the laws of logic describe the natural functioning of the human mind and that they prescribe its correct functioning. But how can they play both roles at once?

In spite of these shortcomings, Wolff's early elaborations on the logic of mental processes form the background of his mature understanding of the relationship between logic and psychology as detailed in his Latin works. The novelty after *Deutsche Metaphysik* is that—perhaps in an attempt to solve the problems mentioned above—Wolff resorts to the traditional distinction between natural and artificial logic. The next sections investigate the consecutive formulations of this doctrine.

8.6 Natural Logic in 1724

As mentioned above, by the late 1710s Wolff is already emphasizing the natural character of all inferential procedures and affirming the identity of the geometrical order of demonstration and the natural order of ordinary thought. However, he does not yet use the distinction between natural and artificial logic, nor does he clarify the relation between the logical rules that the mind unknowingly follows and the rules formalized by logic as a discipline. The final step is taken in the mid-1720s. Wolff adopts the distinction between natural and artificial logic in his 1724 *Anmerckungen* (Annotations) on *Deutsche Metaphysik*, then uses it consecutively in his *Ausführliche Nachricht*, in a brand new chapter added to the fifth, 1727 edition of *Deutsche Logik* (Wolff, 1754/1965a, ch. 16), in the *Prolegomena* to the 1728 Latin *Logica* (Logic), and finally in the first volume of *Ethica* (Ethics), published in 1750. Actually, a passing reference to the *Prolegomena* in the third, 1725 edition of *Deutsche Metaphysik* (Wolff, 1751/1983b, §.191) suggests that Wolff was already working on them in the mid-1720s. Thus, this appears to be the period of Wolff's first but also most intense engagement with the traditional issue of natural vs. artificial logic.

From the very beginning, Wolff's treatment of the two logics appears to be closely related to his previous investigations into the syllogistic structure of thought. The first occasion for mentioning natural and artificial logic is provided by Wolff's analysis of the categorization process—the process he later calls *vindicare nomina rebus* (ascribing names to things) (Wolff, 1738/1968, §§.354–363). According to *Deutsche Metaphysik*, categorization too is an inferential process whose formal structure is syllogistic. For instance, when I see a dove and judge "This is a dove", my categorization judgement cannot be the mere effect of my visual perception. If seeing a dove were a sufficient reason for judging that it is a dove, we would always form this judgement whenever we see that bird, which Wolff denies to be the case. Thus, my judgement is actually the conclusion of a syllogism whose premises are the empirical judgement (elicited by my attention to the perceptual content) "This being has this form" and the universal judgement (retrieved by memory) "What has this form is a dove" (Wolff, 1751/1983b, §.333, p. 189). The dove example is paradigmatic of categorization processes, just as Tom's example is paradigmatic of all the inferential operations that consist in applying a universal judgement to a particular case. The 1724 annotation to this paragraph points out that these two types of syllogistic reasoning are based on two different logical principles. Whereas Tom's reasoning is based on the principle that what belongs to a certain genus or species also belongs to the subordinate species or individuals (Wolff, 1740/1983c, §.104), categorization is always based on the principle that *Cui competit definitio, illi competit definitum* (If the definition applies to a certain thing, the defined name also applies to it) (§.103, p. 173). But how can a principle of scholastic logic, known only to educated people, explain such a basic, "natural" activity as categorization? To justify the recourse to logical principles in the analysis of ordinary cognitive processes, Wolff evokes for the first time the relation between natural and artificial logic:

> What this principle has to say in artificial logic—which is but a distinct explanation of natural logic—coincides precisely with the principle of judgments in natural logic by which the name is ascribed to a certain thing when the thing occurs to us, except that we use a confused concept instead of the definition. (1740/1983c, §.103, pp. 173–174)

Natural logic works with confused concepts whereas artificial logic works with definitions, i.e. distinct concepts, but both follow the same inferential patterns, whose validity rests on the same principles. The "difference between natural and artificial logic" is thus expressly reduced to "the difference between confused and distinct cognition" (1740/1983c, §.103, p. 174). Clearly, Wolff's substantial (albeit not formal) identification of natural with artificial logic aims to provide a theoretical foundation for his key claim that the rules of syllogistics are but explicit formulations of the universal rules of thought and can therefore help us investigate the workings of the mind. On the other hand, since natural logic is intrinsic to the very nature of the human soul, it is the task of psychology to demonstrate "the main principles of logic from the nature of the soul" (§.104, p. 176), and it is on this basis that one can assess whether the rules prescribed by this or that artificial logic are useful or not.

This embedding of natural logic within the soul's nature helps solve the circularity problem raised above concerning the relation between logic and psychology. Since the rules of logic are embedded in the cognitive faculty, it makes sense to maintain both that psychology provides the foundations of logic by deriving its principles from the soul's powers and that logic supports psychology in analysing our cognitive processes. Furthermore, we are now in a better condition to explain why logic appears to have both a descriptive and a normative side. Artificial logic turns the natural logical laws—which are inscribed in our mind and described by psychology (at least as far as their basic principles are concerned)—into explicit prescriptions. Thus, natural logic provides both a description of how we think (via psychology) and (via artificial logic) the norm that we must follow to think correctly, that is in conformity with our nature.

8.7 Natural Logic and the Laws of Thought

Whereas the 1728 *Prolegomena* introduce the terms *logica naturalis* (natural logic) and *logica artificialis* (artificial logic) by means of definitions and further divisions, the earlier works use their German counterparts without defining their meaning, as though they were current expressions in the philosophical jargon. However, Wolff gradually approaches a definition of these terms precisely by evoking two German idioms, *Mutterwitz* (mother wit) and *Schulwitz* (school wit).

In *Ausführliche Nachricht*, after claiming once again that his logic is but a distinct explanation of the "natural manner of thinking", Wolff observes that his position defeats those who deny the usefulness of logic on the pretext that they prefer "a pinch of mother wit to one hundred pounds of school wit" (Wolff, 1733/1973a, §.67, pp. 206–207). Wolff rejects the very opposition between innate and learned skills:

> As I have shown that school wit is the same as mother wit, i.e. that artificial logic is the same as natural logic (if one attains the truth in this matter), it is no longer possible to draw from this an argument against artificial logic. (1733/1973a, §.67, p. 207; see also 1750/1970, §.108)

This passage may have inspired later identifications of *Mutterwitz* and *Schulwitz* with natural and artificial logic, respectively (see Sánchez Rodríguez, 2012). However, the new, 1727 chapter of *Deutsche Logik* advances a more refined view. Here, Wolff criticizes again the idea that natural logic has absolute priority over artificial logic or is even "sufficient by itself for all the functions of the intellect" (Wolff, 1754/1965a, ch. 16, §.2, p. 244), as though one knew how to reason without having to learn the rules of reasoning. Wolff defends the usefulness of studying logic, but without renouncing the idea that the basis of our logical skills is inborn. To do so, he tries to dispel the confusion surrounding the concept of natural logic:

> That is how it is, the human being has a natural aptitude for the functions of the intellect, and the intellect conforms to the rules prescribed to it even if it does not understand them, just as bodies move according to certain rules, and the human being himself observes certain rules in walking and standing which he does not understand. These rules that are prescribed by God to the intellect, along with the natural aptitude for acting accordingly, constitute natural logic. In particular, the latter constitutes the so-called mother wit. (1754/1965a, ch. 16, §.3, p. 244)

Here, *Mutterwitz* appears to designate only the natural human ability to abide by the rules that the intellect must follow, whereas natural logic includes both this ability and the rules themselves. This formulation paves the way for the 1728 definition of natural logic (see below). As far as I know, this is also the only Wolffian passage that expressly affirms that the rules of logic are "prescribed by God". Although this reference to divine legislation might suggest that the laws of logic are positive laws, namely obligations depending on God's will, the immediately preceding comparison with the physical laws of motion and statics suggests the contrary. How could the rules of logic be merely positive norms, if the intellect acts according to them even without knowing them, just as bodies move according to the laws of mechanics? According to Wolff, even these latter laws are prescribed by God to the nature of bodies. However, it seems to me that the status Wolff ascribes to the laws of logic is even more similar to the status of the (moral) law of nature, whose sufficient reason lies in the essence and nature of the human being (see Wolff, 1738/1971, §.135). The (moral) law of nature is dictated by God insofar as He is the author of nature (§.136n, §.273), but unlike the laws of physics it does not depend on God's free choice.

The existence of rules for mental operations is the starting point of the extended 1728 discussion on natural logic. Wolff takes it to be empirically evident that the exercising of our cognitive faculty follows certain rules (1740/1983a, vol. 2, §.3). Against those who believe that the mind has no rules simply because they ignore them, he maintains that the mind respects its own rules in cognizing "even if it does not understand them" (§.4, p. 108), just as all organic bodies follow certain rules in moving, such as the rules formulated by G.A. Borelli (1608–1679) in his *De motu animalium* (On the Motion of Animals). Our mind is not free to direct its inferences to any conclusion whatsoever, just as our body is not free to move in any direction whatsoever. The logical constraints that our thought experiences in reasoning are like the biomechanical constraints that our body experiences even in voluntary motions.

However, this set of mental rules does not by itself constitute natural logic, which is rather characterized as the inborn disposition to follow those rules: "There is a natural disposition of the mind to direct its own cognitive operations in conformity with these rules" (1740/1983a, vol. 2, §.5, p. 108), and this natural disposition "is called *natural logic*" (1740/1983a, vol. 2, §.6, p. 109). In order to achieve knowledge by means of its cognitive operations, the mind need not follow any instructions; it simply has to apply its inborn powers. However, this is not to say that the use of the cognitive faculties (and especially of reason) requires no learning at all, for at least some practical training is necessary to "actualize" the mere disposition and develop it into a real *habitus* (ability). The natural disposition that is called natural logic "is not brought to actuality, let alone developed into an ability, without prior exercise" (1740/1983a, vol. 2, §.5, p. 108). Whereas natural dispositions are inborn, abilities are acquired (Wolff, 1738/1968, §§.426–430). Thus, Wolff distinguishes between *logica naturalis innata* (innate natural logic), the inborn disposition, and *logica naturalis acquisita* (acquired natural logic), "the ability to direct the operations of the mind in cognizing truth, acquired by practice and without rules" (Wolff, 1740/1983a, vol. 2, §.6, p. 109).

8.8 Logical Dispositions and Abilities

The distinction between disposition and *habitus* is a cornerstone of Wolff's psychology and practical philosophy (Wolff, 1740/1983a, vol. 2, §.6n), but it turns out to be relevant to the foundations of logic as well. An acknowledgement of these two components of natural logic, the inborn disposition and the acquired *habitus*, is the hallmark of Wolff's mature elaboration. Of course, Wolff is not the first to introduce this distinction (nor does he claim to be); he merely observes that the distinction is "not yet generally accepted" (§.6n, p. 109).[6] Indeed, the issue is key to understanding Wolff's position with respect to the traditional accounts of natural logic. Since he deplores that "philosophers generally ignore that there is something acquired even in natural logic" (§.8n, p. 111), one of his polemical targets appears to be the view that natural logic is a purely inborn ability that belongs to all humans qua rational beings and cannot therefore be the object of teaching or training. Although Wolff mentions no names, he is possibly referring to Johann Andreas Schmidt (1652–1726) (see Favaretti Camposampiero, 2017, p. 118), who characterized natural logic as "the *potentia disserendi* [power to reason] that pertains to the human

[6] Thus, I do not agree that Wolff "conveys the impression that it was he who invented this distinction" (Hoenen, 2010, p. 105).

being as a rational animal" (Schmidt, 1702, §.1, p. 2). Against such a restriction of the natural to the inborn (i.e. to the "purely" natural), Wolff maintains that "this faculty of the soul to direct the operations of the mind in investigating the truth without the rules that we have been taught is not purely natural; rather, it is an ability acquired by virtue of some pre-existing natural disposition" (Wolff, 1740/1983a, vol. 2, §.8n, p. 111).

As a consequence, the distinction between *logica docens* (theoretical logic) and *logica utens* (practical logic), traditionally treated as a subdivision of artificial logic, must find a place also within natural logic, insofar as this includes an acquired component (Wolff, 1740/1983a, vol. 2, §.8n). Besides having an inborn disposition to follow the rules of logic, the mind may acquire by mere practice both a confused acquaintance with those rules, which constitutes the *logica naturalis docens* (theoretical natural logic), and the *habitus* of putting this knowledge into practice, or *logica naturalis utens* (practical natural logic) (§§.8–9). To the objection that natural logic cannot be taught, and hence should not be called *docens*, Wolff replies that "we teach not only by words but also by deeds", that is by offering concrete examples of (cognitive) conduct that serve as models for our disciples to imitate (§.8n, p. 111; see also §.10n). Thus, "acquired natural logic" is not an oxymoron; this logic is natural insofar as it is acquired by imitation of similar cases, without the explicit cognition and formal teaching that characterize artificial logic (§.7).

On the other hand, Wolff reports that when people speak of "natural logic", they usually refer to the *logica naturalis utens* alone (Wolff, 1740/1983a, vol. 2, §.9n). This reveals the other target of Wolff's criticism, namely the view that natural logic is merely an ability acquired by practice, which neither presupposes any inborn rules or any disposition to follow them nor involves any awareness, however confused or implicit, of such rules. This radically empiricist view is the opposite of the radically innatist view of someone like Schmidt, who reduces natural logic to *logica naturalis innata* alone.

Given these shifts in meaning, one better understands Wolff's complaint that "judgments on natural logic are ambiguous and arguable from both sides of the contradiction", unless it is specified whether one is referring to the innate or the acquired component of natural logic (Wolff, 1740/1983a, vol. 2, §.6n, p. 109). Wolff might well have had the following case in mind. If one says that natural logic belongs to every human being, the truth value of this proposition varies according to the intended meaning of "natural logic". For, although every human being has an innate disposition to properly use the cognitive faculty (which makes the proposition true in the innatist reading), not everyone develops the relevant abilities (which makes the proposition false in the "acquisitional" reading). Consider the two case stories that Wolff regularly adduces as evidence for the distinction between having a certain faculty and having the use of that faculty (Favaretti Camposampiero, 2008, 2009, pp. 617–646). In 1694, hunters captured a feral man allegedly brought up by bears. In the early 1700s, a deaf from birth suddenly recovered hearing. Once released from their previous state of isolation, both men started to develop higher cognitive and linguistic abilities, which they previously lacked entirely. Materialists and anti-innatists took advantage of these cases to argue that reason is not essential to humans but rather acquired through training like any other skill. Against such subversive conclusions, Wolff invokes the distinction between reason as an inborn, essential faculty and the use of reason, which depends on the acquisition of language and the imitation of rational agents, and thus requires a social context in order to develop (Wolff, 1740/1972b, §.461). The same distinction separates inborn natural logic from the acquired use of it:

> Both examples [the feral man and the deaf from birth] teach us that humans do not make use of the mental operations required to know the truth unless they have imitated what they have seen and heard from other humans, in whose society they live. (Wolff, 1740/1983a, vol. 2, §.5, p. 109)

8.9 Wolff's Mechanization of the Mind

A recurring theme in Wolff's accounts of natural logic is the analogy mentioned above between the laws of thought and the laws of bodies (see also Wolff, 1738/1968, §.394n). The intuition that even the mind must have its own laws has deep roots in Wolff's philosophy. In the *Ausführliche Nachricht*, he traces this idea back to the earliest stages of his career:

> Long ago in my very young years, when in Leipzig I began to lecture on mathematics and philosophy, I realized that the soul has its own laws, according to which its changes occur in conformity with its essence, just as bodies have the laws of motion, according to which their changes happen in conformity with their essence. (Wolff, 1733/1973a, §.93, p. 258)

As a reference, Wolff cites a corollary of his 1705 *Methodus serierum infinitarum* (Method of Infinite Series):

> Notwithstanding the immateriality and freedom of the mind, we can affirm that even the operations of our mind are performed in a mechanical manner. The knowledge of this mechanics and its laws contributes a great deal to moral science and rational science, which still lack almost every perfection. Thus Leibniz, the most ingenious of mortals, very aptly compares mind and body sometimes to two pendulums, sometimes to two clocks, in order to explain his system of pre-established harmony. (Wolff, 1755/1974, section II, pp. 318–319)

In this early dissertation, acknowledgement of the fact that the mind operates according to laws goes hand in hand with the idea that the mind is a sort of mechanical device. In the same year, 1705, Wolff writes to Leibniz that "the thoughts in the mind follow one another no less necessarily than do the motions of the gears in a machine" (Gerhardt, 1860, p. 47). The following year, he explains to Leibniz the *mechanismus mentis* (mechanism of the mind)—that is the analogy between mind and machine sketched out in the 1705 corollary—as follows:

> Just as in machines there are several parts, each one of which, when set in motion or caused to move in a certain way, necessarily influences the motion of the other according to fixed laws of motion, so in the mind there are several faculties or potencies, each one of which, when caused to think, necessarily influences the thought of the other according to fixed laws of thinking. (Gerhardt, 1860, p. 54)

The laws of thought that will later structure Wolff's psychology and ground his account of natural logic initially emerge from the mechanistic analogy between mental and mechanical operations. However, the problem quickly dismissed in 1705 ("Notwithstanding the immateriality and freedom of the mind...") reappears as a serious threat in the 1720s, when Wolff faces the charges of Spinozism, necessitarianism and materialism. In 1724, Wolff tries to purge his mechanistic view of the laws of thought of any necessitarian consequences:

> Although the soul observes certain rules in thinking which it cannot transgress any more than bodies in their motion can deviate from the rules of motion, it does not follow from this that the soul must necessarily have all the same thoughts; thus, one cannot pass off the series of thoughts as something absolutely necessary [...]. (Wolff, 1740/1983c, §.105, p. 177)

To save mental freedom, Wolff claims that the mind is free both to direct its attention and to infer a conclusion or not. Moreover, he argues that the existence of mental laws is essential to make our thoughts correspond to the actual world. If the soul were free to determine the course of its thoughts with no rule to respect, then "all the representations in it would be arbitrary, hence their similarity to things in nature would be lost along with all certainty of whether we possess the truth or not" (Wolff, 1740/1983c, §.106, p. 180). This is even truer if one subscribes to the idea of pre-established harmony, which rejects psychophysical causation in favour of internal mental causation, so that each perceptual state of the soul is entirely determined by the series of the previous states but also represents the corresponding state of the external world. To preserve such a parallel or correspondence without interaction, both the soul and the body (or even the entire physical world) must work predictably like machines, that is, move from one state to another according to their own specific laws.

8.10 Conclusion

In the wake of Wolff's reflections, I suggest the following analogy to clarify his views on the relation between logic and human thought. Using the cognitive faculty is like using a technological device. The device works according to the laws of its functioning, which are determined by its particular structure and the physical laws of this world. These laws of functioning are explained in a booklet that contains instructions for the correct use of the device. Thus, the descriptive laws that govern the device are the basis for the prescriptive instructions that we should follow to obtain the best results from our device and avoid malfunctions. Now, we can use the device even without reading the instructions; we can simply rely on our intuition or on previous experiences with similar devices and proceed by trial and error. Still, the device will necessarily work according to its laws of functioning. In this sense, it always follows the rules embedded in its own nature, even though its user may ignore them.

Analogously, a set of rules is embedded in our mind, just as the laws of nature are embedded in any material device, and this is what Wolff calls natural logic. It consists of the invariable laws of the mind's functioning along with its disposition to work accordingly. By contrast, artificial logic is the set of instructions that prescribe the correct use of the cognitive faculty, a use that perfectly conforms to its natural functioning and thus makes the most of it. Handbooks of logic are like the owner's manuals usually supplied with appliances; they are the user guides for our cognitive faculty. Studying them is useful, since it makes us aware of how this faculty works and how we should use it so as to avoid fallacies; but it is not absolutely necessary, since we may learn to judge and reason simply by practising our natural disposition and imitating the cognitive (and especially inferential) behaviour of our fellow human beings. Some logic books are even harmful, as reading them may negatively affect our intellectual development. This is because the rules they dictate do not reflect the genuine natural logic, so that it is actually impossible to think in the way they prescribe. As examples, Wolff cites the logic books written by J. Lange (1670–1744), J. P. de Crousaz (1663–1750) and J. J. Syrbius (1674–1738). In order to think in the manner they prescribe, a human being "should receive a soul entirely different from the one he has" (Wolff, 1740/1983c, §.55, p. 127).

According to our analogy and the resulting picture, it is true that artificial logic is but a distinct explanation of natural logic, but it is also true that the former is prescriptive, whereas the latter is descriptive. Thus, the distinction between natural and artificial logic effectively clarifies the *prima facie* ambiguous character that Wolff ascribes to logic and its laws. Furthermore, it illuminates the relation between logic and psychology as being actually threefold. From the Wolffian point of view, psychology can provide the principles of *artificial* logic only insofar as it brings to light the *natural* logic that is connatural to the human mind.

References

Engfer, H.-J. (1992). Von der Leibnizschen Monadologie zur empirischen Psychologie Wolffs. In S. Carboncini & L. Cataldi Madonna (Eds.), *Nuovi studi sul pensiero di Christian Wolff* (pp. 193–215). Olms.

Favaretti Camposampiero, M. (2008). Homo inter bestias educatus: Langage et raison à partir du 'Discursus praeliminaris'. *Lumières, 12*, 29–39.

Favaretti Camposampiero, M. (2009). *Conoscenza simbolica: Pensiero e linguaggio in Christian Wolff e nella prima età moderna*. Olms.

Favaretti Camposampiero, M. (2016). Christian Wolff's philosophy of medicine: An early functional analysis of health and disease. *Quaestio, 16*, 75–94.

Favaretti Camposampiero, M. (2017). Christian Wolff on error and ignorance. In A. Pelletier (Ed.), *Christian Wolff's German logic: Sources, Significance and Reception* (pp. 99–121). Olms.

Gerhardt, C. I. (Ed.). (1860). *Briefwechsel zwischen Leibniz und Christian Wolff*. Schmidt.

Goubet, J.-F. (2004). In welchem Sinne ist die Wolffsche Psychologie als Fundament zu verstehen? Zum vermeintlichen Zirkel zwischen Psychologie und Logik. In O.-P. Rudolph & J.-F. Goubet (Eds.), *Die Psychologie Christian Wolffs* (pp. 51–60). Niemeyer.

Harman, G. (1986). *Change in view: Principles of reasoning*. MIT Press.

Hoenen, M. J. F. M. (2010). From natural thinking to scientific reasoning: Concepts of *logica naturalis* and *logica artificialis* in late-medieval and early-modern thought. *Bulletin de philosophie médiévale, 52*, 81–116.

Sánchez Rodríguez, M. (2012). *Logica naturalis*, healthy understanding and the reflecting power of judgment in Kant's philosophy. *Kant-Studien, 103*, 188–206.

Schmidt, J. A. (1702). *Logica positiva sive dialectica et analytica* (2nd ed.). Süstermann.

Wolff, C. (1965a). Vernünftige Gedanken von den Kräften des menschlichen Verstandes und ihrem richtigen Gebrauche in Erkenntnis der Wahrheit. In H. W. Arndt (Ed.), *Christian Wolff: Gesammelte Werke* (Deutsche Logik) (Vol. 1, Pt. I). Olms. (Original work published 1754).

Wolff, C. (1965b). Mathematisches Lexikon. In J. E. Hofmann (Ed.), *Christian Wolff: Gesammelte Werke* (Mathematisches Lexikon) (Vol. 11, Pt. I). Olms. (Original work published 1716).

Wolff, C. (1968). Psychologia empirica, methodo scientifica pertractata, qua ea, quae de anima humana indubia experientiae fide constant, continentur et ad solidam universae philosophiae practicae ac theologiae naturalis tractationem via sternitur. In J. École (Ed.), *Christian Wolff: Gesammelte Werke* (Psychologia empirica) (Vol. 5, Pt. II, 2nd ed.). Olms. (Original work published 1738).

Wolff, C. (1970). Philosophia moralis sive ethica, methodo scientifica pertractata. Pars prima. In J. École (Ed.), *Christian Wolff: Gesammelte Werke* (Philosophia moralis sive ethica) (Vol. 12, Pt. II). Olms. (Original work published 1750).

Wolff, C. (1971). Philosophia practica universalis, methodo scientifica pertractata, pars prior. In J. École et al. (Eds.), *Christian Wolff: Gesammelte Werke* (Philosophia practica universalis) (Vol. 10, Pt. II). Olms. (Original work published 1738).

Wolff, C. (1972a). Ratio praelectionum wolfianarum in mathesin et philosophiam universam et opus Hugonis Grotii de jure belli et pacis. In J. École (Ed.), *Christian Wolff: Gesammelte Werke* (Ratio praelectionum) (Vol. 36, Pt. II, 2nd ed.). Olms. (Original work published 1735).

Wolff, C. (1972b). Psychologia rationalis, methodo scientifica pertractata, qua ea, quae de anima humana indubia experientiae fide innotescunt, per essentiam et naturam animae explicantur, et ad intimiorem naturae ejusque autoris cognitionem profutura proponuntur. In J. École (Ed.), *Christian Wolff: Gesammelte Werke* (Psychologia rationalis) (Vol. 6, Pt. II, 2nd ed.). Olms. (Original work published 1740).

Wolff, C. (1973a). Ausführliche Nachricht von seinen eigenen Schrifften. In H. W. Arndt (Ed.), *Christian Wolff: Gesammelte Werke* (Ausführliche Nachricht) (Vol. 9, Pt. I, 2nd ed.). Olms. (Original work published 1733).

Wolff, C. (1973b). Kurtzer Unterricht, von der mathematischen Lehrart. In J. E. Hofmann (Ed.), *Christian Wolff: Gesammelte Werke* (Anfangs-Gründe aller mathematischen Wissenschaften I) (Vol. 12, Pt. I, 7th ed.). Olms. (Original work published 1750).

Wolff, C. (1974). Meletemata mathematico-philosophica. In *Christian Wolff: Gesammelte Werke* (Meletemata mathematico-philosophica) (Vol. 35, Pt. II). Olms. (Original work published 1755).

Wolff, C. (1983a). Philosophia rationalis sive logica, methodo scientifica pertractata et ad usum scientiarum atque vitae aptata. In J. École (Ed.), *Christian Wolff: Gesammelte Werke* (Philosophia rationalis sive logica) (Vol. 1–3, Pt. II, 3rd ed.). Olms. (Original work published 1740).

Wolff, C. (1983b). Vernünfftige Gedancken von Gott, der Welt und der Seele des Menschen, auch allen Dingen überhaupt. In C. A. Corr (Ed.), *Christian Wolff: Gesammelte Werke* (Deutsche Metaphysik) (Vol. 2, Pt. I, 11th ed.). Olms. (Original work published 1751).

Wolff, C. (1983c). Der Vernünfftigen Gedancken von Gott, der Welt und der Seele des Menschen, auch allen Dingen überhaupt, Anderer Theil, bestehend in ausführlichen Anmerckungen. In C. A. Corr (Ed.), *Christian Wolff: Gesammelte Werke* (Anmerkungen zur deutschen Metaphysik) (Vol. 3, Pt. I, 4th ed.). Olms. (Original work published 1740).

Chapter 9
Image Composition as an Aesthetic–Epistemological Problem in Wolff's Empirical Psychology

Márcio Suzuki and Mario Spezzapria

9.1 Introduction

Although it is not possible to say that Christian Wolff developed a proper and complete aesthetics, we can trace in his philosophy a discourse on art in a broad sense, as a "technical" activity of production. In fact, within our philosopher's reflections on production, we find the principle of pleasure arising from knowledge about the conformity of an object with the rules of construction and from a judgement about creation (i.e. a judgement on the artist's ability to obtain the desired effect, starting from the aims posed by him). Such a discourse, combining reflection on the emergence of pleasure, on art as technology (production) and on teleology, is made possible by considering *representation* as a productive activity and *imagining* not merely a present activity but as an anticipation of a future desired perfection.

In the present chapter first part, we take into consideration the Leibnizian presuppositions of Wolff's psychological reflections, which are relevant to the way in which Wolff understands the active, productive and expressive nature of the soul as a representative force. In the second part, we show how our author uses reflections on artistic praxis (in particular painting and architecture) in order to think about the physiological and cognitive patterns that can explain the formation of representations. Supposing that the soul's representative force produces *like* the artist does, Wolff characterizes the psychological *facultas fingendi* in terms of *artistic production*. Finally, in the third part, we show the psychological play between the representative and the appetitive faculties, as well as the role of the intuitive knowledge of perfection and the subjective desire for it, in aesthetic pleasure.

M. Suzuki (✉)
University of São Paulo/CNPq, São Paulo, SP, Brazil
e-mail: marciosuzuki@usp.br

M. Spezzapria
Federal University of Mato Grosso, Cuiabá, MT, Brazil
e-mail: mariospezzapria@ufmt.br

S. d. F. Araujo et al. (eds.), *The Force of an Idea*, Studies in History and Philosophy of Science 50, https://doi.org/10.1007/978-3-030-74435-9_9

9.2 Leibniz and the Theory of Expression

The increased interest in Christian Wolff's work has yielded some particularly good results in aesthetics, a field to which the German philosopher does not appear to have dedicated much attention. Whereas at the beginning of the twentieth century Alfred Baeumler (1887–1968) could write that the Silesian philosopher was the "grandfather" of German aesthetics, having made possible the famous Baumgarten's (1714–1762) work (Baeumler, 1923, p. 45), in 2009, Frederick Beiser asserted that Wolff was Baumgarten's "father", since "Baumgarten's conception of the arts is essentially Wolffian" (Beiser, 2009, p. 48). Beiser's book is relevant for proposing that the aesthetics deriving from Leibniz (1646–1716) and Wolff provide a consistent alternative to the idea of artistic autonomy, the latter having predominated since Kant's *Critique of the Power of Judgment*. The main advantage of a rationalist aesthetics would consist in keeping a certain distance from formalism and giving value to the expressive and substantive aspects of art. In addition, Beiser touches on a capital point in Wolff's thought on the arts, that is the reformulation of the Aristotelian idea of "imitation" (Beiser, 2009, p. 50) into "representation", as Joachim Krüger (1980, pp. 40–41) and Pietro Pimpinella (2006, p. 13) very astutely observe.

In order to understand the way in which Wolffian empirical psychology deals with notions related to the arts and aesthetics, the concept of representation is central. For sure, there is no proper autonomous aesthetics in Wolff's philosophy: in his system, aesthetic satisfaction is merely an integral component of the complex gear linking the representative force with the appetitive one. The Wolffian explanation for the connection between representation and desire has its origin in the reflections contained in Leibniz's *Principles of Nature and Grace*, when he says that monads have nothing else but *perceptions* and *appetitions*. As a "principle of change", the latter are "tendencies to pass from one perception to another" (Leibniz, 1925, p. 407). Moreover (as explained in the *Monadology*), this "change or passage from one to another" is led by an "activity of the inner principle" called *appetition* (Leibniz, 1925, p. 226).

In a passage of a letter to Wolff, in which he explains his doctrine of pre-established harmony, Leibniz says that when one compares the parts of the body with the different faculties of the mind, one can understand that the soul has a two-fold relation with the body: it expresses either the body's present state in relation to external objects (*expressionem praesentis externorum status*)—a state in which the soul is in agreement with its body (*Animae convenientem secundum corpum suum*)—or it can be conceived as a tendency towards a new expression, "which represents a tendency of the bodies (or of the external things) to the future state" (Gerhardt, 1860, p. 56). When the pre-established harmony between soul and body occurs in the present state, the soul has a *perception* (*perceptio*); if it anticipates a future state of the body, this perceptive anticipation is called *percepturitio*. Such an explication is fundamental for the Leibnizian "aesthetics" and is centred not on the passive or imitative but on the *expressive* way in which a monad is related to the world. In order to show how such an *expressivity* manifests itself, Leibniz appeals to the Platonic conception of the unity in multiplicity (a very important conception for much of western aesthetics, too): the way in which a soul expresses the external world is an "expression of the multiplicity in unity" (*expressio multitudinis in Unitatem*) (Gerhardt, 1860, p. 56).[1]

Wolff adopted this Leibnizian distinction between a perception addressed to a present state and one addressed to a future state, as well as the relationship between perception and appetition, or—in his own words—between a representative force and an appetitive one. This distinction is quite interesting from an aesthetic point of view, since both the representation of the present state of the body and that of its future state are conceived by means of a comparison with art. Although Wolff did not adopt either pre-established harmony or monadic theory from Leibniz, his "aesthetics" nonetheless has an affinity with an aesthetics of expressivity, since it is not based on imitation.

9.3 Representation as Picture

Unlike Leibniz, Wolff accepts that external bodies cause modifications in the sensitive organs: this is, in fact, why there are representations of what happens in the outer world. In order to explicate how external objects affect sensitivity, philosophy can make use of knowledge from the sciences, such as optics, which teaches how physical objects placed on a straight line to the eye *paint* (*abmahlen*) their image on the back of its interior. Thus, the way in which we become conscious of such objects would depend on the *painting* (*Gemälde*) that they—so to speak—imprint on the fundus of the eye, and therefore, when these "painted images" are obscure or clear, confused or distinct (to employ the Leibnizian criteria used by Wolff), the corresponding ideas are likewise clear or obscure, confused or distinct (Wolff, 1720/1999, §.219). For Wolff, then, art demonstrably plays an exemplary role in explaining the birth of representation already in its physiological stage. The paintings that things produce in the fundus of the eye become exemplars for understanding what happens in the other sensitive organs (Wolff, 1720/1999, §.219).

However, the optical account is not sufficient to elucidate how a representation originates or the different ways in which we become conscious of it. In fact, the eye may receive only a vague impression of something, without necessarily seeing the parts composing the shape of this "something". The soul has a *clear* representation of an item only when it recognizes some parts that make it possible to distinguish the item's shape from the shape of other things, and it is only by recognizing its characteristic parts one by one that we can have a *distinct* representation of the object (Wolff, 1738/1968, §.37). The eye's physiological structure, then, is not

[1] The doctrine that a substance is an expression of the whole universe, and particularly that the soul is not affected by the body but has with it an expressive relation, is presented by Leibniz in the letters 56 and 57 to Arnauld from September and October 1687 (Leibniz, 2009, p. 238s).

enough to provide a clear or a clear and distinct idea of something. A figure's acquisition presupposes the capacity to simultaneously gather *in space* the parts comprising the figure; moreover, if such an acquisition has to be a clear and distinct one, these parts necessarily have to be captured sequentially *in time* (each one of them occupying a moment sequentially, separated from other moments). It is the *ordering* of these parts in space and time that constitutes a unified representation.

Time, space and order are therefore the three basic conditions that make possible the acquisition of any external object, this object always being something *compound*. It is for this reason that there must be a faculty in the soul in charge of the compositional figuration, which Wolffian empirical psychology calls the *facultas fingendi* or *Dichtungsvermögen*. This faculty of composition presupposes the work of the imagination in recollecting sequentially, one by one, something's marks (or parts), and of memory holding onto those characteristic marks, made available by the imagination. Finally, the soul's composing power combines all these marks in a single unified composition.

Here, we can find a trace of the Leibnizian heritage: although, from a physiological point of view, we must say that when a sensation occurs, it is the object that "paints its image" in the eye, from a cognitive point of view there is in no way passivity in cognition, because such an image is produced by the soul's faculties. Wolff seems to believe that all previous epistemologies have erred in their excessive simplification by not having considered the entire process involved to produce an image of something. These epistemologies were conceived to acquire the external objects as if they were simple, whereas they are not at all. Only the soul is simple, and so the difficulty lies in explaining precisely how it is capable of representing bodies, which are composite entities.

In principle, an image derived from sensitivity is no different from one derived from the imagination, except for being generally stronger or more intense. With regard to what interests us here, namely, *image production*, the two are entirely the same. In order to be generated, images from sensitivity need the very same faculties as images from imagination because they are both pictures of something compound (Wolff, 1720/1999, §.750). They share this characteristic with paintings and sculptures:

> But both sensations and imaginings correspond to images like paintings and sculptures in that they are representations of a compound, and for this reason, the representations of physical objects [*körperliche Dinge*] are called *images*, too. In fact, an *image* is a representation of a compound item. However, sensations and imaginings are different from paintings and sculptures because they occur as simples, whereas the latter occur as compounds. An *image* produced by art is a representation of a compound within a compound; specifically, a *picture* is a representation of the compound on a surface, while a sculpture, or a *sculpted* or *embossed image*, is a representation of the compound in a physical space [*körperliche Raume*]. (Wolff, 1720/1999, §.751, p. 598. Our translation.)

Sensitive or imaginative, any picture the soul produces is comparable to a plastic or pictorial image in the sense that it is a unity of the multiplicity within the simple, whereas the image produced by a sculptor or painter presents a unity of the multiplicity in the composite. This analogy is by no means a weak one: what is at stake here is that the soul's powers involved in the ideation of a sculpture or a painting are the same as the faculties needed for image production in general. The comparison is meant to show the independence of the representative forces from the matter provided by sensibility. This imaginative power is also to be found in geometry, when we design a curved line that has never been seen before (Wolff, 1720/1999, §.241).

In fact, the imaginative–compositional power frequently slips up in such cases when it combines objects in an unregulated way. This may happen because of the *associative* capacity of the imagination, which—even in its unruliness—nevertheless obeys the rules of similitude. This would explain those crazy images that painters, sculptors and other artists create when they lose full control over their pictures (Wolff, 1720/1999, §§.243–244).

When imagination obeys the principle of reason (i.e. when the selection of the composition's parts produces "images in which there is some truth"), we follow another, completely different path of creating (Wolff, 1720/1999, §.245). Now, according to our author, there is a (transcendental) truth in something when it is produced according to the order in space and time. The creations of a sculptor "worthy of the name", as well as those of an unregulated artist, are brought into the light by the very same faculties. Nonetheless, only the "good" sculptor has control over his own creation: you won't see simply what has popped into his head but rather the beauty he has seen and carefully observed in other works of art. His compositional faculty uses the associative imagination to identify models of beauty that can serve the representation he wishes to realize. To achieve this, the imagination must already be well supplied with images of all kinds; that is, a good sculptor needs to have studied carefully the existing examples in order to be able to create a new ordered and beautiful composition. In the *Empirical Psychology*, the way in which the *facultas fingendi* operates is described precisely in terms of the architect's practice:

> *If, from what he sees in various buildings, the architect composes the idea of a building under the principle of sufficient reason, the building is made according to the architecture's rules.* And, in general, *if, from what he sees related to his art in various different artificial bodies, the artist composes the idea of some artificial body of his own art, under the principle of sufficient reason, this body [follows] the art's rules.* As a matter of fact, if, from what he sees in various different buildings, the architect composes the building's idea under the principle of sufficient reason, he will not admit in it anything of which he cannot give a sufficient reason for which it must be more present than absent, [or] why it must be more in one way than in another (§.70 *Ontol.*). Since in civil architecture the singular parts' reasons are chosen for their purpose (§.5 *Arch. civil.*) and ultimately all the particular reasons resolve themselves in the founder's scope, which is the whole building's purpose (§.2 *Archit. civil.*), he [the architect] will investigate the reasons for those things he sees having being made in other buildings, and judges [*judicat*] whether those things are in agreement with the purpose of the building whose idea he must conceive in his mind [...]. Therefore, a building follows the architectonic rules if it is built according to the idea of what the architect saw in various different buildings (and) composed it according to the principle of reason. (Wolff, 1738/1968, §.150, p. 103. Our translation.)

This excerpt on architecture provides a general account of Wolff's artistic vision: for him, architecture is modelling. In the architectonic art, it becomes particularly evident that a composition can assemble in one "artificial body" the construction rules identified in many other buildings, but only if the architect has a clear understanding of how those parts apply: doors, windows, corridors, rooms, etc., observed elsewhere are like many individual rules that the faculty of composition must make compatible with a common purpose. The architect's art is paradigmatic for the procedure of composition because it is complex, implying knowledge of various crafts, such as masonry, carpentry, joinery and hydraulics (it is certainly not by chance that Kant uses the idea of an "architectonic of pure reason" to talk about the system of sciences). Much as in architecture, all other crafts are based on a process of composing aimed at covering a specific field; this is true not only of the so-called useful or servile arts but also of the "liberal arts" or fine arts (Wolff, 1738/1968, §.150, p. 103). As a matter of fact, Wolff does not distinguish between the "fine arts" and the useful arts, treating them all under the umbrella term "technology". At first sight, this seems to distance his thought from any autonomous conception of works of art, but his effort to obtain a philosophical knowledge of art by means of an explication of the rules, according to which every single craft operates, makes his "philosophy of all the arts" (*philosophia artium quarumcunque*) a forerunner of the attempt by Diderot and D'Alembert's encyclopaedic project to be—as the work's title says—a *reasoned dictionary of the sciences, arts and trades*.[2]

The lack of distinction between the liberal and servile arts does not make Wolff a utilitarian *avant la lettre*; rather, one can say that he is the introducer of a very particular conception of *homo technologicus*. Man improves the arts as a means for his own self-development. The perfectibilian use of techniques can be targeted, as in the invention of machines to improve our vision (glasses, telescope, microscope) or of calculation tools, to expand the capacity of our understanding. However, the most valuable service that a specific art has for our improvement can be indirect and less obvious. The architect plans a building while aiming at a particular purpose—it may be a dwelling, say—or with a more general sense of the functional role of the building. In a work planned under rules, the objective utility of the construction is accompanied by "subjective" gratification, both of the architect and of those who live in the building or who just contemplate its accomplishment. To a great extent, Wolff's thought pivots on such a tightly convergent relationship between the human artefact and its producer and/or user. All more or less well-finished creations generate satisfaction with the (manual or mental) ability to create it. For this reason, as a philosophy of art, Wolffian philosophy is a discourse on human aptitudes and abilities (see Krüger, 1980, p. 31). It is for this reason, too, that the same faculties of the soul are translated in terms of technical or artistic production, as happens when paintings and sculptures are brought to mind for the elucidation of the mental or physiological genesis of an image or a sensitive representation. Contrary to what one may think (and of what Kant himself believed, too), Wolffian dogmatic philosophy is no scholastic knowledge; instead, it is addressed to practice.

[2] For Wolff on technology, see his *Preliminary Discourse* (Wolff, 1740/1983a, §.71, p. 3). On philosophy of art in this broader meaning, see Krüger (1980, p. 29).

For Wolff, art is far away from all spontaneity and voluntarism. Sculptors and architects are educated through the study of the most comprehensive possible repertory of fine works, contemporary or traditional. Without making it explicit, the philosopher shares the neoclassical belief that the artist must attempt to approximate the maximum of ideal beauty, which is feasible only by means of exercise and contact with works of art. To really do so, artists need to attend art academies, which are to be promoted by the state's authority (Wolff, 1736/1975, §§.310–311; see Pimpinella, 2006, p. 12).

9.4 Representation and Desire

The Wolffian theory of composition as an order in time and space puts the formation of representation in a relationship of similitude and difference with the formation of painting/sculpture. The "mental" image is an inner sculpture or painting, composed as a result of the major or minor ability of each individual's inner paintbrush or chisel; still, the composition does not remain within the compound material, but rather within the soul's unity. This productive capacity is no small thing, and yet it could be said that such explications are only given from a cognitive and theoretical point of view and do not offer any indication of the entire context in which artistic representation is involved in the Wolffian system. Returning to Leibniz's distinction, we can say that mental images are perceptions of the present state of things, but they do not account for the perception of their future state yet. The image of a building designed by an architect gives us a better representation of a future state.

Wolff adopted the distinction between *perception* and *percepturitio* that he learned from Leibniz. Like him, Wolff thinks the soul makes an effort (*Streben*) to pass from one representation to another. This *conatus* (inclination) is defined using the same Leibnizian term: "In every present perception there is an inclination to change the perception" (Wolff, 1740/1972, §.480, p. 395. Our translation). This engagement to move on to a future perception is called *percepturitio* (Wolff, 1740/1972, §.481, p. 396). Clearly, the displacement of the focus from the present perception to the perception of a future something changes the entire game, since we leave a pretty much neutral or "theoretical" relationship with the representation's object and start to have a relationship of desire with it; in other words, the representation now involves not only the representative faculty but also the volitional or appetitive one. The appetite does not originate from nowhere; instead, it comes from the "cognition" of something (*appetitus nascitur ex cognitione*). There is a lexical care in this expression that must be pointed out: at stake here is not the *object's cognition* but the cognition of the fact that the object is desirable, and more precisely that changing the present perception is desirable in order to satisfy this wish (*connatus mutandi perceptionem*). To identify this kind of "consciousness of something", as opposed to the "theoretical" consciousness of it, Wolff introduces the term *appercepturitio*, coined from the Leibnizian words *apperceptio* and *percepturitio*. The *appetitus* is not a neutral representation but an *inclinatio* towards the object (Wolff, 1738/1968, §.509).

The key concept in the representation-desire device is the notion of perfection, as used by Leibniz but of Platonic descent. The inclination that constantly makes the soul depart its present state for "another" one is a constant quest for perfection. The soul seeks to achieve a state of higher perfection: when such a state is reached, the soul feels *pleasure* (the Latin term used by Wolff is *voluptas*; the corresponding German term is *Lust*). The soul is always looking for objects closer to perfection, giving greater satisfaction than those it already knows. The background of Wolff's arguments (though he might not have been completely aware of it) is the idea of a *soul's inquietude*, originally an Augustinian conception and strongly present in the reflections of Pascal, Locke and Malebranche, eventually filtered and mitigated by Leibniz. The soul must always be in expansion; it must avoid objects that keep it where it is or that cause repugnance and displeasure (*taedio*, *Unlust*) because these states of lower perfection are likely to bring it to "atrophy" and imperfection.

As Wolff himself explains, the statement that perfection is the cause of pleasure is not one of his own: Descartes had already expressed it distinctly (*distincte*) in his letter to the princess Elisabeth on 6 October 1645.[3] The French philosopher's explanation would have brought Wolff (according to what he himself said) to define pleasure as intuition of perfection—or intuitive knowledge of perfection (Wolff, 1738/1968, §.511; 1720/1999, §.404). In the Wolffian representative–appetitive–affective system, this implies that it is by means of representation that we glimpse something good we want to reach, and it is the representation of such a glimpsed perfection that gives pleasure. Pleasure is therefore prior to and independent of the realization or achievement of what is represented. With this, Wolff means that *voluptas/Lust* is *already* given *in the mere representation of what is perfect*. Such a conception, entirely in agreement with the central role of the *vis repraesentativa* in his system, gives autonomy to the representation, rather than conceiving of it as dependent on the desired thing. This point deserves a further discussion because it was fundamental for the development of the subsequent aesthetics.

Aesthetic (and architectonic) representation works, once again, as a paradigm to explain the correlation between pleasure and perfection. "If I see a picture", Wolff states, "that is similar to the object it represents, and I consider its similitude, I derive pleasure [from this painting]. Now, a painting's perfection consists in similitude" (Wolff, 1720/1999, §.404, p. 344. Our translation). A painting in which we recognize its similitude with what is represented is a source of a pleasure, perceptible by the recognizance of the similitude. In his *Empirical Psychology*, Wolff mentions a similitude between the image and the prototype of perfection (*similitudine imaginis* cum *prototypo perfectio ejus consistit*) (Wolff, 1738/1968, §.512). This characterization is problematic, however, because it seems to consider the artistic creation again from the perspective of imitation. The more complete the mutual

[3] According to Pietro Pimpinella (2005, p. 253), when Wolff reconciled his own theory of the passions with that of Leibniz, he was also aware of Descartes' efforts in the *Passions of the Soul*.

relationship between representation and that which is represented, the more perfect the imitation will be: "In fact, there is nothing in the image that does not represent something in the prototype, and there is nothing in the prototype that is not represented in the image" (Wolff, 1738/1968, §.512, p. 390. Our translation). Similar statements can also be found in the text *De voluptate ex cognitione veritatis percipienda*, one of the principal sources enabling us to trace an "aesthetic theory" in Wolff. In any case, the importance of imitation in Wolff can be evaluated once we understand with more precision what he meant by *similitude*.

With regard to aesthetics, one of the interesting points in the Wolffian argument is the assertion that the "intuition" of perfection does not necessarily need to be true. Pleasure can arise from the vision of a false perfection, too: "[...] it is not necessary that pleasure be grounded on a true perfection; it is sufficient for it to have an appearance of perfection" (Wolff, 1720/1999, §.404, p. 344. Our translation). Being merely a first approximation (the glimmer of something of a future we do not yet know), the apprehension of something good can be founded on a mere appearance, on an illusion or mirage:

> It is of the highest importance that we learn that pleasure is perceived no less by the apparent perception than by the true one [*non minus percipi ex perceptione apparente, quam ex vera*], although the seed of all moral evil resides in that fact. (Wolff, 1738/1968, §.511, p. 389. Our translation).

This sentence shows the risk that adherence to perfection, considered equivocally, as such may cause; the other side of the coin, even so, is that this adherence itself proves to be testament to the representative power and of the power of the representation itself. To a large extent, we inevitably fall victim to error at this level, because it is just the first contact with something unknown, and above all because intuitiveness is deprived of reflection (intuition in essence does not involve any judgement as to whether something corresponds to the truth). This intuition only *provides* something to be known, a gift acting like a fuse lighting up the desire for something; it is in no way the accurate knowledge of this something it suggests. Anachronistically, we can say that Wolff achieves a phenomenological description of the manifestation of a heretofore unconscious desire. The Wolffian process does indeed have something phenomenological about it, given that its purpose is to describe and distinguish, in its different nuances, *how this knowledge of something shows up*. As in the case of the Leibnizian discernment of the ideas, the intuition of perfection can be obscure, clear, confused or distinct. In the mistaken intuition of perfection, the conscious mind is not capable of seeing the characteristic marks of the represented object correctly. Another case, however, is more interesting: the distinct perception of something must be accompanied by an indistinct perception to be able to arouse pleasure.

Some examples may help us to better understand this. Comparing side by side the demonstrations of the infinitesimal calculus of Newton and Leibniz, the clearest order in which the Newtonian sequence is presented patently provides the soul with much more pleasure than does the Leibnizian sequence. This means that, regardless of its content, Newton's exposition is more elegant. Being mathematical knowledge, no doubt the arguments of both the English scientist and the German philosopher are irreproachable and not lacking in clarity and distinction. However, one of them has (or it has in major proportion) something the other does not: in the exposition of the Newtonian sequence, the soul feels delighted by the beautiful order (Wolff, 1729/1983b, §.5). It can be said comprehension of the infinitesimal calculus runs on two levels: the first is knowledge of the truth; the second is the cognition of intellectual perfection in the order of its presentation (far better in Newton than in Leibniz). This example makes clear that pleasure is thus somehow independent of the represented "content" (the infinitesimal calculus), being foremost a pleasure derived from representation.

This double way of considering the same facts remained unperceived by Descartes and Locke. It was spotted but not exactly explained by Leibniz: when the mind knows, it also has—so to say—a gaze fixed on itself; it observes what happens in itself, although usually without noticing it (Wolff, 1720/1999, §§.727–730). Without doubt, one of Wolffian psychology's greatest discoveries is the elucidation of this dual, objective–subjective addressing of consciousness. Wolff was led to this result by, *inter alia*, this conclusion: clear and distinct knowledge has, by definition, to be conscious knowledge, and moreover, by being conscious (i.e. "knowing that it knows"), consciousness must be at once knowledge of the difference between itself and what it is conscious of. Now, it happens that consciousness cannot be simultaneously a clear and distinct consciousness of the object and of itself, and for this reason, when it has knowledge (of something) with clarity and distinction, it can only have a confused knowledge of itself. Wolff knows that the possibility of concomitance between these two views implied the contrast in clarity between them. A converse example may help us to understand the point: the soul can have clarity and distinction about the fact it has obscure knowledge of something. Such technicalities should not allow us to lose sight of the fundamentals, however; that is, that even the perfection we know with the highest possible clarity and distinction cannot provoke an equally clear and distinct representation in cases where it might generate pleasure. An explication in the *German Metaphysics* on why science and discovery provide pleasure elucidates this point. A deep (*gründlich*) knowledge and new discoveries give a greater and more sensible (*so viel empfindlicher*) pleasure:

> [...] the greater the effort we made before to understand it or to find them out. In fact, we have then an intuitive knowledge of our understanding's perfection and at the same time [*zugleich*] of the object we know distinctly [*deutlich*], as well as of the discoverer's perfection, if we learn to comprehend what someone else has found out. Now, the more demanding it is to comprehend or to find something out, the greater is the knowledge we get of our perfection, particularly if we call to mind all the other things that we have already come to know with less of a struggle; in that case, the pleasure is undoubtedly greater. (Wolff, 1720/1999, §.412, p. 350. Our translation.)

It becomes clear that the focus of the Wolffian reflection here is not the knowledge gained or the new discoveries but the difficulty and the struggle to obtain them, the pleasure felt being proportional to the effort. This point is fundamental: it is as though the reflection puts into parentheses the objective side of the cognition, concentrating only on its subjective activity. This is essential to understanding the problem of artistic production and representation in Wolff.

As we have seen, the pleasure we derive from a well-executed painting originates from the existing similitude between the image and its prototype. When a painter sketches an image (e.g. when he paints a grape), he does not pretend to be doing anything other than representing on the canvas the prototype of which he has made an image. That is, it is necessary to distinguish in the image what we distinguish in the prototype (Wolff, 1738/1968, §.512). All this seems to reinforce the impression that imitation is essential to appreciating a picture. Indeed, this does matter but—much as we saw in the case of knowledge and discoveries—the role of imitation has to be put into parentheses in order to be able to identify the source of the true pleasure. It cannot be said that pleasure comes from the represented object: the poor grape is nothing in comparison to the artist's power to imitate it. The admiration a picture generates does not come from the comparison between the real object and the represented object but from the capacity of the artist to achieve the effect of representing something. The same grape projected into a mirror, with the same play of shadows, would be a mere product of chance, and he who observed it could, at most, admire the casual artistic potential of such a sight, but no intentional proposal behind the representation.

Wolffian reasoning has its fulcrum, in fact, in the creator's intention, and judgement of his creation must be addressed to his ability to achieve the desired effect, starting from the aim he established beforehand. The principle of reason commanding the execution of any work of art is final, although the execution itself is the result of mechanical actions; that is, it is subject to rules of efficient causality. We admire a clock insofar as it is efficient at correctly showing the hours: "pleasure consists here, too, in an intuitive knowledge of perfection", and likewise if a connoisseur of architecture (*Bau-Kunst Verständiger*) contemplates a building made according to architecture's rules, he knows its perfection (on the basis of such rules) (Wolff, 1720/1999, §.404, p. 344. Our translation). Pleasure originates, then, from such knowledge of the conformity of something with its principles of construction: put another way, with its *rules*. This "finalism" might be expanded to a global perspective, in which the universe is conceived according to the rules of a great artificer—a great architect—which is God. However, the ingeniousness of Wolffian thought is surely not addressing theological proof, since its concern, as we have said, is the human capacity for knowledge and refinement. Every well-executed work generates pleasing approval because the observer recognizes its manufacture according to the standards required by the relevant art.

We can now better understand the status of painting and imitation in Wolff. The *purpose* of a pictorial work is to imitate, just as any other type of art (liberal or utilitarian) must produce an effect starting from its own specific finality. What matters, however, is not the imitated content, but whether the imitation is well done. The relevance of Wolffian "aesthetics" consists in the fact that intuitivism is the paradigm of the relationship between representation and perfection—an "image" anticipates something that we do not yet know but that we desire in our quest for perfection. In this sense, it is necessary to understand that *intuitivism is not a synonym for visibility* but for something broader, called *representativeness*. The obvious fact that it is somehow much easier to talk in terms of images and similarity cannot prevent us from recognizing the broader significance of the representative faculty; it embraces, besides painting, complex objects like Newton's mathematic calculation, architecture and clockwork mechanisms. As an anticipation of perfection, painting comprises only one of the classes of the wider genre named *technique*, involving representation. The same thing must be said, retrospectively, of pictorial and plastic imagery as a paradigm of present representation. These kinds of art are not the present perception's only means of representation (if this were the case, it would imply as a consequence the exclusion of the non-visual senses), but they are probably the best means of explication of what a present perception is (the unity of the composite in the simple) because of their capacity to demonstrate *that the mental representative force produces its images like an artist*.

It is the representative capacity, in fact, that is at stake. Someone who follows a demonstration can admire any order contained in it and thus distinguish it from a disordered demonstration (order is the expression of the principle of reason). In the case of a very long demonstration, it is always useful to divide it and place it within a framework that can be intuitively understood at a single glance (*uno obtutu*) (Wolff, 1729/1983b, p. 382). In the same way, architects (or connoisseurs) who contemplate a building made under the best rules of their particular architectonic craft are not able to hold in their mind individually (i.e. distinctly) all the rules used to make the building; rather, they perceive the good order implicit in its construction, consisting, for instance, in the good rhythm (*eurithmia*) of its parts, that is in the similarities between them and in relation to the building as a whole (Wolff, 1729/1983b, p. 381). What is important, once again, is that although the architect may be aware of all the rules separately, in the moment in which he contemplates the work of art he is not able to keep them all in his mind with clarity and distinction.

9.5 Conclusion

Wolff brings all the arts together under a singular principle of finalistic explication (in which what is at stake is observing whether the realization of the purpose is well executed), and nothing would seem to be farther from the idea of the autonomy of the aesthetic object than such an explanation. However, this finalistic vision is constructed in parallel with a highly innovative conception of *representativity*. It was along this path that his followers—A. G. Baumgarten (1714–1762), G. F. Meier (1718–1777), M. Mendelssohn (1729–1786)—proceeded. Kant knew, directly or indirectly, such Wolffian ideas. It was without doubt the autonomy of representation that linked Wolff to aesthetic thought; it was for this reason that Joachim Krüger very appropriately affirmed that the dogmatic philosopher got very near to the modern idea of exposition (*Darstellung*), a topic dear to Kant and to German idealism (Krüger, 1980, p. 41).

References

Baeumler, A. (1923). *Kants Kritik der Urteilskraft. Ihre Geschichte und Systematik.* Max Niemeyer.

Beiser, F. C. (2009). *Diotima's children: German aesthetic rationalism from Leibniz to Lessing.* Oxford University Press.

Gerhardt, C. I. (Ed.). (1860). *Briefwechsel zwischen Leibniz und Wolff.* Schmidt.

Krüger, J. (1980). *Christian Wolff und die Ästhetik.* Humboldt-Universität.

Leibniz, G. W. (1925). *The Monadology and other philosophical writings.* Oxford University Press.

Leibniz, G. W. (2009). *Sämtliche Schriften und Briefe* (Philosophischer Briefwechsel (1686–1694)) (Vol. 2, Reihe II). Akademie Verlag.

Pimpinella, P. (2005). *Wolff e Baumgarten. Studi di terminologia filosofica.* Leo S. Olschki.

Pimpinella, P. (2006). La théorie wolffienne des arts à l'origine de l'esthétique. *Revue Germanique Internationale, 4*, 9–22.

Wolff, C. (1968). Psychologia Empirica, methodo scientifica pertractata, qua ea, quae de anima humana indubia experientiae fide constant, continentur et ad solidam universae philosophiae practicae ac theologiae naturalis tractationem via sternitur. In J. École (Ed.), *Christian Wolff: Gesammelte Werke* (Psychologia Empirica) (Vol. 5, Pt. II). Olms. (Original work published 1738).

Wolff, C. (1972). Psychologia rationalis, methodo scientifica pertractata, qua ea, quae de anima humana indubia experientiae fide innotescunt, per essentiam et naturam animae explicantur, et ad intimiorem naturae ejusque autoris cognitionem profutura proponuntur. In J. École (Ed.), *Christian Wolff: Gesammelte Werke* (Psychologia rationalis) (Vol. 6, Pt. II, 2nd ed.). Olms. (Original work published 1740).

Wolff, C. (1975). Vernünftige Gedanken von dem gesellschaftlichen Leben der Menschen. In H. W. Arndt (Ed.), *Christian Wolff: Gesammelte Werke* (Deutsche Politik) (Vol. 5, Pt. I). Olms. (Original work published 1736).

Wolff, C. (1983a). Philosophia rationalis sive Logica, methodo scientifica pertractata et ad usum scientiarum atque vitae aptata. Praemittitur Discursus praeliminaris de philosophia in genere. In J. École (Ed.), *Christian Wolff: Gesammelte Werke* (Discursus praemlinaris) (Vol. 1.1, Pt. II, 3rd ed.). Olms. (Original work published 1740).

Wolff, C. (1983b). De voluptate ex cognitione veritatis percipienda. In J. École (Ed.), *Christian Wolff: Gesammelte Werke* (Horae subsecivae Marburgenses anni MDCCXXIX) (Vol. 34.1, Pt. II). Olms. (Original work published 1729).

Wolff, C. (1999). *Metafisica Tedesca.* Rusconi. (Original work published 1720).

Chapter 10
In-Between Psychology and Moral Philosophy: Christian Wolff's Principle of Natural Obligation

Dieter Hüning

10.1 Introduction

In what follows, I hope to illuminate some aspects of the relationship between practical philosophy and empirical psychology in the thought of Christian Wolff. To that end, I will focus my investigations on Wolff's concept of *obligatio naturalis* [natural obligation] and its psychological presuppositions. Wolff believed that this conception of obligation would allow him to remedy certain problems of justification he encountered in natural law theory. Given its novelty, his concept of natural obligation and the closely related issue of free will formed a major point of contention in Wolff's dispute with the Halle theologians. In this context, Joachim Lange in particular charged Wolff with fatalism a charge that has been investigated thoroughly by Bruno Bianco (1989, pp. 116ff.). Bianco shows that Lange developed his charge of fatalism primarily against the background of the ontological and cosmological presuppositions of the Wolffian system and its doctrine of pre-established harmony (Bianco 1989, p. 117). In contrast, my contribution will focus on the aspects of Wolff's doctrine of obligation that specifically concern the will, and on Lange's criticism thereof.

It will do so in four parts: First, I will examine Wolff's criticism of Pufendorf's conception of obligation (Sect. 10.2). I will then discuss Wolff's concept of natural obligation with the aim of showing what its novelty consists in and how it relates critically to the natural law conception of obligation dominant till then, as it derives from Pufendorf (Sect. 10.3). Subsequently, I will briefly address Wolff's project of systematically grounding practical philosophy on the results of empirical psychology (Sect. 10.4). I will only be able to do so in outline, however, since a full account of these psychological presuppositions would not only require going into Wolff's empirical psychology but also necessitate discussing the metaphysical basics of his

D. Hüning (✉)
University of Trier, Kant Research Center, Trier, Germany
e-mail: huening@uni-trier.de

S. d. F. Araujo et al. (eds.), *The Force of an Idea*, Studies in History and Philosophy of Science 50, https://doi.org/10.1007/978-3-030-74435-9_10

philosophy and his relationship to Leibniz. Finally, I will illustrate the difficulties with Wolff's conception of the will in relation to his treatment of the problem of indifference (Sect. 10.5).[1]

10.2 Wolff's Critique of Pufendorf and the Proponents of a Voluntarist Account of Obligation

Wolff's account of the foundations of practical philosophy is set in notable opposition to Pufendorf's and Thomasius's natural law theory. In particular, this opposition manifests itself in Wolff's revision of two central concepts of practical philosophy and natural law theory, viz., the concept of law and the concept of obligation. Pufendorf and Thomasius had defined the concept of law as an obligating commandment issued by a superior authority. In this vein, Pufendorf writes: "In general, law seems best defined as a decree whereby a superior obligates someone subject to himself to conform his actions to the superior's prescription" (Pufendorf, 1672/1998, p. 71).[2]

Wolff empathically rejects this definition because, on his view, it abrogates the concept of natural obligation:

> One usually defines the law as a superior's commandment promulgated to an inferior and obligating him; but this is not the definition of the law in general. It is the definition of those who abrogate natural obligation by negating the intrinsic goodness and badness of actions and who hold that all actions as such are indifferent before they are related to God's will as a superior being. Since we have already established the intrinsic goodness and badness of actions, which is confirmed by the old philosophers and theologians; we also define the law in general as the way in which an obligation respectively demands that certain actions be done and other actions be omitted. (Wolff, 1738/1971, I, §.131 nota)[3]

He contrasts Pufendorf's definition with his own, which defines the concept of law as follows: "A law is called that prescription according to which we are obligated to arrange our actions" (Wolff, 1738/1971, §.131; cf. also Wolff, 1754/1980, §.39). With this entirely formal definition, Wolff characteristically eliminates any reference to an external, commanding will and instead merely highlights the obligatoriness of the law (Hartung, 1999, p. 133).

A similar opposition to Pufendorf's natural law theory is detectable in Wolff's account of obligation. Following Roman law, Pufendorf understands *obligatio* as a *vinculum juris* (legal bond), "by whose necessity we are constrained to do something" (Pufendorf, 1672/1998, p. 72). Elsewhere, Pufendorf defines obligation as an operative moral quality (*qualitas moralis operativa*) that arises from an authority's

---

[1] Part of Sects. 10.2 and 10.3 have been adapted from Hüning (2002). For the translation, I have to thank Sascha Settegast (University of Trier).

[2] In this chapter, Pufendorf's citations have been translated by Michael Seidler (Carr, 1994). Cf. also Christian Thomasius's definition (Thomasius, 1718/1963, I, 5, §.3).

[3] Regarding Wolff's critique of Pufendorf, see also Wolff (1738/1971, I, §.63 nota).

law and "by which someone is bound to do or suffer something" (Pufendorf, 1672/1998, p. 72). While Pufendorf thus conceives of obligation as a moral necessity arising from law, Wolff's concept of obligation highlights its psychological character: Obligation, if considered as an act, which we will call active obligation (obligationem activam), is the connecting of a motive to an action, whether it is [an action] to be done or to be omitted (Wolff, 1754/1980, §.35).

In the *Deutsche Ethik* a quarter-century earlier, Wolff had formulated it succinctly: "To oblige someone to do or to refrain from doing something, is nothing but to connect a motive of willing or not willing to it" (Wolff, 1733/1976, §.8, p. 9). Thus, the moral necessity that constitutes the obligatory character of the law does not rest on the law's relationship to a law-giving will, but rather depends on the psychological conditions that determine the will's decision at any one time, that is, it depends on the presence of relevant motives (Hartung, 1999, p. 129).

On Wolff's view, the human will is constituted such that a recognition of the good connected to an action functions as a "motive of the will [...], such that we will [the action]" and, conversely, that a recognition of its evil functions as a "motive of not-willing, or of eschewing a thing" (Wolff, 1733/1976, §§.6, pp. 7–8).[4] In this respect, the motivation of the will through a representation of the good or evil connected to an action is identical to its obligatoriness. As Clemens Schwaiger has emphasized, psychologizing the concept of obligation in this way, with the aim of "combating a purely positivistic conception of obligation," indeed represents a "profound innovation" in the context of the early modern theory of obligation: "Obligation equals motivation—this, in short, is Wolff's solution to the problem of obligation" (Schwaiger, 2000, pp. 251f.). Yet, it also raises the question whether the specifically normative character of obligation—that it represents an ought, a moral kind of necessitation rather than a merely natural one—is not thereby being dissolved into mere psychology.

In this respect, Wolff introduces a distinction between *obligatio activa* (active obligation) and *obligatio passiva* (passive obligation).While the former is generated from the will of the one who institutes the obligation by connecting an action to a motive (Wolff, 1738/1971, §.118), *obligatio passiva* refers to the necessity conferred upon the action through this act of instituting the obligation (*actus obligatorius*) (Wolff, 1738/1971, §.121). Active obligation necessitates an action by connecting it to a particular motive. Connection between motive and action (*connexio* [...] *motivi* cum *actione*): this is Wolff's formula for obligation.[5] In contrast, passive obligation consists in the moral necessity of "determining an action in such

[4] Elsewhere, Wolff clarifies that he understands the will "in a narrow sense" like the Scholastics as "rational desire" (*appetitus rationalis*) (Wolff, 1740/1983, §.155). See also the definition of the will in Wolff (1738/1968a, §.880). Regarding the systematic problems connected to this intellectualist conception of the freedom of the will, cf. H. Wolff (1949, pp. 109ff.).

[5] "Obligation [...] is the connection of a motive to an action [...]" (Wolff, 1754/1980, §.35). "To oblige someone to do or to refrain from doing something, is nothing but to connect a motive of willing or not willing to it" (Wolff, 1733/1976, §.8).

and no other way" (Wolff, 1740/1968b, §.57 nota), which is instituted by the *obligatio activa*.[6]

With this definition of obligation, Wolff reacts to the decisive weakness in Pufendorf's voluntarist account of obligation. This weakness consists in the problem that, in the words of Julius Ebbinghaus, "the principle of the divine will [as assumed by Pufendorf] itself plainly lies beyond anything that could take on the character of law as conceivable by human beings" (Ebbinghaus, 1986, p. 311). In other words, for human beings subject to such arbitrary legislation, Pufendorf's voluntarist account of obligation entails that they are incapable of comprehending the divine legislation, including the reasons why God commands or prohibits the things he does. In consequence, the idea of submission to such arbitrary legislation immediately abrogates the very possibility of the will's morality. If the obligation to follow the natural law only rests on God having commanded it, then there is no properly moral ground for following it and submitting to God's will. For in this case, there is no ground of compliance that is intrinsic either to the natural law or to the obligations it imposes as such. Consequently, the reason why people follow the natural law as commanded by God can then only consist in that divine attribute "which I am able to conceive in independence of the determinations of His will, namely in His omnipotence" (Ebbinghaus, 1990, pp. 398f.).

A further point of difference between Wolff and the preceding natural law tradition concerns the systematic location or place within the system of philosophy at which the concept of obligation is to be treated. While for Pufendorf the issue of obligation represents a problem that belongs to natural law theory proper, Wolff places his treatment of this concept within *Philosophia practica universalis* (universal/general practical philosophy), that is, within that discipline which provides the normative foundations for both natural law theory and ethics more narrowly construed.

A particular punch line of Wolff's polemic lies in his charge that Pufendorf's moral-positivistic conception ultimately abets the amoralism of the atheists. For Pufendorf's disavowal of any necessary connection of human nature and the essence of things to natural obligation on the one hand and his systematic connection of the concept of obligation to the will of God on the other make it easier for the atheist to abrogate all natural obligation together with the belief in God (Wolff, 1738/1971, §.245). This is the reason why the concept of natural law is at the heart of Wolff's engagement with Pufendorf. In this vein, Wolff declares in his *Philosophia practica universalis* that the natural law differs from all positive law in that its *ratio cognoscendi* (ground of cognition) lies in human nature and the essence of things, and thus ultimately in their natural teleology.[7] The obligatoriness of the positive law, in contrast, depends on an alien will (Wolff, 1738/1971, §.147).

---

[6] See also Wolff (1738/1971, §.118).

[7] "One calls a natural law that which has its sufficient ground in the nature of man and of things itself" (Wolff, 1754/1980, §.39). See also Wolff (1738/1971, §.135).

10.3 Wolff's Account of the *Obligatio Naturalis*

In the *Philosophia practica universalis*, Wolff refers to the *obligatio naturalis* as that obligation which arises from the natural law (Wolff, 1738/1971, §.141) and has its sufficient ground in the nature and essence of human beings and of things (*ipsa hominis rerumque essentia*) (Wolff, 1738/1971, §.143). Hence, the *obligatio naturalis* represents a special case of the general concept of obligation discussed above, that is, of the connection of a motive with an action, and in this sense expresses the moral necessity of the natural law, according to which we are obligated by this law, that is, by our own rational nature, to perform actions that are good as such and avoid actions that are bad as such (Wolff, 1738/1971, §.153; also cf. §.128).

On Wolff's view, a natural obligation "obtains when a sufficient motive of action results from the natural constitution of human beings" (Hartung, 1999, p. 131). Accordingly, Wolff defines the *obligatio naturalis* as that obligation "which has its sufficient ground in the proper essence and nature of man and the other things" (Wolff, 1754/1980, §.38; cf. also Wolff, 1738/1971, §.129). What is striking about this definition is that Wolff here too aims at grounding the obligatoriness of the natural law without recourse to any kind of alien will. Thus, if all morality is grounded immediately in human nature, then a rational human being does not seek the good and avoid the evil "from consideration of reward and fear of punishment." Rather, he gives the law to himself, that is, he lets his actions be determined solely by the recognition of the potential good or evil following from them and therefore needs no further motive of action beyond the natural obligation itself (Wolff, 1733/1976, §.38). Hence, for Wolff, as much as already for Grotius, the fear of punishment is "in itself not yet a constitutive attribute" of morality but rather signifies a "practical means of producing norm-conforming behavior, which is supposedly able to guarantee an indispensable minimum of moral conduct" (Grunert, 2000, p. 104).

Incidentally, it is important to note that, in the early modern natural law theory, the concept of obligation (*obligatio*), or of being obligated, is distinguished strictly from the concept of duty (*officium*), although both terms largely overlap in contemporary colloquial use, particularly in German (cf. *Verpflichtung* and *Pflicht*). For Wolff, (passive) obligation refers to the moral necessity of a type of action or omission (Wolff, 1738/1971, §.118) as it arises from a law, while duty signifies the particular act demanded by the law (Wolff, 1738/1971, §.224). In the Preface to the second edition of his *Deutsche Ethik* (German Ethics), Wolff himself regarded the introduction of a novel concept of *obligatio naturalis* as his proper contribution to the field of practical philosophy (Wolff, 1733/1976). As he writes in the *Ausführliche Nachricht*:

> I have given a general concept of obligation, suchlike as did not exist heretofore, and, since like all true and distinct concepts it is fruitful such that all that may be cognized of obligation can be deduced from it, I have demonstrated from it that in the nature of man and the constitution of free action there is grounded an obligation that I call *natural*, and which even he must recognize who either does not recognize what kind of being GOD is or even denies that a GOD exists. Even though I have asserted together with Grotius and our theologians that even *in hypothesi impossibili athei*, i.e. under the impossible condition that no GOD

> should exist, a law of nature must be conceded, in order to persuade those of their folly who are suited by atheistry because it would, in their opinion, allow them to live as they will; yet I have ascended further and shown that the originator of this natural obligation is GOD and that through this He binds man in yet another way to direct his actions such that they redound to his perfection and even to that of the whole human race and of the entire world. Yet insofar as GOD obligates us, we have to recognize him as the legislator of natural law. (Wolff, 1733/1973a, §.137, pp. 395–396)

In what follows, I will highlight some important aspects of Wolff's theory of obligation. First of all, it is notable that Wolff rehabilitates the classically Scholastic doctrine of *perseitas* [perseity], that is, the doctrine of the self-sufficient being of good and evil, which finds clear expression in his claim that, through his novel concept of natural obligation, he has "demonstrated that man's actions are in themselves necessarily good or evil, and by no means first become good or evil through a superior authority's command or prohibition" (Wolff, 1733/1976, Preface to the second edition).[8] Naturally, Wolff is aware that the primary target of this claim is Pufendorf's natural law theory, as Pufendorf follows Hobbes in his strict opposition to the doctrine of *perseitas*. According to this doctrine, particular actions are morally good or bad as such, that is, independently of the law-giving will of a superior, so that, e.g., theft, adultery, or incest are bad "as such and according to their nature," even apart from any legal stipulation.[9] Pufendorf rejects the doctrine of *perseitas* because, on his view, it implies an artificial and misleading distinction between the *ius naturale* and the *ius divinum positivum*:

> Some lay down as the object of the natural law those acts which are morally necessary or base of themselves, and which are therefore in their own nature either owed or illicit, and for this reason understood as necessarily prescribed or forbidden by God. This feature, they maintain, distances natural law not only from human law but also from divine voluntary or positive law, which does not prescribe or forbid things that are of themselves and by their own nature owed or illicit, but makes things illicit or owed by forbidding or prescribing them. For the things forbidden by the natural law are not base because God has forbidden them, they say, but God forbade them because they were base in themselves. Similarly, those which are prescribed by that same law are not made honorable or necessary because they are prescribed by God, but they are prescribed because they are honorable in themselves. (Pufendorf, 1672/1998, pp. 132)

Instead, Pufendorf maintains that the *ratio formalis* (formal ground) of our moral judgment about the goodness or badness of an action is conditioned on a law, so that an action is called good when it accords with the law and called bad when it departs from it (Pufendorf, 1672/1998, I, 7, §.3).[10]

[8] That this conception of objective morality (*moralitas objectiva*) or the intrinsic goodness or malice of an action (*bonitas ac malitia intrinseca actionum*) (Wolff, 1738/1971, §§.55ff.) derives from Scholasticism, as emphasized by Wolff himself (Wolff, 1733/1973a, §.137). For his views on *perseitas*, also cf. Wolff (1738/1971, §.172; 1754/1980, §.15).

[9] Cf. Arist. Nic. Eth. II.6; Thomas Aquinas, *Summa Theologica* II, 2, qu. 57 a, 2 ad 2; Grotius, *De jure belli ac pacis* I, 1, §.10,5. For a critique of this view, cf. already Hobbes, *De Cive* VI, 19.

[10] Also cf. Welzel (1962, p. 137) and Behme (1995, pp. 54f.).

Yet, for Wolff, the ultimate ground of validity for the *obligatio naturalis* does not lie in its promulgation or imposition by a superior authority, but in nature conceived as teleological and aimed at the realization of perfection, and hence it is our nature itself, on Wolff's view, which imposes the natural law on us as the binding norm that governs our actions (Wolff, 1738/1971, §.129; 1733/1976, §§.9, 12; 1754/1980, §§.36, 38). In the context of the present contribution, I cannot discuss in detail the problems connected to this ontological concept of nature and to the principle of perfection as the highest practical principle.[11] Here it must suffice to take note of Kant's insight that, while the principle or requirement of perfection does indeed represent a principle of ethics, it is of no help as a juridical principle for determining the external use of freedom and therefore does not belong within the doctrine of right (Geismann, 1974, 41f.).

Thus, in contrast to Pufendorf's voluntarist doctrine, Wolff determines the moral quality of action in recourse to the concept of perfection: "That which makes both our inner as well as our outer condition perfect, is good (§422 Met.); that, however, which makes both less perfect, is evil (§426 Met.)" (Wolff, 1733/1976, §.3, p. 6; cf. 1738/1971, §.55; 1754/1980, §§.13–15).[12] In this vein, Wolff also characterizes "the perfection of ourselves and our condition" as the "ultimate intent of all our free actions, and the main intent of our entire life" (Wolff, 1733/1976, §.40, pp. 29–30).

Wolff himself identifies the concept of perfection as the source of his practical philosophy (*fons philosophiae meae practicae*) (Wolff, 1753/1973b, Preface).[13] In itself, this concept signifies an ontological category, yet here we are concerned with "its application for practical use" (Schwaiger, 1995, p. 94). This ontological anchoring of the concept of natural obligation serves to demonstrate that the difference between good and evil actions is "independent of human convention and independent of a specifically divine authority" (Schröer, 1988, p. 144), and hence to guarantee the autonomy of practical philosophy.

A second aspect of Wolff's self-assessment regarding his introduction of the concept of *obligatio naturalis*, as quoted above from the *Ausführliche Nachricht*, concerns the relationship of atheism to morality, a highly contentious issue during the seventeenth and eighteenth centuries, and Wolff's handling of the so-called *hypothesis impossibilis athei* (impossible assumption of God's nonexistence) (Hüning, 2002, pp. 239f.). With his conception of *obligatio naturalis*, Wolff takes on the heritage of natural law theory in the tradition of Grotius. In contrast to Pufendorf, Wolff is able to appropriate the thesis of hypothetical atheism, as expressed in Grotius' "even if we were to say" (*etiamsi daremus*) formula (Wolff, 1733/1976, §§.5 and 20)[14] and to emphasize that the norms of natural law are eternal and immutable and

[11] On the moral principle of perfection, cf. Schröer (1988, pp. 84ff., 91ff., 107ff., 114ff.), Winiger (1992, pp. 192f., 243ff.), and Schwaiger (2001, pp. 317–328).

[12] See also Wolff (1738/1968a, §§.554, 564, and 565).

[13] On Wolff's concept of perfection, cf. Schwaiger (1995, 93ff.), who particularly emphasizes the influence of Leibniz's critique of Wolff's early conception of ethics.

[14] On the concept of hypothetical atheism, cf. Schröder (1998, pp. 162f.), who shows that Grotius' formula was already used in the seventeenth century to justify atheist doctrines.

hence cannot be altered even by God (Wolff, 1738/1971, §.282).[15] On Wolff's view, the specific advantage of the concept of *obligatio naturalis* lies in the fact that:

> every rational man, even an atheist, must allow a place for it, and suchlike instructed how to make use to that end of the motives of the disgracefulness and harmfulness of the vices and, on the opposite, of the excellence and advantage of virtue: whereby those of the atheists are put to shame who would like to live according to their appetites and desires and thus opine that, if there were no GOD, there would be difference no more between virtues and vices. (Wolff, 1733/1976, Report on the Third Edition, §.4)

By detaching morality from the divine will and from moral theology, Wolff makes an essential contribution to a new understanding of morality, which for him—as much as subsequently for Kant—does not consist in the conformity of our actions with the natural law, but in the conformity of our will with that which the natural law demands. Thus, for Wolff, morality is the orientation of one's own will in accordance with the natural law. In this vein, Wolff writes in the *Deutsche Ethik*: "Since we recognize through reason what the law of nature wants to have [i.e. requires]; thus a rational man needs no further law [than the natural law], but by means of his reason he is a law unto himself" (Wolff, 1733/1976, §.24, pp. 18–19).[16]

In regard to Wolff's ethical legislation, some interpreters have taken the view that it already represents a conception of autonomy, that is, of self-legislation in the Kantian sense, since Wolff too proceeds on the assumption that the moral human being recognizes no laws other than those he gives to himself (Schmucker, 1961, p. 40). And it is in fact impossible to overlook the progress that Wolff's concept of morality represents in this regard. However, from the autonomy of moral philosophy, an ethics of autonomy does not immediately follow. It is indeed correct that Wolff's practical philosophy rests on the principle of the autonomy of moral philosophy, i.e., of its independence from theological presuppositions.[17] Yet, neither this autonomy of moral philosophy nor Wolff's concept of morality as an inner determination of the human will is sufficient to turn his moral philosophy into an ethics of autonomy, at least if understood in the Kantian sense. Wolff's establishment of the principle of the autonomy of morality and his determination of morality as obedience to a self-imposed law have nothing to do with Kant's formula of the autonomy of the will as the highest principle of morality. For Wolff's, the highest moral principle is a material principle, which from the standpoint of Kantian ethics belongs to the heteronomy of the will.[18] Rather, Wolff's ethics rests on the principle of self-obligation, by means of which human beings in virtue of their own reason

[15] Also cf. Wolff (1733/1976, §.29).

[16] See Wolff (1738/1971, §.268). "The key to the heart of Wolff's justification of morality thus lies in the thesis that the rational man is, in virtue of his reason, a law unto himself and requires no further laws beyond this" (Schröer, 1988, p. 213). Also cf. Joesten (1931, pp. 27ff.).

[17] And in this sense, Clara Joesten (1931, pp. 26ff.) speaks of the "autonomy of morality" in Wolff.

[18] That the establishment of an autonomous morality (in the sense of combating other theological moral principles) must be differentiated from Kant's positing of the principle of the autonomy of the will as the sole principle of morality has been stressed empathically by Klaus Reich (1989, pp. 86f.).

subject themselves to the natural law. In this respect, it is true that Wolff too considers the determination of the will by one's own reason as the essence of morality. Yet, on his view, reason does not determine the will to the conformity of its maxims with "the will's own universal legislation" (Kant, 1911, p. 431) but only to the conformity of its maxims with a law that, as its universal norm, is antecedent to the will itself.[19] Thus, Wolff's moral philosophy does not rest on the principle of self-legislation through which human beings qua practical reason subject their willing and acting to a law. Rather, it rests on the very different principle of self-obligation in the face of a law that is systematically antecedent to the will and whose validity is not grounded in practical reason but in the teleological constitution of nature and the world (Röd, 1984, p. 252)

However, Wolff's attempt to introduce the idea of self-obligation, that is, of the individual imposing an obligation on itself, into practical philosophy (Joesten, 1931, pp. 26ff.), is afflicted with a number of problems. First of all, we may note the Kantian objection that it is impossible for the highest principle of morality to be material in nature. Second, one may query what it means to say that nature is the source of natural obligation and as such obligates us to perform particular actions. In this respect, the problem does not only consist in how we are capable of recognizing what nature prescribes for us; rather, the very idea that nature as such could exhibit a law-giving or normative will is itself problematic (Winiger, 1992, pp. 271ff.). Finally, Wolff's determination of the relation between practical philosophy and moral theology seems to me equally unsatisfactory. While Wolff's purely rational justification of norms within the bounds of *Philosophia practica universalis* derives the obligatory force of the natural law exclusively from "the essence of man and of things," his *Theologia naturalis* [natural theology] considers the natural law to emanate from the divine will. Even though Wolff asserts that the ground of the obligatoriness of the natural law does not lie in God's will but has "its sufficient ground in the essence and nature of man and of things" (Wolff, 1754/1980, §.38; Winiger, 1992, p. 179), he always held on to the view that it is possible to also consider the norms of natural law as divine commandments and therefore God as the author and legislator of the natural law (Wolff, 1738/1971, §§.273ff.; 1754/1980, §.41). In his *Theologia naturalis*, Wolff correspondingly discusses an *obligatio divina* [divine obligation], which differs from the *obligatio naturalis* and obligates man to "determine his free actions not from his arbitrary will but in accordance with the will of God" (Wolff, 1739/1978, §.974).[20]

The reason for Wolff's recourse to God as originator and legislator of the natural law lies in his conviction that the existence of human beings, and of the world overall, is not necessary in itself but, as a contingent phenomenon, points to God as its necessary cause (Bissinger, 1983, p. 153). Thus, he writes in the *Deutsche*

[19] For this *distinction between self-obligation and autonomy*, which is decisive for understanding the difference between the practical philosophies of Wolff and Kant, cf. Geismann (2000, pp. 441f.). On the relation of Wolff's principle of perfection to Kant's concept of autonomy, also cf. Schröer (1988, pp. 196–206).

[20] See also Wolff (1739/1978, §.975 nota, 944).

Metaphysik: "If God did not exist, neither would men nor the law of nature" (Wolff, 1740/1983, §.364, p. 590; cf. Bissinger, 1983, p. 154). On Wolff's view, both ways of justifying obligation, viz., the philosophical and the moral-theological, are not rival theoretical options but merely differ in their respective emphasis. Yet, bracketing the ontological problem of the contingency of the world, practical philosophy retains its character as an autonomous discipline. In regard to the issue of the obligatoriness of the natural law, this implies that recourse to God as the originator of the world is superfluous from a systematic point of view (Wolff, 1739/1978, §.975 nota).

10.4 On the Connection Between Psychology and Moral Philosophy

A distinctive feature of Wolff's practical philosophy is its close connection to empirical psychology. More precisely, this connection consists in the fact that Wolff makes comprehensive and systematic use in his practical philosophy, and particularly in his theory of obligation, of concepts and doctrines he previously developed within empirical psychology, which he considers part of metaphysics. Notably, his accounts of the will and its freedom belong here, which are essential both for ethics and law. Wolff himself emphasizes this grounding function at various locations in his oeuvre. In this respect, he indicates the importance of the *Psychologia empirica* (empirical psychology) in the *Ausführliche Nachricht von seinen eigenen Schriften* (Detailed report on his own writings), which effectively represents the apex of his German language writings. From this part of psychology, which treats of that "which one may learn of the soul from experience," it is possible on Wolff's view to derive "important truths," and not merely about "the rules of logic, whereby the intellect is directed in its recognition of truth, but also the rules of morality, whereby one directs the will of man toward the good and holds it back from evil." Wolff expressly declares that this grounding function of empirical psychology is "something novel [...] which one is not yet used to" (Wolff, 1733/1973a, §.89, p. 252; cf. Wolff, 1738/1968a, §.509). He similarly emphasizes in the *Anmerkungen zur Deutschen Metaphysik* [Remarks on the German Metaphysics] that "the benefit of that" which is taught in empirical psychology shows "itself for the main part in morality" (Wolff, 1740/1983, §.55, p. 122; cf. §§.69 and 131). Moreover, in the *Psychologia empirica* itself, Wolff highlights the exceptional yet so far unrecognized usefulness of empirical psychology for investigating and developing the concepts of natural law and obligation, and in particular its usefulness in relation to the duties concerning the soul, which it is impossible to establish adequately without knowledge of the capacities of the soul (Wolff, 1738/1968a, p. 14*). Wolff declares *Psychologia empirica* to be the "foundation of moral philosophy and of natural and civil law" (Wolff, 1738/1968a, §.945, p. 709). And finally, even Wolff's definition of the *Philosophia practica universalis* as an "affective and practical science for directing free actions by means of most general rules" (Wolff, 1738/1971, §.3, p. 2; cf.

Winiger, 1992, p. 132) gives us a decisive clue: for, as an affective science (*scientia affectiva*), the *Philosophia practica universalis* relies on the results of empirical psychology in discussing how the will is determined toward the good.

What is the reason for this close connection between psychology and moral philosophy? At first sight, it seems to be nothing new. Both Hobbes and Pufendorf had made use of psychological presuppositions in various parts of their natural law theories. Hobbes, for instance, preceded his doctrine of the citizen (*de cive*) with a doctrine of man (*de homine*), while Pufendorf developed an extensive account of the freedom of the human will in the context of his discussion of juridical imputability. Nevertheless, I maintain that the foundational role of psychology for practical philosophy is significantly more important in Wolff than in his predecessors. The reason for this, on my view, lies in Wolff's understanding of ethics in its broadest sense, namely, as a practical science that enables human beings to devise their free actions in accordance with the natural law (Wolff, 1753/1973b, Prolegomena, §.1). Recourse to empirical psychology then becomes necessary to answer the question which motives need to be present in order to determine the will "to the exercise of virtue and the prevention of vices" (Wolff, 1740/1983, §.55, p. 122).

Yet, in order for practical philosophy to fulfill this steering or directive function, two psychological presuppositions must be given: First, that the intellect has primacy over the will, and that the latter must therefore be understood, as in Scholastic psychology, as rational desire (*appetitus rationalis*), that is, a desire determined by reason (Wolff, 1738/1968a, §.880; 1740/1983, §.155).[21] And second, that for this reason the will is determined through a representation of the good and evil consequences connected to an action: "The motive of the will is that represented by the intellect" (Bissinger, 1983, p. 153).[22]

10.5 The Problem of Indifference

In Wolff's intellectualist account of the will, the will appears as "the inclination of the mind toward a thing on account of the good that we deem to perceive in it" (Wolff, 1751/2003, §.492, p. 299). Correspondingly, not-willing is the "withdrawal of the mind from a thing on account of the evil that we deem to perceive in it" (Wolff, 1751/2003, §.493, p. 300). Thus, on Wolff's view, the will is oriented

[21] Also cf. Lange's critique in this respect: "It is also wrong when it is said in §.520 that, since intelligence notes the interconnection of things, reason is the ground of freedom. For the will is the ground of freedom: and reason directs the free will to the right use of freedom" (Lange, 1724/1998, §.8).

[22] In the context of the present contribution, I cannot go further into some distinctions important to empirical psychology, e.g., that between the higher and the lower part of the soul (Wolff, 1726/1985, §.35), between *appetitus sensitivus* (Wolff, 1738/1968a, §§.579ff.) and *appetitus rationalis*, and between the *idea boni confusa* (confused idea of the good) and *idea boni distincta* (distinct idea of the good) (Wolff, 1738/1968a, §.890).

essentially by means of a representation of the good that lies in the desired object. Now, a crucial problem with this account pertains to the question of the soul's relation to the motives of its actions, which are given to it and precede the will's coming to a decision.

Wolff maintains that the will is determined by a representation of the good. We are faced here only with the psychological causation of action, that is, with an action's psychological or motivational necessity pursuant such a representation, which still leaves open how such psychological necessitation differs from whatever form moral necessitation may take. An adequate account of this issue would require a comprehensive development of Wolff's empirical psychology, which is beyond the confines of the present contribution. Instead, I will limit myself to a particular problem for Wolff's psychology, namely, the problem of indifference, which puts some core difficulties with his views into stark relief. As Bruno Bianco (1989) has shown, this problem also formed a major point of contention in Wolff's dispute with the Halle theologians because it renders their differences regarding the will's relation to the motives preceding it and regarding the possibility of the will's coming to a decision when these are indifferent particularly salient. Concerning the problem's relevance for the issue of free will, Wolff's interpreters for the most part agree that, all of his reassurances to the contrary notwithstanding, Wolff could not dissolve the contradiction between the principle of the universal determination of all things in the real order of the world on the one hand and his concurrent claim that all the acts of our will are spontaneous on the other.[23] I would like to add that there was no way for him to dissolve it given his basic metaphysical assumptions. In what follows, I will attempt to show that the objections of his opponents, the religious dogmatism that finds expression in them notwithstanding, actually hit a weak point in the systematics of Wolff's practical philosophy.

At the heart of the contemporary polemic were the following, closely related claims by Wolff, viz.:

1. That the will must be understood as an "inclination of the mind toward a thing on account of the good that we deem to perceive in it" (Wolff, 1751/2003, §.492, p. 299)

[23] "Spontaneity is the intrinsic principle of self-determination to act" (Wolff, 1738/1968a, §.933, p. 702). The consensus among interpreters is paradigmatically expressed by Arndt (1976, p. viii): "On the ontological level, it may be the case that contingency negates the absolute necessity connected to things' essences, which is the same in all possible worlds. Yet, this 'contingency' is merely another name for the 'hypothetical' necessity everything existing in this world is subject to, including everything that happens, which is determined by the real order of succeeding states of the world as premised on a first state of the world created by God. Wolff was not able to overcome, any more than Leibniz, the discrepancy in his philosophy between his assertion of the spontaneity of all our acts of will and actions on the one hand and the complete determination of all events and states of things in the real order on the other."

2. That there is no action without motive and "that we at all times need motives why we will something" (Wolff, 1733/1973a, §.96, p. 265)[24]
3. That nevertheless the soul is not necessitated by the presence of relevant motives (Wolff, 1740/1983, §.165)
4. That in the presence of equally strong motives, no decision of the will is possible since there is then no motive that could tip the scales either way (Wolff, 1738/1968a, §.889)[25]

The true point of contention between Wolff and his opponents is the application of the metaphysical principle of sufficient reason to the problem of voluntary actions. Wolff himself emphasizes in various places that there is a connection between the requirement of a motive for a decision of the will and the ontological principle of sufficient reason. For such an account of the will, as Wolff endorses it in following Leibniz, the old question whether there can be a decision of the will in the state of indifference is of particular interest.

It is certainly no accident that Wolff's harshest critic, the theologian Joachim Lange, chose the problem of indifference, which is particularly suitable for highlighting the metaphysical presuppositions of Wolff's psychology, as the central object of his denunciation.[26] In contrast to Wolff, Lange considers the claim that some sort of motive has to precede the will's decision as its determining ground the epitome of a deterministic psychology, which cannot but issue in a "mechanical morality." On his view, the mark of the freedom of the human will precisely is the independence of its decisions from external motives:

> It is one thing to will or not will without any motives, and another thing to will or choose one thing out of two when motives are indifferent. For in the former case there are no motives at all: in the latter, however, there are and remain motives, in that the motives for one thing are not in themselves abrogated by the equal validity [Gleichgültigkeit] of the other thing. And in such indifference, the soul owes it to see whether it cannot, by means of impartial deliberation or consultation with others, bring it about in itself to be swayed more toward the one thing rather than the other. If it cannot bring this about; then it avails itself of its freedom and chooses one of both; and does so according to its liking, since it could also have chosen the other: but still does so not without reasons, but from those it finds on the side of the thing chosen. Now, if one denies to the soul the capacity to, as it were, overcome its indifference and choose one thing over the other according to its liking, then one

[24] Cf. Leibniz's letter to des Bosses on February 8, 1711 (Leibniz, 1890a, p. 420). Hence, Wolff's criticism of what, on his view, is a false concept of freedom as the overcoming of indifference: "Therefore, those are in error who explain freedom as a capacity to choose, from two conflicting things, either one or the other, without the presence of a motive why one chooses the one over the other. As has been extensively discussed, such a capacity is contrary both to reason (§.369) and to experience (§.325)" (Wolff, 1751/2003, §.511, pp. 312–313).

[25] "Just as there can be no tipping of the scales, due to the principle of sufficient reason, if one of the weights is not increased by adding to it, so the soul cannot choose either of the two if nothing is added on one side to the already existing motives" (Wolff, 1751/2003, §.510, p. 311).

[26] Cf. Schröder (1938, pp. 87f.). The roots of the problem of indifference reach as far back as Antiquity. For Leibniz, this problem is tied closely to the position of the Skeptic Carneades (Leibniz, 1885, pp. 307–308).

> denies it its freedom. Which is what our author §.508 [of the *Deutsche Metaphysik*] does. (Lange, 1724/1998, §§.134f.)

> [...] Freedom is that capacity of the soul from which it chooses without inner necessity and external constraint and from which it acts as it likes, so that what it has done it could have omitted to do, and what it has omitted it could have done. (Lange, 1724/1998, §.138)

In the philosophical literature of the seventeenth and eighteenth centuries, the problem of indifference is illustrated by two allegories. One of them is the famous allegory of Buridan's ass, which Leibniz used to discuss the problem.[27] Thus, a donkey that is standing in equal distance between two bundles of hay, which are of equal size and indistinguishable in all respects, would have to starve since there could be no motive that would determine its will in one direction rather than the other. Since the bundles of hay as objects of the will cannot present differentiated motives due to their indistinguishability, the donkey's will is said to be paralyzed in this case. For neither of the two bundles of hay is capable of presenting a motive to the will for preferring one to the other.

As Andreas Dorschel has pointed it, it may seem suspicious that the dilemma of indifference is illustrated by means of the animal that is considered in received opinion as the dumbest animal (Dorschel, 1992, pp. 92f.). That a donkey is made to play this part is due to the allegory's function of thwarting the claim that the rational will is oriented according to the quality of the object desired. For the result that follows if one takes this premise seriously is not particularly rational: the will's paralysis in cases where two identical objects cannot give sufficient determining grounds for the will's decision. Thus, the allegory of Buridan's ass calls the assumption into question that the will is oriented according to the quality of its contents and ultimately determined by them in its decisions. It denies the soundness of the premise by deriving an absurd conclusion from it. Leibniz, on the other hand, discusses the allegory of Buridan's ass in his *Théodicée* with the aim of rescuing it from the problem formulated by it. Yet, his rescue attempt ultimately affirms the allegory's critical point that the will is determined by the content of its willing and merely disputes the inference to the will's paralysis because situations of complete indifference are impossible in reality (Leibniz, 1885, pp. 129–130; cf. Leibniz, 1882, pp. 105–106).[28] The assertion of complete indifference ultimately rests on the fact that the

---

[27] On what follows and especially on the role of Buridan's ass in deterministic psychology, cf. Dorschel (1992, pp. 92ff.). On Leibniz's treatment of indifference, cf. Platz (1973, pp. 127ff.).

[28] Leibniz disputes that "a complete indifference between moral actions, as in those of Buridan's ass" is possible at all. Such an assertion of complete indifference ultimately rests on the fact that those "impressions which are capable of tipping the balance" are *insensible*, that is, unnoticeable. "For even if I do not always see the reason for an inclination that makes me choose between two things that appear equal, there is always some impression, however imperceptible, which determines us" (Leibniz, 1885, pp. 297–298). Also cf. a view of Leibniz followed by Wolff in this context: "Between things absolutely indifferent, there simply is no choice, and consequently no selecting or deciding, since a choice requires some reason or principle" (Leibniz, 1890b, p. 371).

impressions capable of tipping the scales can be utterly unnoticeable or *insensible*, as Leibniz puts it.[29]

In contrast to Leibniz, Wolff does not illustrate the effect of motives on the will by means of Buridan's ass, but with the allegory of the scale, which is equally common in the psychological literature of the seventeenth and eighteenth centuries (Dorschel, 1992, pp. 86, 96f., 104). In general, this allegory served to elucidate the thesis that the will is determined by its content, and in particular to clarify the principle of sufficient reason's validity in regard to decisions of the will. Just as a scale can tip only toward one side or the other if there is a sufficient weight, so the will as an "inclination of the mind toward a thing on account of the good that we deem to perceive in it" (Wolff, 1976, §.492) can only incline toward one act or another if there is a sufficient motive. As Wolff puts it in the *Deutsche Metaphysik*:

> [...] For [...] if we will; then our mind is inclined toward the thing: if we eschew it, it is withdrawn from it: if we omit willing; then it remains, as it were, upright and immovable, neither inclining toward the thing nor withdrawing from it. This can be elucidated by analogy to a scale. When the scale is in balance; then this is the state resembling the state of the mind as we neither will nor eschew. When the balance is tipped to one side and the scale inclines over to it, then this is akin to us willing something. Yet the other side, from which the scale withdraws, we imagine as the state that we eschew. From this are taken our everyday expressions when we talk of the will. (Wolff, 1754/1980, §.494)

The difficulty in using this allegory consists in the fact that motives, and particularly the strongest motive that is supposed to determine the will, are represented in this allegory as real grounds, that is, causes of the will's decision. Hence, this allegory, just as much as that of Buridan's ass, is suggestive of a deterministic account of the will, in which the contents of the will are considered to be externally caused by the object. As Hobbes (*De cive*, XIII, 16), Bayle (1727/1966, pp. 782–785), and Leibniz[30] before him, Wolff is of the view that our practical subjectivity can be conceived in analogy to the functioning of a scale:[31]

> As long as the weights in both trays of the scale are equal; the scale remains still and cannot tip over to either side. If it is to tip; then we have to add to the weight on one of the sides. In this allegory, the scale represents the soul, and the weights are to be interpreted as the motives. (Wolff, 1751/2003, §.509, p. 310)

[29] "Yet I see that among those who speak of freedom there are those who do not take account of those insensible impressions, which are capable of tipping the balance, and imagine a complete indifference between moral actions, as in the case of Buridan's ass which is torn between two meadows" (Leibniz, 1882, p. 105). Wolff similarly argues: "For it is well comprehensible that specific motives do not always catch everyone's eye but are sometimes so hidden that they are only discernible by the astute, who are used to examine things thoroughly, as has been recalled above (§. cit [498])" (Wolff, 1751/2003, §.508, p. 310).

[30] "It is true that *Reasons* have in the mind of the sage, and *Motives* in some mind that has them, which is the same, the effect that weights have on a scale. One objects that this notion tends toward necessity and fate" (Leibniz 1890b, p. 389).

[31] "The core of the deterministic reasoning [that is articulated in the allegory of the scale] is [...] this: Just as the scale cannot incline more to one side than the other when the opposing weights are equal, so a human being cannot come to a decision when the differing motives are of equal value" (Dorschel, 1992, pp. 86, 96f.).

Wolff's main opponent, the Halle theologian Joachim Lange, objects to this action-theoretical account because it implies a deterministic conception of the will's relation to its antecedent motives:

> While the allegory of the scale seems to give a good elucidation; and in a certain way it can be used sensibly to that end: Yet, if one extends it too far, it is no longer appropriate but falsifies the whole matter. And this happens particularly when one applies, with the same notion of necessity, to the soul that which happens necessarily in the case of the scale. For while the scale is not a thing that is at freedom and can act to come into motion when the weights are equal and tip over to one side, but will have to remain in balance; or be pulled or steered according to the greater weight: the soul, however, is at freedom not only, when it contemplates the motives for two different actions, to incline in a state of indifference to where it wills; but can also in misusing such freedom turn to that which has the worst and misshapen reasons speaking for it: thus, one well comprehends that the allegory of the scale is not at all appropriate for the free actions of the soul. Yet if the soul should not have aforesaid freedom to knowingly and intentionally turn to that side which has no valid reasons, the whole ground of morality would be overthrown and no imputation would take place. And this is where the author, with this dearly beloved allegory, heads to or leads. (Lange, 1724/1998, §§.134f.)

Wolff's remarks in §.510 of the *Deutsche Metaphysik* can be understood as a reaction to the objection, as formulated by Lange and others, that the allegory of the scale implies a determination of the will by its motives and is thus incompatible with the doctrine of the freedom of the will (Wolff, 1738/1968a, §.941). In his reply, Wolff takes up a Leibnizian distinction between two kinds of necessity, according to which a motive, although it inclines the will toward one side of the scale, does not necessitate it (Leibniz, 1885, p. 127). Wolff says:

> I am well aware that some harbor the thought that the allegory of the scale is inappropriate to the will. For the scales moves with necessity; the soul, however, is at freedom to will and to not will. For that reason one cannot infer from what is necessary to what is free. But really! Who infers from what is necessary to what is free? Those who believe that do not understand the allegory. The comparison of the scale's tipping to the will does not go farther than that both require sufficient reason (§.30). For as long as both weights are equal, there is no reason why the scale should tip more to the right than to the left. And likewise is the case with the will. As long as the motives for both parts are of equal weight, there is no reason why one should choose one rather than the other. Just as due to sufficient reason the scale cannot tip if one of the weights is not increased by addition; so the soul cannot choose either of both if something is not added to the already existing motives on one side. This is how far the comparison goes, and no one worries much whether there is necessity to the tipping of the scale: the soul, however, is not necessitated by its motives. For the question is not whether the motives exert compulsion but whether one of them is stronger than the other. This is why the allegory of the scale is well appropriate here after all, since the will has been explained above as an inclination toward a thing on account of the good that we perceive in it (§.492). For this form of expression is taken from a body that is being inclined by a force from the vertical line toward the horizontal line on one side; which also happens in the case of the scale's tipping, as was shown in detail above (§.494). Notwithstanding that this word must indeed have a different meaning when used of the soul, since the concepts of corporeal things are not fit for it; this does not abrogate the similarity between that which we find in the soul and the corporeal, which is the reason for the appellation. Yet what the

> inclination of the soul consists in can only be shown once I have explained the nature of the soul. (Wolff, 1751/2003, §.510, pp. 310–312)[32]

The reason that Wolff nonetheless holds on to the allegory of the scale and asserts the doctrine that a motive is necessary as the starting point for the will's decision, results systematically from his ontology, and in particular from his *application of the principle of sufficient reason*—"everything has its sufficient ground why it is" (Wolff, 1751/2003, §.31, p. 18)—to the doctrine of the human will (Dorschel, 1992, 138f.):

> For since everything requires its sufficient ground for why it is rather than is not; so it must have its sufficient ground why we will or do not will something, just as it is impossible for a scale to tip over if there is no weight present to cause it. These grounds of willing and not-willing we usually call motives. (Wolff, 1751/2003, §.496, p. 302)

[32] "By rejecting the *Indifferentiam perfecti aequilibri*, I make use of, as have others before me, the allegory of the scale. But to avoid the thought that in doing so one attempts to infer from material things to the soul, I have explicitly raised this objection in detail and shown how this allegory can serve to elucidate the will without committing such an inference, which I do not at all condone. Hence it comes across as strange when one comes up with this objection against me as an issue I supposedly have not seen and that others had to hold up to me in the first place. Even more fantastical does it come across, however, when Mr. D. Lange does this and draws dangerous consequences from it, since he explains this allegory in his *Medicina mentis* in the objectionable sense that he wants to burden me with" (Wolff, 1733/1973a, §.96, pp. 265–266). Also cf. Karl Ferdinand Hommel's critique of using the allegory of the scale for illustrating free decisions of the will: "I have no idea what spirit of quackery drove the scholars to represent the scale as an image of freedom. It is true, if no weights lie in it, I would say: the scale hangs freely. For it is able to stir or not to stir. Yet alone it does not have a capacity to move itself, which is after all all that matters. For if I place a weight on it, the scale necessarily has to move. In consideration of this, I know of no better model to indicate compulsion and necessity than the scale, which one miraculously deploys as an example of freedom. [...] Nothing is easier to move than a scale, and the lowering of the tray so unavoidable that its opposite is not even thinkable. Moreover, it is not a self-moving thing, as is the soul according to common opinion, of which one believes that it has its principle of movement within itself. Thus, the scale is a completely inept symbol. This tool, which cannot even resist a downy feather, is supposed to signify freedom. Compulsion, fate, necessity can be signified by it, but not freedom. [...] What is the weight in the case of a scale, is in the mental realm and in the case of the will a certain representation of the mind that something is good to do; or even more often the constitution of my body, namely animal drives, as they derive from the humors and blood. [...] The true concept of freedom therefore does not consist in the ability to will or not to will something, to move or not to move, to desire or not desire something, but in the capacity to will something without an inlaid weight and to self-move without external drives, in short without all cause. The former I call passive, the latter active freedom [...]" (Hommel, 1772/1970, §.20). In contrast to Wolff, Hommel consistently interprets the allegory of the scale in terms of determinism while recognizing the principle of sufficient reason: "Thus, the true question is: whether our soul has a capacity to act from itself without any representation caused in it from outside? Such a capacity cannot be found in it, for otherwise something would happen without sufficient reason; but just as the scale would forever stand still without an inlaid weight, so the human will would be eternally dead if not certain representations originating from close by things would enliven it from outside. [...] I ask, however, whether there exists apart from God one self-moving thing in the world, i.e. one, that has the principle of movement solely in itself?" (Hommel, 1772/1970, §.22) Our "feeling of freedom" Hommel (1772/1970, §§.22f.) considers a mere epiphenomenon.

A decision of the will that was not determined by some sufficient reason relating to its subject matter would amount to a violation of the universality of the causal law, which leads Wolff to think the will's process of decision in analogy to the relation of cause and effect in processes of nature. Similarly, it is the basic idea of a continuous determination of the world in the sense of a continuity of grounding relations, that is, of the universal validity of the principle of sufficient reason, whose application here forms the actual foundation of Wolff's psychology—and thus not, as Wolff intends, some insight won from experience. Wolff declares a decision of the will without motive to be impossible, since that would abrogate the universal validity of the principle of sufficient reason.

While Wolff's use of the allegory of the scale is at least suggestive of a deterministic account of the will, he certainly attempts to weaken and obscure its deterministic implications. On the one hand, he disputes that motives are attended by an "unavoidable necessity":

> Actions are as such not necessary but merely contingent; their motives do not render them necessary either but merely certain; and the soul has the ground of its actions within itself. [...] It is also free from all inner compulsion, since motives do not have unavoidable necessity, but the soul can depart from it, as in fact frequently happens. (Wolff, 1751/2003, §.883, pp. 547–548)

On the other hand, he claims at the same time that the predominance of one motive as such already effects a decision for one or the other option, and even independently of whether the subject making the choice is distinctly aware of the motives' respective weight. This is clear from Wolff's example that someone will choose, between two otherwise identical ducats lying on the table, the one more conveniently placed, since "this convenience" supposedly acts as the determining "motive." Yet, of this motive, Wolff explicitly declares that it "remains hidden" to the agent (Wolff, 1751/2003, §.498, p. 304).

10.6 Conclusion

Wolff's psychology suffers, to summarize the preceding considerations, from a continuous "confusion of causes with reasons," in that the thoughts an agent has about the possible objects of their decision, and which insofar constitute the reasons for their action, are mistaken for a causality operating on the will, as if their choice coincided with the causes supposedly acting on their will (Dorschel, 1992, p. 139). The insight that many eighteenth-century accounts of the will rest on an "*impermissible application* of the causal nexus to matters [...] of the life of the spirit" belongs to a later period in the history of philosophy. To unfold it, we would have to talk about Hegel, who formulated this criticism (Hegel, 1978, §.400), and about his *philosophy of subjective spirit*, which among other things promises a solution to the difficulties that attend rationalistic psychology (Wolff, 1992). That, however, is a different topic.

References

Arndt, H. W. (1976). Einleitung zur *Deutschen Ethik*. In J. École (Ed.), *Christian Wolff: Gesammelte Werke* (Vol. 4, Pt. I). Olms.

Bayle, P. (1966). Response aux questions d'un provencal. In P. Bayle (Ed.), *Oeuvres diverses* (Vol. III). Olms. (Original work published 1727).

Behme, T. (1995). *Samuel von Pufendorf: Naturrecht und Staat. Eine Analyse und Interpretation seiner Theorie, ihrer Grundlagen und Probleme*. Vandenhoeck & Ruprecht.

Bianco, B. (1989). Freiheit gegen Fatalismus. Zu Joachim Langes Kritik an Wolff. In N. Hinske (Ed.), *Zentren der Aufklärung I: Halle* (pp. 111–155). Schneider.

Bissinger, A. (1983). Zur metaphysischen Begründung der Wolffschen Ethik. In W. Schneiders (Ed.), *Christian Wolff 1679–1754. Interpretationen zu seiner Philosophie und deren Wirkung* (pp. 148–160). Meiner.

Carr, C. (Ed.). (1994). *The political writings of Samuel Pufendorf*. Oxford University Press.

Dorschel, A. (1992). *Die idealistische Kritik des Willens. Versuch über die Theorie der praktischen Subjektivität bei Kant und Hegel*. Meiner.

Ebbinghaus, J. (1986). Über die Idee der Toleranz. Eine staatsrechtliche und religionsphilosophische Untersuchung. In H. Oberer & G. Geismann (Eds.), *Julius Ebbinghaus, Gesammelte Schriften* (Vol. 1). Bouvier.

Ebbinghaus, J. (1990). Mensch und Ratio im Europa der Neuzeit. In H. Oberer & G. Geismann (Eds.), *Julius Ebbinghaus, Gesammelte Schriften* (Vol. 3). Bouvier.

Geismann, G. (1974). *Ethik und Herrschaftsordnung*. Mohr.

Geismann, G. (2000). Sittlichkeit, Religion und Geschichte in der Philosophie Kants. *Jahrbuch für Recht und Ethik/Annual Review of Law and Ethics, 8*, 437–531.

Grunert, F. (2000). *Normbegründung und politische Legitimität. Zur Rechts- und Staatsphilosophie der deutschen Frühaufklärung*. Max Niemeyer.

Hartung, G. (1999). *Die Naturrechtsdebatte. Geschichte der Obligatio vom 17. bis 20. Jahrhundert* (2nd ed.). Alber.

Hegel, G. W. F. (1978). Wissenschaft der Logik. Erster Teil. Die objektive Logik (1812/13). Zweites Buch: Die Lehre vom Wesen. In Rheinisch-Westfälische Akademie der Wissenschaften (Ed.), *G. W. F. Hegel, Gesammelte Werke* (Vol. 11). Meiner.

Hommel, K. F. (1970). *Über Belohnung und Strafe nach türkischen Gesetzen*. Schmidt. (Original work published 1772).

Hüning, D. (2002). Die Grenzen der Toleranz und die Rechtsstellung der Atheisten. Der Streit um die Verbindlichkeit des natürlichen Gesetzes in der neuzeitlichen Naturrechtslehre. In L. Dannenberg, S. Pott, J. Schönert, & F. Vollhardt (Eds.), *Säkularisierung in den Wissenschaften seit der frühen Neuzeit* (Vol. 2, pp. 219–264). De Gruyter.

Joesten, C. (1931). *Wolffs Grundlegung der praktischen Philosophie*. Meiner.

Kant, I. (1911). Grundlegung zur Metaphysik der Sitten. In Preußische Akademie der Wissenschaften (Ed.), *Kants Gesammelte Schriften* (Vol. 4, pp. 385–463). Georg Reimer.

Lange, J. (1998). Bescheidene und ausführliche Entdeckung der falschen und schädlichen Philosophie in dem Wolffianischen Systemate Metaphysico von Gott, der Welt, und dem Menschen; und insonderheit von der so genannten harmonia praestabilita des commercii zwischen Seel und Leib: wie auch der auf solches gegründeten Sitten-Lehre. In J. École (Ed.), *Christian Wolff: Gesammelte Werke* (Vol. 56, Pt. III). Olms. (Original work published 1724).

Leibniz, G. W. (1882). Nouveaux Essais sur l'Entendement humain. In C. I. Gerhardt (Ed.), *Leibniz. Die philosophischen Schriften* (Vol. 5, pp. 39–509). Weidmann.

Leibniz, G. W. (1885). Essais de Théodicée sur la bonté de Dieu, la liberté de l'homme et l'origine du mal. In C. I. Gerhardt (Ed.), *Leibniz. Die philosophischen Schriften* (Vol. 6, pp. 1–436). Weidmann.

Leibniz, G. W. (1890a). Brief an des Bosses vom 8. Februar 1711. In C. I. Gerhardt (Ed.), *Leibniz. Die philosophischen Schriften* (Vol. 2, pp. 419–420). Weidmann.

Leibniz, G. W. (1890b). Streitschriften zwischen Leibniz und Clarke (1715–1716). In C. I. Gerhardt (Ed.), *Leibniz. Die philosophischen Schriften* (Vol. 7, pp. 347–440). Weidmann.

Platz, B. (1973). *Fatum et libertas. Untersuchungen zu Leibniz' 'Theodizee' und verwandten Schriften sowie Ciceros 'De fato.'* PhD Dissertation. University of Köln.

Pufendorf, S. (1998). De jure naturae et gentium. In W. Schmidt-Biggemann (Ed.), *S. Pufedorf, Gesammelte Werke* (Vol. 4.1). Akademie Verlag. (Original work published 1672).

Reich, K. (1989). Kant und Rousseau. *Neue Hefte für Philosophie, 29*, 80–96.

Röd, W. (1984). Die deutsche Philosophie im Zeitalter der Aufklärung: Christian Wolff. In *Geschichte der Philosophie* (Vol. 8, pp. 233–294). Beck.

Schmucker, J. (1961). *Die Ursprünge der Ethik Kants in seinen vorkritischen Schriften und Reflexionen*. Anton Hain.

Schröder, K. (1938). *Das Freiheitsproblem bei Leibniz und in der Geschichte des Wolffianismus*. PhD Dissertation. University of Halle.

Schröder, W. (1998). *Die Ursprünge des Atheismus. Untersuchungen zur Metaphysik und Religionskritik des 17. und 18. Jahrhunderts*. Frommann.

Schröer, C. (1988). *Naturbegriff und Moralbegründung. Die Grundlegung der Ethik bei Christian Wolff und deren Kritik durch Immanuel Kant*. Frommann.

Schwaiger, C. (1995). *Das Problem des Glücks im Denken Christian Wolffs. Eine quellen-, begriffs- und entwicklungsgeschichtliche Studie zu den Schlüsselbegriffen seiner Ethik*. Fromman.

Schwaiger, C. (2000). Ein 'missing link' auf dem Weg der Ethik von Wolff zu Kant. Zur Quellen- und Wirkungsgeschichte der praktischen Philosophie von Alexander Gottlieb Baumgarten. *Jahrbuch für Recht und Ethik/Annual Review of Law and Ethics, 8*, 247–261.

Schwaiger, C. (2001). Vollkommenheit als Moralprinzip bei Wolff, Baumgarten und Kant. In M. Oberhausen & R. Pozzo (Eds.), *Vernunftkritik und Aufklärung. Studien zur Philosophie Kants und seines Jahrhunderts* (pp. 317–328). Frommann.

Thomasius, C. (1963). *Fundamenta Juris Naturae et Gentium ex consensu communi deducta*. Aalen. (Original work published 1718).

Welzel, H. (1962). *Naturrecht und materiale Gerechtigkeit* (4th ed.). Vandenhoeck & Ruprecht.

Winiger, B. (1992). *Das rationale Pflichtenrecht Christian Wolffs. Bedeutung und Funktion der transzendentalen, logischen und moralischen Wahrheit im systematischen und theistischen Naturrecht Wolffs*. Duncker & Humblot.

Wolff, C. (1968a). Psychologia empirica, methodo scientifica pertractata, qua ea, quae de anima humana indubia experientiae fide constant, continentur et ad solidam universae philosophiae practicae ac theologiae naturalis tractationem via sternitur. In J. École (Ed.), *Christian Wolff: Gesammelte Werke* (Psychologia empirica) (Vol. 5, Pt. II, 2nd ed.). Olms. (Original work published 1738).

Wolff, C. (1968b). Jus naturae, methodo scientifica pertractatum (Pars I). In M. Thomann (Ed.), *Christian Wolff: Gesammelte Werke* (Vol. 17, Pt. 1). Olms. (Original work published 1740).

Wolff, C. (1971). Philosophia practica universalis methodo scientifica pertractata (Pars Prior). In J. École (Ed.), *Christian Wolff: Gesammelte Werke* (Philosophia practica universalis) (Vol. 10.1, Pt. II). Olms. (Original work published 1738).

Wolff, C. (1973a). Ausführliche Nachricht von seinen eigenen Schrifften. In H. W. Arndt (Ed.), *Christian Wolff: Gesammelte Werke* (Ausführliche Nachricht) (Vol. 9, Pt. I, 2nd ed.). Olms. (Original work published 1733).

Wolff, C. (1973b). Philosophia moralis sive Ethica (Pars V). In J. École (Ed.), *Christian Wolff: Gesammelte Werke* (Philosophia Moralis) (Vol. 16, Pt. II). Olms. (Original work published 1753).

Wolff, C. (1976). Vernünfftige Gedancken von der Menschen Thun und Lassen. In J. École (Ed.), *Christian Wolff: Gesammelte Werke* (Deutsche Ethik) (Vol. 4, Pt. I). Olms. (Original work published 1733).

Wolff, C. (1978). Theologia naturalis, method scientifica pertractata (Pars I). In J. École (Ed.), *Christian Wolff: Gesammelte Werke* (Theologia Naturalis) (Vol. 7.2, Pt. II). Olms. (Original work published 1739).

Wolff, C. (1980). Grundsätze des Natur- und Völckerrechts. In J. École (Ed.), *Christian Wolff: Gesammelte Werke* (Vol. 19, Pt. I). Olms. (Original work published 1754).

Wolff, C. (1983). Der Vernünfftigen Gedancken von Gott, der Welt und der Seele des Menschen, auch allen Dingen überhaupt, Anderer Theil, bestehend in ausführlichen Anmerckungen. In C. A. Corr (Ed.), *Christian Wolff: Gesammelte Werke* (Anmerkungen zur deutschen Metaphysik) (Vol. 3, Pt. I, 4th ed.). Olms. (Original work published 1740).

Wolff, C. (1985). Oratio de Sinarum philosophia practica. In M. Albrecht (Ed.), *Rede über die praktische Philosophie der Chinesen*. Hamburg. (Original work published 1726).

Wolff, C. (2003). Vernünftige Gedancken von Gott, der Welt und der Seele des Menschen, auch allen Dingen überhaupt. In C. A. Corr (Ed.), *Christian Wolff: Gesammelte Werke* (Deutsche Metaphysik) (Vol. 2, Pt. I). Olms. (Original work published 1751).

Wolff, H. M. (1949). *Die Weltanschauung der deutschen Aufklärung in geschichtlicher Entwicklung*. Francke.

Wolff, M. (1992). *Das Körper-Seele-Problem. Kommentar zu Hegel, Enzyklopädie (1830), §389*. Klostermann.

Chapter 11
The Relation Between Psychology and the Other Parts of Metaphysics: Ontology, Cosmology, and Theology

Jean-François Goubet

11.1 Introduction

Psychology, be it empirical or rational, forms part of metaphysics. Ontology, cosmology, and theology are the other parts of metaphysics, according to Wolff. In what order do these disciplines appear? And does this order have any significance?

In the first section of this chapter, I will show that deductive order matters a great deal in Wolff's system, and, therefore, that psychology has to take the right place in this order.

After these general preliminary considerations, I will follow a historical common thread in order to reconstruct the evolution of Wolff's thinking. His system was first sketched in a prospective writing, the *Ratio praelectionum* (Lecture Plan). The second section will analyze Wolff's first metaphysical insights, while the third will consider the German works, particularly in logic and metaphysics. Stress will be laid on the order in which metaphysical disciplines appear.

The fourth section will look in more detail at the links between psychology, ontology, cosmology, and theology as they appear in the Latin works. I will begin with another prospective work, the famous *Discursus praeliminaris de philosophia in genere* (Preliminary Discourse on Philosophy in General), and then comment on the different treatises Wolff wrote on the particular metaphysical topics. The fifth and final section will use a case study on pleasure to show how separate disciplinary considerations can conspire: pleasure does not just belong to psychology also but includes ontological, cosmological, and theological features. A joint examination is required in order to understand it more completely.

J.-F. Goubet (✉)
University of Lille, Lille, France
e-mail: jean-francois.goubet@inspe-lille-hdf.fr

S. d. F. Araujo et al. (eds.), *The Force of an Idea*, Studies in History and Philosophy of Science 50, https://doi.org/10.1007/978-3-030-74435-9_11

11.2 The Systematic Order of the Metaphysical Disciplines in Wolff's Works

Wolff was the creator not just of empirical and rational psychologies as metaphysical disciplines, but also of general cosmology as an important part of metaphysics. Prior to Wolff, the term "general cosmology" was not used in metaphysics schools (Wolff, 1731/2001), and the discipline had not been systematically and consistently treated as a part of metaphysics. Although the German works are commonly claimed to be an *adumbratio* or *esquisse* (sketch) of the more developed and complete Latin works (Wolff, 1736/1962, 1737/1964, 1738/1968, 1739/1978b, 1740/1972, 1741/1981), neither of them is really isomorphic, as if the latter were a mere reproduction, with more commentaries and precisions, of the former. A shift took place between the two sets of works as Wolff decided to change the order in which they had to be released. While in *German Metaphysics* cosmology follows empirical psychology, the Latin works display a different order by placing cosmology before empirical (and rational) psychology. In this section, I will discuss the order of metaphysical disciplines, showing the importance of systematicity in Wolff's thinking.

Wolff was of the opinion that metaphysics had to receive new parts for it to be exhaustive or, at the very least, more complete than before. But completeness was not the only aspect that was important for Wolff: the ordering of metaphysical content was equally important, if not more so. The deductive framework in Wolff's writings cannot be ignored. But what can the subsequent disciplines borrow from those that are presented first? What was the content a student had to be taught and understand before he could turn to psychology? Two factors will be important in this regard: first, what Wolff says in general about the need to borrow certain concepts and topics, and second, what Wolff actually does when he writes his metaphysical works. It is not enough to simply say anew what the author has already told us in his forewords and prolegomena. Following certain doctrinal expressions sometimes teaches us more about the actual relevance of cosmological themes, for instance, than the general preliminary statements.

Although I have only discussed psychology and cosmology, Wolff's philosophy also includes two other metaphysical disciplines, both of which are of major importance: ontology and theology, each occurring at the opposite extremities of the range—ontology comes first, theology last. Ontology has to occupy the first position because it gives all the other disciplines their concepts, including, for instance, force (*vis, Kraft*). This is employed both in cosmology and psychology, meaning that it has to be defined beforehand. Ontology appears in this regard as the foundation of further metaphysical disciplines, including cosmology, psychology, and, at last, theology. One cannot decide what the major issues of psychology are without first putting ontology on solid ground. The generality of its concepts explains why they must be defined in advance: what can be said of a genus is also predicated on its species and, ultimately, on the individual *par excellence*, who is God. But generality in itself is not enough to explain ontological notions. In addition, these concepts have to be directive and fruitful (Wolff, 1729–1730/1983a, 1730–1731/1983d). With the use of clearly defined primary concepts, new truths can be gleaned in further doctrines. Ontological priority is therefore not just a matter of classifying an already definite stock of truths; rather, the use of ontological concepts also allows this stock to grow.

Let us now take a brief look at the relationship between psychology and theology. Psychology functions as a ladder to God in many respects. In Wolff's view, there are numerous ways to prove the existence of God, many of which appear to be fallacious. But there are only some correct proofs for this existence, as Jean École points it (1990). Wolff prefers to approach the doctrine of God by employing arguments that refer to his powers and soul, ascending to the ultimate source, and not descending from it to its consequences. This a posteriori approach makes the previous stabilization of some psychological truths necessary: it has to be proved that men are spirits—and not just mere creatures—that are destined to improve their knowledge and behavior. In other words, they are born to become wise. The link between psychology and theology in Wolff's philosophy refers to the pneumatology that already existed before him in metaphysics. As we shall see, Wolff discusses metaphysical issues dealing with this tradition, for instance, in the *Ratio praelectionum* and, later, the *Discursus praeliminaris de philosophia in genere*. God and mind are thereby related through a theory of spirits. Indeed, the demonstrative way of moving from the soul to God is a particular mark of Wolff's approach.

11.3 Wolff's Early Insights Before the Completion of the Metaphysical Works

What is metaphysics? In the first version of the *Ratio praelectionum*, Wolff gives an overview of its common meaning. For him, it is the science of the principles of God and the human mind; but it also can be thought of as the conjunction of *philosophia prima* (first philosophy), i.e., ontology, and pneumatic science, i.e., the science of the soul. The difference is just a difference of words, not of subject matter, and Wolff declares that he is at ease with words. More important is the distinction to be given to notions and the evidence to be bestowed on propositions (Wolff, 1718). In contrast to the Scholastics, Wolff intends to obtain determined propositions in metaphysics, which calls for the distinction of the first concepts, the ontological ones, and demonstrative strength to infer new propositions in subsequent disciplines such as cosmology, psychology, or theology. The philosopher wants to break with the former metaphysical tradition while relying on a scientific, deductive method. Wolff's allusions to Goclenius' (1547–1628) metaphysical dictionary (Goclenius, 1613/1980) are to be understood in the same way. This subject matter cannot be tackled without deductive order, and entries arranged in alphabetical order alone, regardless of any logical connection, are not sufficient. Determined propositions can only be obtained in all metaphysical chapters when the first notions are in themselves clear and fruitful, and capable of giving birth to further truths in subsequent disciplines.

The first version of the *Ratio praelectionum* is interesting too, because it gives a picture of Wolff's early philosophical development. We learn that the mind/body problem was crucial for him from the start, as he wrote on human language, and that Leibniz's (1646–1716) *Theodicy* tried to solve the problem using theological principles. Unwilling to become part of any philosophical sect, Wolff developed his own system. In metaphysical matters, Wolff suggested inquiring about how modifications in the mind take place in analogy with physical changes that occur as a result of the natural force. The analogical link in the treatment of theological and psychological matters is alleged next (Wolff, 1718). If one has distinct notions of God and the human mind, one knows that thought can be predicated of these two substances, and knows the force they both have and the laws they follow. In order to explain some modifications in the mind, an inquiry into theology-related subjects such as materialism and idealism is required. And then we arrive at the conclusion concerning the ascension from psychology to theology: "The doctrine of the human mind for the best precedes the doctrine of God, which follows, in so far as one arrives at this latter knowledge a posteriori" (Wolff, 1718, §.16, p. 146).

It was not the dignity of the object, but the "order of dependence" of the disciplines, that was the decisive criterion for Wolff. It is also an order based on knowledge—not a value hierarchy or an order between beings (God existing before any creature)—that is crucial in Wolff's eyes. The attributes and existence of God are to be inferred from the knowledge of the human mind.

Knowledge about the human mind, on the other hand, allows a development of general notions such as slumber, wakefulness, time, space, continuity, eternity, and truth. We can see here that psychology occupies a central place in Wolff's metaphysical construction, linked to theology at one pole and to ontology at the other. But cosmology, although not explicitly named in this paragraph, is not forgotten. When Wolff says that truth is the "order of phenomena" (Wolff, 1718, §.18), he also implies a cosmological account, as we shall see later when considering the Latin works of the late period. Truth also has to do with connections of real things, not just with the phenomena present in someone's mind. However, the text of the *Ratio praelectionum* does not go further: it simply indicates the hope that the German text on God, the human mind, and the world will contain some expositions of the primal, i.e., ontological, notions (Wolff, 1718).

Wolff further develops his ideas on psychology in this chapter before proceeding to theology and ontology. Here, we can see how psychological and cosmological considerations intermingle, in particular concerning the mind/body problem and the place from which one's soul sees the world. The link with Descartes is relatively clear in this context, the mind/body hypotheses of modern philosophy being sequels of his metaphysical doctrine. Although Wolff may be considered a follower of Descartes (1596–1650) regarding his methodical positioning, as Corr points out (1997), and his dualism between two sets of substances, he declares that he is not satisfied with the Cartesian definition of *substantia cogitans* (thinking substance). The latter is unable to give an account of what occurs in the human mind, and the difference between spirits, as higher rational creatures, and other creatures endowed with a soul (such as animals), is not inferable from it. Here, Wolff's purpose, as before, is to clarify notions that common thought does not perceive clearly enough and to give the reasons why modifications in the human mind occur (Wolff, 1718). These reasons, as I have already pointed out, are to be drawn from the former metaphysical disciplines, and they will function anew in theology, for which psychology paves the way.

11.4 The Link Between the Metaphysical Disciplines in the German Works

11.4.1 A Brief Look at German Logic

German Logic repeats the importance of method in all philosophical matters, including metaphysics. Logic is important because it covers all inquiries into God, the human mind, and the world, and it gives them their form. In the book on the forces of the human mind and their correct use, Wolff lays stress on the first chapter, related to concepts, and the fourth, related to syllogisms. "Because, where solid knowledge is indeed loved, it is above all a matter of distinct concepts and ordered proofs" (Wolff, 1754/1978a, p. 110).

Wolff, for whom solidity is a key word, also has ontology in mind: the first distinct notions must be taken there, and the correct reasoning procedures will then arrive at new results in all subsequent disciplines. Although the deductive order means so much to Wolff, in the short sketch he gives of the organization of *Weltweisheit* ("world wisdom" or philosophy), Wolff does not begin with ontology but with logic, which sounds normal in a foreword to a logical work. Here we will make some remarks about the connection of this latter discipline and psychology before moving on to the core of this chapter: the link between the subject matters of metaphysics. Let us have a look at Wolff's deductive starting point, which involves logic, psychology, and ontology: "When we pay attention to ourselves, so we are convinced that there is a power in us thinking about what is possible, which we use to call understanding" (Wolff, 1754/1978a, §.10, p. 117).

Wolff's deductive starting point is a *cogitamus* (we think), a revised Cartesian *cogito* (I think) with the help of which the standard of all correct reasoning may be established (Arnaud, 2001). A psychological experience is at stake: the consciousness of thinking, the awareness of being a thinking subject that is in relation to external objects. And logic, understood as the correct use of mental powers, grounds itself in empirical psychology, which gives an account of man's faculties. Wolff's entire systematic project can in this sense be related to psychology. Psychology makes it possible to understand the status of the first inner experience, and it allows an individual to gain power over his own mind in order to make the correct use of his understanding and reason. Science itself, like philosophy and metaphysics in particular, is defined as a competence of understanding (Wolff, 1754/1978a) and is therefore rooted in psychology.

Logic is related to the possible, as something thinkable, but subsequent parts of philosophy, i.e., metaphysical disciplines, have a connection to it too, as something that cannot be in all cases inexistent. First of all, God cannot not exist, and subsequently bodies and souls may or may not exist: theology is in consequence named first, and then the other two particular sciences. It is interesting that both are given a specific appellation. What is possible through the *Kräfte* (powers) of bodies is called physics or the science of nature, while what is possible through the powers of *Geister* (spirits) is called the doctrine of spirits or pneumatology. Psychology, then, is not very accurately related to souls in general (Wolff, 1754/1978a). Here we see that Wolff first thinks in a traditional way, placing human psychology in the vicinity of God rather than animals, which certainly have a soul but neither reason nor free will. Empirical and rational parts of psychology are not precisely distinguished. Cosmology and psychology, furthermore, are presented in an analogical way, as two particular sciences, whereas ontology, or fundamental science, deals with being in general. Something like a transcendental cosmology, which shares the same generality in terms of singular intuitions as ontology (Hinske as cited in Carboncini, 1991), does not appear in these lines, as will be the case in the later Latin works. The parallel presentation of the science of nature and the science of mind conceals another aspect: natural bodies are compounds, while the human soul is simple and not made up of separate parts. Therefore, the complex articulation of metaphysical doctrines cannot be inferred from the short presentation Wolff himself gives of his philosophical enterprise in 1712. It will take the other German writings to give a more accurate overview of metaphysics. Still, the methodical, systematical way of writing, with roots in psychological considerations about the *cogitamus*, is already present in this first philosophical German treatise.

11.4.2 About the Organization of German Metaphysics

German Metaphysics, published 7 years after *German Logic*, reaffirms the same methodological framework. Although not explicitly mentioned, systematicity can be clearly seen in the foreword. Wolff still criticizes the shortcomings of traditional metaphysics, but without mentioning the names of any individuals, such as Descartes, or any movements, such as scholasticism. A comparison with other passages would confirm, however, that Wolff's own metaphysical attempt continues to be distinguished from previous achievements: "There was a lack of distinct concepts, solid proofs and connection between truths" (Wolff, 1751/1997, p. [3]).

Ontology has to give firsthand notions that break with the obscurity of common language. As Wolff himself will say later, an *ontologia artificialis* (artificial ontology) must spring from an *ontologia naturalis* (natural ontology). Ontology as a science has as its foundation something sensitive, and it originates in sensibility (Paccioni, 2006). With these notions, when they are made explicit, one can obtain solid proofs that cannot be denied. A link between truths will then appear, and a *nexus veritatum* (connection of truths) will prevail. Hence, science must be thought of as a chain of reasons that are internally coherent and form a web: the German word *zusammenhängen* (to be connected with) used by Wolff (1751/1997) refers to the idea of a system. Truth and system go together, and science is not just a matter of understanding and thoughts about what is possible, but a question of reason, a systematic connection between all things. As we shall see later in the Latin works, in Wolff's account of reason, science, and transcendental truth, ontological, cosmological, and psychological themes conspire.

Wolff begins *German Metaphysics* with a brief chapter dedicated to the knowledge of ourselves before going on to examine the first principles of our knowledge and all things in general. Who would be foolish enough to deny he is conscious of himself and of other things? It is impossible to say that and, at the same time, experience it (Wolff, 1751/1997, §.1). An indubitable experience ("We think") and an ontological principle ("A cannot be and not be at the same time") enter in a canonic syllogism. I say canonic because all chapters will begin the same way, by referring to it or some similar reasoning. In §.45 Wolff explains what thinking of oneself and thinking many other things outside of us means, and it will serve again and again throughout *German Metaphysics*. All metaphysical sciences have deep roots in a first reasoning, in which the terms are distinct, which proves fruitful for the rest of philosophy.

Untrügliche Erfahrungen (infallible experiences) (Wolff, 1751/1997, §.191) are at the basis of the empirical part of psychology, and distinct concepts of the soul's powers can be gained through them. The scheme of the brief first chapter of *German Metaphysics* repeats itself in the third chapter, dedicated to what we know from experience of the soul in general. Rational psychology, the discourse on the essence of the soul and of the spirit in general, will be supported by the previous, empirical discourse on the soul. Truths from empirical psychology will function as touchstones for new truths concerning the unity of the soul, endowed with a single representative force (Wolff, 1751/1997, §.727). In this introductory paragraph, Wolff does not explain why he included a reference to spirits in the chapter's title. It is plain, however, that he concentrates his pneumatical reflections at the end of the chapter, after discussing the meaning of reason and free will, two distinctive features of man and higher beings. A ladder to the doctrine of God will then be set up, which is to be climbed up to the final metaphysical truths, which, to be understood, require all previously established truths.

In *German Metaphysics*, the first paragraphs of the theological chapter refer to ontology and cosmology, not psychology. God's existence and his independence regarding the world, its eternity or incorruptibility, are to be understood using distinct first notions and a correct view of the world's essence and structure. Psychology enters the stage when God's qualities are explained. Beginning with §.954 (Wolff, 1751/1997), divine attributes are presented. First of all, God possesses an infinite understanding, which allows him to perceive at once and perfectly everything that is possible. His lack of senses and imagination (§.959) will follow. Free will follows later, beginning from §.980. With the concepts of reason and free will, God's wisdom can be examined. What use does God make of his infinite power? Why is he a perfect philosopher (Schneiders, 1983)? Such questions can only be addressed when spirits, not just souls, have been examined in advance. The community that God enters with human beings constitutes the realm of spirits, the unity of beings sharing reason and free will (although not to the same degree).

11.4.3 Wolff's Further Explanations About His German Metaphysics

Wolff never tired of commenting on his own work. After completing his *German Metaphysics*, he penned a large volume in his mother tongue in 1726 to justify himself against the criticism he faced: the *Ausführliche Nachricht von seinen eigenen Schriften* (Detailed report on his own writings). As the title indicates, Wolff wanted to be exhaustive in his clarifications regarding his previous work. It cannot be denied that some extra light is shed on the reason why he treated certain metaphysical disciplines before others.

Wolff's disciples Thümmig (1697–1728) and Bilfinger (1693–1750) made compendia in Latin that candidly took over the master's teachings. They divided their treatises into many parts, leaving some empty space between sections, while Wolff's *German Metaphysics* appeared at first sight in the form of a continuous chain of reasons. This material arrangement was not the only change Thümmig and Bilfinger introduced. In general, they decided to tackle the whole of psychology, be it empirical or rational, after cosmology. Wolff had then to explain this change, first introduced by his disciples.

What does Wolff say about the order of psychological disciplines?

> *Psychologia empirica* [Empirical Psychology] is indeed a story of the soul and may be known without all remnant disciplines; on the contrary, *Psychologia rationalis* [Rational Psychology] supposes cosmology to be already familiar. Hence, if someone wants particularly to treat disciplines in their order, so ontology follows cosmology, and psychology follows the latter (Wolff, 1733/1996, §.79, p. 231).

Wolff justifies two different things here: his previous choice in *German Metaphysics* and his future choice in his Latin works. Empirical psychology can be understood with the help of ontology alone. Accordingly, it was no problem when the former was treated just after the latter; and it will be no problem to address it just after cosmology, since material things and the soul are two species of substances that are to be developed one after the other. But rational psychology, because it deals with the difficult mind/body problem and the question of pre-established harmony, cannot be understood without any knowledge of the *nexus rerum* (the connection of (material) things). In the future, when the two parts of psychology will combined, the correct order will be ontology/cosmology/psychology. The former order was correct too, because empirical psychology does not draw its principles from cosmology. In fact, we shall see that this is not so simple in the Latin works, in which the *voluptas* (pleasure) derived from truth, perfection, and order is an empirical psychological notion whose definition is connected to cosmological matters. But in 1726, Wolff did not have the details of his later doctrine so clearly in mind, so he could say without any doubt that empirical psychology does not refer to cosmology.

Wolff treated empirical psychology in second place also for a didactic rather than a logical reason:

> I have placed one part of psychology, namely, the *empirica*, before *cosmology*, because it is easier than the latter and more attractive for beginners, eliminating their frustration with *ontology* caused by having to pay more attention to various things than they are used to. (Wolff, 1733/1996, §.79, p. 232)

The order of study is also an important factor in Wolff's eyes, as confirmed by the final chapter of the *Ausführliche Nachricht* and Chap. III of *Discursus praeliminaris* (1963, §§.55–113). The relative ease of empirical psychology (École, 1990) was a factor in teaching it early on in the metaphysical curriculum. This factor will also be important for Kant (1724–1804) as he will teach empirical psychology before cosmology and ontology (Kant, 1765/1912). Wolff, like Kant, was a philosophy professor, for whom the order of disciplines mattered. He was not a lonely writer working at his desk, surrounded by books and distant from students. After all, disciplines, as such, are doctrinal bodies to teach someone.

Theology, which takes its principles from all previous metaphysical sciences, comes naturally in the last position, however tedious it may prove itself to be. The order of reasons precedes all other considerations in this case, in particular those linked with the order of study.

11.5 The Concrete Links Between Metaphysical Themes in the Latin Works

As Wolff shifted from German to Latin to write his entire metaphysical work anew, he did not just translate himself but also introduced some organizational changes and new material. I have focused my attention thus far on the general structure. Now I would like to say some additional words about one of Wolff's famous booklets, the programmatic *Discursus praeliminaris de philosophia in genere*. I will then discuss some specific aspects of the Latin metaphysical works, in which connections between psychology and earlier or subsequent disciplines appear in more detail than previously.

11.5.1 General Considerations About the Succession of Metaphysical Disciplines in the Preliminary Discourse

The subject matter of the third chapter of the *Preliminary Discourse* is the different parts of philosophy. The chapter presents considerations about philosophy in general and philosophical method. It is here that Wolff provides a criterion for dividing the metaphysical sciences according to their subjects and proposes a justification of the demonstrative order to be followed.

Just as he did in the *Ratio praelectionum*, Wolff deals with German *Schulphilosophie* (school philosophy, i.e., previous metaphysics). Goclenius, although not named, is once again a major point of reference, from whom Wolff feels a strong urge to distinguish himself. As Pierre-François Moreau clearly saw:

> The one who says 'I'm introducing a new part' transforms *ipso facto* the whole in so far as he changes the articulation of parts while introducing a new region, and taking a decision on the way these regions command each other. (Moreau, 2002, p. 8).

Wolff named new knowledge areas and, as a result, kept a distance from classical German metaphysics in at least one major concern: he introduced cosmology in a key position in order to deter his readers from accepting Spinozism. World unity and the coherence of the universe were to be maintained; however, they should be compatible with finality. Cosmology, although named very late in the order of metaphysical disciplines, does indeed play a major role because it commands the rest of them: psychology and theology will have to borrow some major concepts from it and not contradict what has been established in it. The first named metaphysical disciplines, theology and psychology—the more traditional ones—have their content changed for this reason. The key role played by cosmology is reinforced by the addition of the adjective "general" (Wolff, 1963, §.78), which brings it close to ontology. The science of being in general and the science of world in general are obviously able, as such, to rule special domains of metaphysics and, thereafter, non-metaphysical philosophical disciplines.

The third chapter of the *Preliminary Discourse* is important too, because it tells us a little more about what Wolff had precisely in mind in stating that ontology and cosmology were to be treated before psychology. §.98 is dedicated to the ontological and cosmological principles in psychology, and it provides some examples of what notions the latter discipline can borrow from the former:

> No one can deny that the soul possesses the force of representing the universe in accordance with the modifications that occur in the sensory organs. Careful analysis will show that this notion is primary and provides the reason for the other modifications of the mental faculties and of the things which they understand. This force is not distinctly understood unless one has acquired the general notion of force from ontology and the general understanding of the world from cosmology [...]. (Wolff, 1963, §.98, p. 50)

Wolff wants to follow a natural order in psychology, beginning with the sensation of the external world and ascending toward reason and free will. Spinozism is not the only danger philosophy may be prey to: idealism, i.e., what we would now call solipsism, threatens it too. Historical knowledge and conscious experience as a matter of fact must be positioned at the beginning. Not just I am, but we are, and we are stuck not only with other minds but also with a well-structured material world. The materiality of the world is present twice, once as the content of external perception and once as the bodily form of perception. Modifications in the mind conspire with modifications in the body; representations accompany "material ideas" that are diffused in the nervous system (École, 1990). Furthermore, the first sensitive matter of perception can be worked out to give content to all subsequent faculties or powers of the mind. Rational issues, such as the mind/body problem, and an empirical issue, such as the deduction of the particular laws (compared to general law, which is related to *vis*) of mind, fall under cosmology.

Where is the concept of force defined? In ontology, since it plays a major role in cosmology as well as in psychology. The opposition between a unique force or principle and a plurality of powers or specific laws (of nature or mind) structures Wolff's metaphysical discourse. It is interesting that Wolff says that cosmology is also required in order to obtain a distinct concept of representative force. Not that physical force and psychical force are the same, but cosmology functions here as a precedent, with its articulate doctrine of a unique force together with multiple laws of motion, in order to investigate another domain. Body and mind are two separate beings; nevertheless, the science of the world is a model for the science of the mind when the question is to derive miscellaneous laws of the mind, such as the law of imagination (Rumore, 2018), from a single representative force.

11.5.2 Ontology, Cosmology, and Empirical Psychology

It is sometimes difficult in Wolff's doctrine to separate what is derived from his logical and methodological positioning, what comes from his ontology, and what is rooted in his cosmology. Many chains of reason tend to confirm themselves when applied to a new subject. Let us take an example found in Wolff's theory of truth—developed in a chapter dedicated to intellectual competences—to see this more clearly. There is a connection between all universal truths. And what is the proof?

> The connection of universal truths is founded in the connection of things established in *Cosmologia*, part. I, sect. I, c. I, in so far as logical truth is founded in the transcendental truth of things (§.499, *Ontol.*). And there is no such connection, nor is it possible to build genuine systems outside of what I described in *Horae subsecivae* [Leisure Hours] anno 1729, fall, No. 3. (Wolff, 1738/1968, §.482, p. 371)

Reason, i.e., systematic understanding, the link between all possible mundane things, and truth, perfection, and order go hand in hand. The logical determinability of a predicate through a subject conspires with reality and ontological connections. Reason is not just a separate power of mind, with no relation to the world, and this explains the fact that ontological statements have an influence on psychological statements.

Ontology is related to our perception of the world. Indeed, without the first ontological principle—the principle of contradiction—we would live in a world of wonder, where Titus would perceive as hot the same thing Maevius perceives as cold. Logical truth is therefore founded in the objective world, which has its own consistence. The second ontological principle—the principle of reason—accounts for the determination of propositions, the fact that predicates can be rightly attributed to a given subject.

At the empirical psychological level, this means that it would be impossible to cultivate reason as an intellection of the connection of truths if no such connection existed and was to be thought. The way men think is not independent of states of things. Not just propositions are true, in this regard, but beings themselves are true as well. Wolff defines anew what the transcendental concepts are, and this shift in his ontology, compared to what it was before, has repercussions on the theory of human intellectual powers.

Here we should remember that Wolff sees science not so much as an objective set of true statements in themselves but as an intellectual disposition that must be cultivated in order to fulfill one's destiny. Wisdom is the reasonable horizon of spirits. The opus Wolff refers to is called *De differentia intellectus systematici et non systematici* (On the Difference Between a Systematical and a Non-Systematical Understanding), and it begins with ontological and cosmological matters before stating the moral duty of a creature endowed with reason: being excellent (for instance, Wolff, 2011, §.18). The notion of system is related to cosmology: before speaking of *systema doctrinarum* (system of doctrines) (Wolff, 2011, §.3), Wolff speaks of the connection between things (Wolff, 2011, §.2). *Nexus veritatum* is akin to *nexus rerum*. The systematic display of propositions expresses the architecture of the world, but a slight clarification must be made in this context. All things, as elements, are connected with each other, but propositions are not like elements, being judgments or relations between two terms. The demonstration of a new true proposition requires the presence of two previous true propositions. Therefore, there is no homology between the way connections are made in logic and the way they are made in cosmology.

Further links between physics (the empirical counterpart of cosmology) and empirical psychology exist at a structural level. Empirical psychology is compared to experimental physics, and it establishes rules and laws for the various powers of mind in the same way as physics does for the rules and laws of motion (Wolff, 1738/1968). This comparison implies that general cosmology is to be paired with rational psychology. The similitude between physical and psychical objects also becomes clear when Wolff sketches the project of psychometrics, introducing measurement into the conception of mind (Feuerhahn, 2003). For him, the analogy between empirical psychology and experimental physics serves as his Ariadne's thread to tackle new problems.

11.5.3 Ontology, Cosmology, and Rational Psychology

The *nexus rerum* question, which is already present in the empirical counterpart, will appear again in *Psychologia rationalis* (Rational Psychology). At first sight, we simply find here a confirmation of truths already established. Sensual ideas and phantasms have a similitude with the external world they represent; accordingly, they embrace, unclearly and sometimes fallaciously, its past and its future state (Wolff, 1740/1972). Still, it is not just inferior faculties but also higher powers of the mind that represent the whole world. Universal truths must be consistent, and it is possible to reject any single assertion that will not fit the established set of valid propositions. The correlate of what is, cosmologically speaking, the entire universe is, psychologically speaking, the unique representative force, whatever its manifestations may be. The soul does not have any parts, and its manifestations, be they passive or active, or lower, middle, or higher, have one force as their source.

But does Wolff really say similar things when he speaks of the *nexus rerum*? Indeed, it seems that there is a shift in the point of view between the two psychological tomes. The first laid emphasis on the genesis of representations in time; it was a story about what occurs in the mind. Apperception or self-consciousness was central here, which led Faustino Fabbianelli to speak of a phenomenological point of view, which is evident in the treatment of the lower faculties (Fabbianelli, 2007). The relation to an objective world appeared more in the treatment of higher faculties, such as reason, and it continues to prevail in *Psychologia rationalis*, which focuses on the real connections, and whose point of view is therefore cosmological. The interpretation supposes that the emphasis on the real connections possibly follows the reformulation of the Leibnizian pre-established global harmony in favor of a local harmony between body and soul (Fabbianelli, 2007). Whether this hypothesis is correct or not, it is clear that Wolff's topic, in his second Latin treatise on psychology, is the representative force in connection not only with the universe as a mere external structure but also with a body, situated in it, and which physical movements correspond to psychic modifications.

Let us sum up: rational psychology has to deal with the relations between the body and the soul, and must therefore refer to the truths which have been previously demonstrated in cosmology. Which is the better hypothesis to account for the soul/body connection? Is there a physical influence? Is the doctrine of occasionalism, defended by Malebranche (1638–1715), more convenient to explain how the body and the mind conspire? Or is the Leibnizian account of this connection, the hypothesis of pre-established harmony (here revised as a local vs. global harmony), the better one? These questions are treated in rational psychology and are supported by cosmological truths (Goubet, 2018).

One of the cosmological truths is the conservation of the quantity of motion or, rather, "living forces" (Simmert, 2018). This metaphysical principle, which was widely used in physics, makes the hypothesis of physical influence (for example) hard to support. If the body has a direct influence on the soul, is not a certain amount of living force irremediably lost? Physics (the empirical counterpart of cosmology) and its metaphysical foundation serve here as a model for the newest metaphysical science named psychology. It is very important for Wolff to show that physical and psychological phenomena occur at the same time but do not mingle. Materialism is a major danger that must be kept at bay: the right to exist of a separate science named psychology has to be obtained at all costs for the German philosopher. Idealism, although equally unilateral because it does not recognize the existence of a dualism of substances (i.e., the existence of both body and soul)—and asserts the mere existence of the soul—is not so dangerous. Indeed, it does not jeopardize the doctrine of immortality, nor does it question the existence of God.

11.5.4 Psychology and Natural Theology

In fact, the community between spirits, in particular between human beings and God, is the most important in Wolff's metaphysics. We humans and God share the fact that we have a soul endowed with reason and free will; still, there is a major difference of degree, limitation, and contingency between us, as creatures, and our creator. We lack the so-called *aseitas* (aseity). Wolff needs to postulate continuity in the scale of beings in order to build arguments in his natural theology. Truths established in both tomes of psychology will help him reformulate some traditional proofs of the existence of God, and define his infinite power of knowledge and action in relation to our own.

The first proof is *a contingentia* (derived from contingency [of beings]). Someone distinct from contingent beings has to exist and to be necessary. Furthermore, it must be shown that this necessary being is the same as the God of the Bible. The latter part of the argument does not concern psychology, but the first part actually includes some psychological considerations. Where the Scholastics displayed the *a contingentia* proof with the only reference to the universe, Wolff relies on the duplicity of contingent substances to assert that God is a necessary being compared to both the world and the souls (Theis, 2018). Parallel chains of reasons, produced in transcendental cosmology and then in empirical and rational psychology, converge in natural theology.

What is linked in particular to empirical psychology here? Human beings have to free themselves from the limitations of their powers of knowledge in order to reach a correct representation of God. This infinite being does not have any body and cannot be perceived by the senses. But empirical psychology begins with consciousness of ourselves as affected by external objects. How can we obtain a distinct idea of God under these conditions? Manuela Mei shows how symbolic knowledge is used to think about this invisible being. We can only reach God through figurative knowledge, through so-called hieroglyphs. Such figures are not copies but aliments for the faculties of imagination and invention. "A triangle, by instance, is not by itself like God, but its likeliness demands the use of *perspicacity* and enables us to conclude from the one (a geometrical figure) to the other (God)" (Mei, 2011, p. 119).

The epistemological possibility of God's knowledge by a limited human mind is, as we see, rooted in the *Psychologia empirica*. And what is more specifically related to rational psychology in this theological matter? That God possesses a unique representative force, able to perceive intuitively all the *nexus rerum*, is consistent with the cosmological (vs. phenomenological) positioning of *Psychologia rationalis*. God is said to be always awake, to perceive distinctly and intellectually all possibilities at once (Mei, 2011). In other words, God never dreams, and the content of his thought can never be a world of wonder. Transcendental truth, a truth relating to a possible state of the world, is once again at stake in such theological reflections.

Psychology is not just present in the first proof of the existence and attributes of God. As Matteo Favaretti-Camposampiero explains, it is possible to speak of a "psychotheological way to God" in general. The first proof invited to compare the human mind to God's soul, to understand the similitude and, at the same time, the difference between finite and infinite. Another proof, developed in the second part of the *Theologia naturalis* (Natural Theology) and based on God's perfection, is indeed a demonstration *ex contemplatione animae* (derived from the contemplation of the soul): divine existence and attributes are not obtainable if one does not consider one's own soul (Favaretti-Camposampiero, 2011).

It follows that psychology is clearly the ladder to God, and that it in fact provides many theorems and notions for grasping the divine existence and attributes. The liberation of our own mind from its limitations, which had already been claimed by Wolff as early as 1717 (Favaretti-Camposampiero, 2011), finds an achievement in the late Latin theological works. The metaphysical system, without being absolutely the same from the beginning to the end of the philosopher's career, is however supported by strong initial decisions, for which the critical interpretation of Leibniz's *Theodicy* was important (Lalanne, 2011).

11.6 The Pleasure Taken in the Perception of Truth: A Transversal Concept in Wolff's System

Wolff's philosophical output, after he was driven away from Halle to settle down in Marburg, was not entirely contained in his major works. It was also expressed—maybe in a more convincing manner—in shorter texts published in the so-called *Horae subsecivae*. Carboncini (1991) raised awareness of one of them, which concerns the pleasure taken in the contemplation of truth. In so far as it requires ontological, cosmological, and psychological notions to be understood, I want to follow some of Wolff's developments to show the entanglement of the chains of reasons in his manner of philosophizing.

Descartes defined pleasure as consciousness of our own perfection. Wolff, without rejecting this idea, introduces a complement: we take pleasure not only in our own perfection but also in the perfection of a given object, such as a clock or building (Wolff, 1729–1730/1983b, §.2). Furthermore, we can take pleasure not only while intuiting, knowing a perfection sensibly, but also while thinking an objective, rational order. Here we see how something rooted in empirical psychology—pleasure as related to certain forms of representation—is linked to something that is found in ontology and cosmology, i.e., transcendental (metaphysical) truth as a connection between all coexistent and successive possible things. Wolff himself mentions a future use of pleasure taken in truth, which is related to teleology, the science of goals (Wolff, 1729–1730/1983b, §.2). When we remember that one way leading to God is the contemplation of order in nature, we clearly see that this pleasure also has a theological function.

Truth, order, and perfection are three ontological notions that directly imply the two major ontological principles: the principle of contradiction and the principle of sufficient reason:

> Who perceives the truth of things knows the order in which, by the force of the principles of contradiction and sufficient reason, they are to be placed as predicates that are either constantly in [their subjects] or follow each other. All order is perfect […]. (Wolff, 1729–1730/1983b, §.4, p. 176)

And these ontological links between concepts have an immediate logical repercussion. Reasonable knowledge is possible because there is something like a transcendental (i.e., metaphysical) truth, and pleasure can be taken in the use of our faculties when we investigate order and perceive its perfection. Doing mathematics or proving a systematic understanding means the same thing as making a correct use of one's reason. Philosophizing can lead to this kind of pleasure too, and Carboncini (1991) points out that Wolff is aware of the pleasure that a student can take in his philosophy when he perceives the accordance between methodological positioning and the constitution of things.

Wolff was aware of the difficult character of his ontology. This metaphysical discipline, although very fruitful, could be tedious for untrained beginners. Rational psychology, as opposed to plain empirical, historical observations (historical knowledge is the lowest grade of knowledge—see Wolff, 1963, §.22,), could, for example, be too hard for blunt minds to understand. Wolff also knew only too well that the mathematical way of thinking could be very demanding, in particular when entering the realm of integral calculus and algebra (on mathematics, see Goubet, 2007). Nevertheless, he managed to find a solution to give a foretaste of perfection to all men, whatever the cultivation of their higher faculties might be, when he sketched the first outlines of what could be an aesthetic pleasure taken in contemplating ordered objects, like well-designed clocks or well-built castles. In the pleasure taken in the revelation, another common channel for enjoying perfection was also indicated, which could help to live a better life, according to Christian principles (Wolff, 1730–1731/1983c) or, more generally, to natural moral laws (Wolff, 1980, §.10). Wolff's philosophy, even if it culminates in the affirmation of an intellectual intuition owned only by God, leaves open human intuitive paths to cultivate oneself and to come closer to one's destiny. This is surely a major feature of Wolff's imposing system: it plants plenty of seeds of truth, each of which will grow for its own sake and find someone to enjoy it.

Furthermore, the fact that we are humans and do not have pure reason but possess mixed knowledge (Paccioni, 2006), and that we cannot have an intellectual intuition of all that is possible, is reflected in the discursive realization of the philosopher Wolff. Writing a whole range of metaphysical chapters or books first in German, and then in Latin, he introduced some shifts in the structure and also implemented some changes in the detail. But the succession of rich treatises was not enough, and Wolff himself felt the urge to publish short articles in which he presented his views in a new light, making new connections, such as between truth, order, and perfection in relation to pleasure in *De voluptate ex cognitione veritatis percipienda* (On Pleasure Taken in Knowledge of Truth).

11.7 Conclusion

Let us sum up the major achievements of Wolff's philosophy regarding psychology and the other parts of metaphysics. As we have seen, a key notion such as transcendental truth—i.e., metaphysical and not just logical truth—makes explicit the connections between different parts of the system. But this paradigmatic case is not isolated, for Wolff's desire to establish links between all parts of metaphysics was constant.

The two major ontological principles are also valid for the psychological realm. To perceive something is to perceive something possible, thus not impossible: non-contradiction is essential to representation too. And a perception of something which would not be the consequence of something else, as if we constantly were asleep and dreaming, makes no sense at all: there must be a reason why something exists rather than not. Our representation of the world must abide by these two major principles. Not just some key concepts, such as those of force or perfection, are taken from ontology, but the supreme laws of the soul refer to the first part of philosophy as well.

What does the soul perceive? Not just scarce things, separate local realities, but a possible world, with its internal connections. Psychology does not just follow a doctrine which could have been called a somatology, or doctrine of the bodies, but is related to something more complex, a cosmology, which as such deals with the idea of a universe. In his *Cosmologia* (Cosmology), Wolff begins to treat the world before considering the bodies. Reason is a very important faculty because it gives insight into the structure of things, the connection of the universe. Moreover, systematicity appears to be very important in Wolff's philosophy because it is the only way to look reasonably at all things in general in so far as they are all connected to each other.

When we are acquainted with empirical psychological notions such as reason and free will, we can begin to have some idea of God. With a rational psychological notion like spirit, we can also understand the proximity between Him and us, and we can act according to our wise destiny. Psychology functions like a ladder to theology. But there is a relation between theology and psychology: when we make use of our reason to investigate the truth of the universe, we take pleasure in its order and perfection. And what is the world, according to Wolff, if not God's own creation? The truth present in the universe, in the interrelation between all things, finally points to God as its architect. The world is the mirror that makes us know something of Him and His supreme wisdom.

There are constant links between different parts of Wolff's metaphysics. Even if he had to write in a discursive order, treating each matter at its appropriate time, it is quite obvious that Wolff's work had a plan, that it was intended to be rich and connected; in a single word, that it was systematic.

References

Arnaud, T. (2001). Le critère du métaphysique chez Wolff. Pourquoi une Psychologie empirique au sein de la métaphysique? *Archives de Philosophie, 65*, 35–46.

Carboncini, S. (1991). *Transzendentale Wahrheit und Traum. Christian Wolffs Antwort auf die Herausforderung durch den Cartesianischen Zweifel*. Frommann-Holzboog.

Corr, C. A. (1997). Introduction. In C. A. Corr (Ed.), *Christian Wolff: Gesammelte Werke* (Deutsche Metaphysik) (Vol. 2, Pt. II, pp. 1*–47*). Olms. (Original work published 1751).

École, J. (1990). La métaphysique de Christian Wolff (2 tomes). In J. École (Ed.), *Christian Wolff: Gesammelte Werke* (Vol. 12, Pt. III). Olms.

Fabbianelli, F. (2007). Tatsachen des Bewußtseins und 'nexus rerum'. In J. Stolzenberg & O.-P. Rudolph (Eds.), *Christian Wolff: Gesammelte Werke* (Part 2. Christian Wolff und die europäische Aufklärung) (Vol. 104, Pt. III, pp. 355–366). Olms.

Favaretti-Camposampiero, M. (2011). Der psychotheologische Weg. Wolffs Rechtfertigung der Gotteserkenntnis. *Aufklärung, 23*, 71–96.

Feuerhahn, W. (2003). Die Wolffsche Psychometrie. In O.-P. Rudolph & J.-F. Goubet (Eds.), *Die Psychologie Christian Wolffs. Systematische und historische Untersuchungen* (pp. 278–292). Niemeyer.

Goclenius, R. (1980). *Lexicon philosophicum quo tanquam clave philosophiae fores aperiuntur*. Olms. (Original work published 1613).

Goubet, J.-F. (2007). Das Verhältnis zwischen mathematischer Methode und Logik. In J. Stolzenberg & O.-P. Rudolph (Eds.), *Christian Wolff: Gesammelte Werke* (Part 2. Christian Wolff und die europäische Aufklärung) (Vol. 104, Pt. III, pp. 141–152). Olms.

Goubet, J.-F. (2018). Rationale Psychologie. In R. Theis & A. Aichele (Eds.), *Handbuch Christian Wolff* (pp. 153–174). Springer.

Kant, I. (1912). Nachricht von der Einrichtung seiner Vorlesungen in dem Winterhalbenjahre von 1765–1766. In Königlich Preussische Akademie der Wissenschaften (Ed.), *Kant's gesammelte Werke (Band II)*. Reimer. (Original work published 1765).

Lalanne, A. (2011). *La recension des Essais de Théodicée dans les Acta Eruditorum par Wolff: Une reconstruction systématique?* Traduction inédite et présentation de la recension à l'occasion du tricentenaire de sa publication. Retrieved June 3, 2019, from http://www.philosophiedudroit.org/lalanne, leibniz.htm

Mei, M. (2011). Notio intellectus divini quomodo prodeat. Eine Untersuchung über die gnoseologische Bedeutung unserer Gotteserkenntnis. *Aufklärung, 23*, 97–122.

Moreau, P.-F. (2002). Wolff et Goclenius. *Archives de Philosophie, 1*(65), 7–14.

Paccioni, J.-P. (2006). *Cet esprit de profondeur. Christian Wolff, l'ontologie et la métaphysique*. Vrin.

Rumore, P. (2018). Empirical psychology. In R. Theis & A. Aichele (Eds.), *Handbuch Christian Wolff* (pp. 175–196). Springer.

Schneiders, W. (1983). Deus est philosophus absolute summus. Über Christian Wolffs Philosophiebegriff. In W. Schneiders (Ed.), *Christian Wolff. 1679–1754. Interpretationen zu seiner Philosophie und deren Wirkung* (pp. 85–92). Felix Meiner.

Simmert, S. (2018). Kosmologie. In R. Theis & A. Aichele (Eds.), *Handbuch Christian Wolff* (pp. 197–218). Springer.

Theis, R. (2018). Theologie. In R. Theis & A. Aichele (Eds.), *Handbuch Christian Wolff* (pp. 219–250). Springer.

Wolff, C. (1718). *Ratio praelectionum Wolfianarum in mathesin et philosophiam universam*. Renger.

Wolff, C. (1962). Philosophia prima, sive Ontologia, methodo scientifica pertractata, qua omnis cognitionis humanae principia continentur. In J. École (Ed.), *Christian Wolff: Gesammelte Werke* (Ontologia) (Vol. 3, Pt. II). Olms. (Original work published 1736).

Wolff, C. (1963). *Preliminary discourse on philosophy in general*. (R. J. Blackwell, Trans. The Bobbs-Merrill Company.

Wolff, C. (1964). Cosmologia generalis, methodo scientifica pertractata, qua ad solidam, inprimis die atque naturae, cognitionem via sternitur. In J. École (Ed.), *Christian Wolff: Gesammelte Werke* (Cosmologia generalis) (Vol. 4, Pt. II). Olms. (Original work published 1737).

Wolff, C. (1968). Psychologia Empirica, methodo scientifica pertractata, qua ea, quae de anima humana indubia experientiae fide constant, continentur et ad solidam universae philosophiae practicae ac theologiae naturalis tractationem via sternitur (New edition emended from the prior, with privileges). In J. École (Ed.), *Christian Wolff: Gesammelte Werke* (Psychologia empirica) (Vol. 5, Pt. II). Olms. (Original work published 1738).

Wolff, C. (1972). Psychologia Rationalis, methodo scientifica pertractata, qua ea, quae de anima humana indubia experientiae fide innotescunt; per essentiam et naturam animae explicantur, et ad intimiorem naturae ejusque autoris cognitionem profutura proponuntur. In J. École (Ed.), *Christian Wolff: Gesammelte Werke* (Psychologia rationalis) (Vol. 6, Pt. II). Olms. (Original work published 1740).

Wolff, C. (1978a). Vernünfttige Gedancken von den Kräfften des menschlichen Verstandes und ihrem richtigen Gebrauch in Ernkänntnis der Wahrheit. In H.-W. Arndt (Ed.), *Christian Wolff: Gesammelte Werke* (Deutsche Logik) (Vol. 1, Pt. II). Olms. (Original work published 1754).

Wolff, C. (1978b). Theologia Naturalis, methodo scientifica pertractata. In J. École (Ed.), *Christian Wolff: Gesammelte Werke* (Theologia naturalis) (Vol. 7, Pt. II). Olms. (Original work published 1739).

Wolff, C. (1980). Christian Wolff's prolegomena to empirical and rational psychology. Translation and Commentary by Robert J. Richards. *Proceedings of the American Philosophical Society, 124*(3), 227–239.

Wolff, C. (1981). Theologia Naturalis, methodo scientifica pertractata. In J. École (Ed.), *Christian Wolff: Gesammelte Werke* (Theologia naturalis) (Vol. 8, Pt. II). Olms. (Original work published 1741).

Wolff, C. (1983a). De Notionibus directricibus et genuino usu philosophiae primae. In Horae subsecivae Marburgenses quibus philosophia ad publicam, privatamque utilitatem aptatur. In *Christian Wolff: Gesammelte Werke* (Horae subsecivae) (Vol. 34.1, Pt. II, pp. 310–350). Olms. (Original work published 1729–1730).

Wolff, C. (1983b). De Voluptate ex cognitione veritatis percipianda. In Horae subsecivae Marburgenses quibus philosophia ad publicam, privatamque utilitatem aptatur. In *Christian Wolff: Gesammelte Werke* (Horae subsecivae) (Vol. 34.1, Pt. II, pp. 167–248). Olms. (Original work published 1729–1730).

Wolff, C. (1983c). De Voluptate ex cognitione veritatis revelatae. In Horae subsecivae Marburgenses quibus philosophia ad publicam, privatamque utilitatem aptatur. In *Christian Wolff: Gesammelte Werke* (Horae subsecivae) (Vol. 34.1, Pt. II, pp. 350–386). Olms. (Original work published 1729–1730).

Wolff, C. (1983d). De Notionibus foecundis. In Horae subsecivae Marburgenses quibus philosophia ad publicam, privatamque utilitatem aptatur. In *Christian Wolff: Gesammelte Werke* (Horae subsecivae) (Vol. 34.2, Pt. II, pp. 150–166). Olms. (Original work published 1730–1731).

Wolff, C. (1996). Ausführliche Nachricht von seinen eigenen Schriften, die er in deutscher Sprache von den verschiedenen Theilen der Welt-Weisheit heraus gegeben auf Verlangen ans Licht gestellet. Die andere Aufl., hin und wieder vermehret. In H.-W. Arndt (Ed.), *Christian Wolff: Gesammelte Werke* (Ausführliche Nachricht) (Vol. 3, Pt. II). Olms. (Original work published 1733).

Wolff, C. (1997). Vernünfftige Gedancken von Gott, der Welt und der Seele des Menschen, auch allen Dingen überhaupt. In C. A. Corr (Ed.), *Christian Wolff: Gesammelte Werke* (Deutsche Metaphysik) (Vol. 2, Pt. II). Olms. (Original work published 1751).

Wolff, C. (2001). Review of his own Cosmologia Generalis. In H. A. Laeven & L. Laeven-Aretz (Eds.), *Christian Wolff: Gesammelte Werke* (Sämtliche Rezensionen in den Acta Eruditorum (1705–1731)) (Vol. 35.4, Pt. II, pp. 2042–2048). Olms. (Original work published 1731).

Wolff, C. (2011). De differentia intellectus systematici & non systematici/Über den Unterschied zwischen dem systematischen und dem nicht-systematischen Verstand. Translation, introduction, and edition by M. Albrecht. *Aufklärung, 23*, 229–301.

Part II
Receptions of Wolff's Psychology

Chapter 12
Development and Diffusion of Wolff's Psychology Through His Disciples and Followers

Sonia Carboncini

12.1 Introduction

Wolff's *Psychologia empirica* is today considered a pioneering discipline not only for the establishment of psychology as an autonomous science but also for the development of metaphysics throughout the eighteenth century. It was not so obvious from the beginning. Among the countless disputes provoked by Wolff's philosophy and the many innovations introduced at a systematic level, the creation of a new discipline seems, at first, to remain in the background. Even if many disciples emphasize the originality of the approach and its usefulness, psychology, while showing some adjustment issues within metaphysics, cannot emancipate itself from it.

A problematic reason may lie in the fact that Wolff introduces two different psychological doctrines—*Psychologia empirica* (1732) and *Psychologia rationalis* (1734)—which investigate the soul in two distinct ways: the former, as consciousness, and the latter, as a science of its faculties. Although he considers psychology, according to its object, a unique discipline treated by two different but complementary points of view (Wolff, 1733/1973, §.90), he is from the very beginning aware of the difficulty of the new approach (Wolff, 1733/1973, §.89, p. 252; Meissner, 1737/1970, p. 464). The relationship between both disciplines and their collocation within metaphysics undergo de facto an evolution, also because of the work of the so-called Wolff's school, whose representatives diverge on which one of the two psychologies represents the major novelty.

The goals of the following chapter are three: (1) to show how the early exponents of Wolff's school elaborate on the topic "psychology"; (2) to highlight the difficulties and problems of this process, especially in relation to the coexistence of two psychological disciplines; and (3) to identify the theoretical areas affected by this

S. Carboncini (✉)
Independent Scholar, San Miniato, Pisa, Italy

S. d. F. Araujo et al. (eds.), *The Force of an Idea*, Studies in History and Philosophy of Science 50, https://doi.org/10.1007/978-3-030-74435-9_12

process, which are relevant not only for the destiny of metaphysics but also for the birth of new disciplines between the end of the eighteenth and the beginning of the nineteenth century. For this purpose, we will provide a brief account of the psychological theories of Wolff's most relevant disciples and followers. That will allow us to assess how the debate on psychology within Wolff's school is widespread and articulate.

12.2 The Topic "Psychology" in Wolff and His School

Carl Günther Ludovici's (1707–1778) entries in Zedler's *Universallexikon* (Ludovici, 1748/2001) represent the first great historiographical reconstruction of Wolff's school. They show us two facts about psychology: (1) Ludovici does not give particular emphasis on this discipline or on its division in empirical and rational, and (2) his detailed list of publications of Wolff's early followers contains only one title on psychology as an independent discipline. The difficulty of the theoretical approach to the new discipline because of its duplicity may be the cause of this apparent flop—a duplicity (maybe dualism) that creates some dissent within the school. Although for Wolff empirical psychology has its foundation in rational psychology, which represents the real challenge (Wolff, 1740/1972), some disciples find empirical psychology more interesting for the following developments.

Certainly, to create such a new discipline as empirical psychology entails significant changes. By placing consciousness as the object of a fresh field of investigation, and by introducing the German term *Bewusstsein* (consciousness), Wolff makes it possible for psychology to emancipate itself from metaphysics. In the same way, the notion of *sensus internus* (inner sense) as a foundation of consciousness develops from the all-inclusive Scholastic concept of *sensus communis* (common sense) toward a "psychologization" of psychology that will eventually lead to the *Bewussteinspsychologie* (psychology of consciousness) of the second half of the century (Galle, 2001).

Bringing experience to the forefront implies, therefore, a break with the Scholastic tradition of pneumatology, understood as the theory of spirits (*Geisterlehre*). "Experimental" psychology means to focus on the soul as an object that can be investigated systematically, like the body in physics (Wolff, 1740/1983a, §.111). The attribution of cognitive dignity to inner states is a relevant step toward an anthropological conception of knowledge. The in-depth study of psychic processes enables psychology to be the preparatory discipline to all the others and, therefore, fundamental for the birth of modern pedagogy.

To sum up, Wolff's psychology introduces and elaborates, through the work of scholars and followers, important suggestions for at least four new disciplines: psychology, aesthetics, anthropology, and pedagogy. The dualism between empiricism and rationalism that pervades Wolff's system becomes not only emblematic but also theoretically relevant in psychology, showing original directions for the post-Wolffian philosophy.

The contemporary disciples seem not to take up the challenge immediately. Engaged as they are in the various disputes unleashed against Wolff's alleged atheistic theses, they consciously avoid, most of the time, lingering on more critical issues. They commit themselves to build a monolithic defense of the teacher's doctrines. By doing so, they manage Leibniz's (1646–1716) legacy in a way that emphasizes the interpretative contribution and originality of Wolff's philosophy, the only one capable of expressing consistently and systematically the metaphysics of the genius of Hanover.[1] At least, this is what emerges from the historiographical vulgate. The overall picture may reserve some surprises.

Wolff himself is at the beginning uncertain about the role of psychology in the systematics. Still influenced by the tradition of German Aristotelianism and particularly by Scharff (1595–1650), he defines it as *Pneumatologie* or *Geister-Lehre* (Wolff, 1713, §.12).[2] It will take 20 years (1713–1733) to elaborate on this issue.

The Latin term *psychologia* first appears in writings around 1520 (Luccio, 2014). Since then, it occurs mostly in two major areas: in medical treatises, concerning the physical nature of the soul, and in theological treatises on the soul as a spiritual substance. Although not used sporadically, between the end of the sixteenth and throughout the seventeenth century, the term seems to be ignored by the philosophers, who investigate cognitive processes (Luccio, 2014). While for Leibniz the word psychology is so difficult to write, "that the pen refuses" (Couturat, 1961, p. 526), Wolff overtakes both traditions, introducing a new doctrine and attributing to it, as to every valid discipline, the status of a science.

Yet Couturat's opinion cannot be fully shared. Leibniz anticipates the conceptual terms of the problem that Wolff and his followers deal with: "the original substance is the Monad and its doctrine *psychologia*. *Psychologia* is twofold. One concerns the percipients (senses, etc.), the other the intelligents or the spirits and can be called *pneumatologia*, for it concerns the minds and principally ours" (Couturat, 1961, p. 526).[3] Eventually, *psychologia* is for Leibniz just another name for his theory of the monads. Since he is not too interested in systematics, the name of the discipline does not matter at all.

[1] See my introductions to Leibniz (1720/2008, pp. 1*–12*) and Leibniz (1740/2010, pp. 5*–16*).

[2] Wolff studied on Johann Scharff and certainly knew his *Pneumatologia* (1629), in which this discipline is defined as "a natural science of what we can know of spiritual substances, their nature, properties and operations" (Scharff, 1629/2014, p. 1). It is an autonomous doctrine, belonging neither to mathematics, nor to physics, nor to metaphysics, since the latter is a universal science and therefore cannot deal with particular things (p. 3). Pneumatolgy is a science, as it uses the demonstrative method, based on the *lumen naturale* (natural light) (p. 4) and is of great utility not only for natural theology, since it investigates the human spirit (*mens*) as the noblest of all research's fields (p. 7).

[3] This fragment from *Encyclopaedia arcana*, dated by Couturat after 1696, concerns the philosophical systematics. We know that the very subject of philosophical systematics was at the center of the last interview between Leibniz and Wolff, which took place in July 1716, about 2 months before Leibniz's death. About this last meeting, see Carboncini (2005, p. 23).

For Wolff, instead, terms matter: they have an instrumental function.[4] Psychology is title, discipline, and method: something new that enables the philosophical process. By introducing an essential part of German philosophical language, Wolff redefines the conceptual apparatus deriving not only from Aristotelianism but also from Cartesianism and Leibniz.[5] If in his first vernacular works, starting from the *German Logic*, he uses the term *Geister-Lehre* to define the matter (Wolff, 1754/1978, §.12), in the *Ausführliche Nachricht* (Detailed Information) (1726), he introduces the German term *Psychologie* and uses it with absolute ease (Wolff, 1733/1973, §.79).

Wolff's idea of a popular work involves method as much as terminology. Metaphysics is a demonstrative course, which begins with the question: how do we know that we exist and what do we need this knowledge for? Self-consciousness is the methodological approach to the various steps, which he sorts out starting from the first foundations of our knowledge of things (ontology), progressing to our experience of ourselves (empirical psychology), to the connection of all things among themselves (cosmology), to the essence of the soul as a spiritual entity (rational psychology), and finally to God (natural theology). Empirical psychology precedes cosmology which is followed by rational psychology. Because of the very nature of the treaty, Wolff recommends reading it entirely (Wolff, 1751/1983b).

Regarding psychology, he is aware of dealing with a discipline that, beyond the systematics, has the conditions to be treated independently. "Empirical psychology is specifically a history of the soul and can be known even without the other disciplines; rational psychology, instead, presupposes knowledge of cosmology" (Wolff, 1733/1973, §.79, p. 231). Such a definition may refer to the distinction between historical, mathematical, and philosophical knowledge (Wolff, 1740/1983a, §§.1–20).[6] According to it, the knowledge achieved through sensible experience should be equivalent to the historical one, a simple description of facts. But "historical," related to psychology, means more. It is an approach that shows *how* the faculties of the soul work and, therefore, have its philosophical dignity (Feuerhahn, 2002). Empirical psychology contemplates through psychometrics even the possibility of the mathematical knowledge of the soul (Wolff, 1738/1968, §.522). It could rise therefore to a scientific level that would make its autonomy plausible, at least from a pragmatic point of view, considering the contribution of psychology to practical life.

[4]This conception is very well expressed in the words of the Wolffian author Georg Heinrich Riebow (1703–1774): "we reason not like grammarians starting from the etymology of words, but as philosophers based on definitions" (Riebow, 1726/2002, §.101, p. 126).

[5]It is worth to remember here that the Latin and German translations of Leibniz's most important metaphysical work, *Monadology*, were managed entirely by Wolff's disciples, in close agreement with the master. See Carboncini (2005).

[6]Johann Nicolaus Frobes (1701–1756) seems to validate this interpretation (Frobes, 1734/1998, pp. 11–26). Ludovici later translates *psychologia empirica* as *Seelen-Geschichte* (soul's history) (Lucovici, 1748/2001, p. 913).

Wolff and his followers constantly refer to the word "experience" in psychological context. But what exactly do they mean? Experience (*Erfahrung*) "is the knowledge we come to when we deal with our sensations and changes in consciousness" (Wolff, 1751/1983b, §.325, p. 181). It differs from experiment (*Versuch*) because in experience sensations are given by themselves, while in the experiment we provoke them. Empirical psychology is not just an experimental science, but it is for Wolff much more. It is a fundamental approach to achieve a purpose in practical life, which is Wolff's primary concern with his German writings.

He declares it openly in *German Metaphysics*: empirical psychology shows us how through the self-awareness exercised by the soul we get concepts on which we base both knowledge and will. Therefore, it is a fundamental discipline for logic, morals, and politics (Wolff, 1751/1983b, §.191). This is a very important step toward the birth of what we call scientific psychology: an empirical science of conscious representations. But on a systematic level, this pragmatic or "historical" (Wolff 1733/1973, §.79, p. 231) approach, however important, is not compatible with the rationalistic framework to which Wolff remains faithful. It does not account for rational psychology. The duplicity of psychology reflects the duplicity of its object. Rational psychology deals with the soul as a set of faculties, while empirical psychology investigates the soul as the thinking subject, identified with consciousness.

Ordo rei (intrinsic order) and *ordo doctrinae* (methodological order) seem not to coincide, as they should. Therefore, the position of psychology within the metaphysical system changes during the transition from German works to the Latin. This mobility is not only due to systematic needs but also to an internal evolution of Wolff's philosophy, a process that will have considerable importance in the subsequent development of psychology as a discipline.

12.3 Ludwig Philipp Thümmig (1697–1728)

The first of Wolff's followers to introduce psychology is Thümmig with his *Institutiones philosophiae Wolffianae* (Institutions of Wolff's philosophy, 1725–1726), a work written in agreement with his master to pursue a cultural strategy. Wolff's German works have a sensational impact on the public not only because they address readers even outside the academic field (Wolff, 1733/1973, §.16) but also because they introduce a new terminology. But once the new philosophy is on everyone's lips, it becomes necessary to translate it into the academic language which continued to be Latin. There are at least three reasons for this change: to clear the field from linguistic misunderstandings in disputes triggered by Pietist adversaries; to secure the primacy of Wolff's philosophy within the academic world, providing material for the various teachings; and finally, to export Wolff's ideas outside of Germany. Probably influenced by Descartes (1596–1650), but much more by Leibniz, who never taught at a university, Wolff brings the outside world into the university with the enormous appeal of his German teachings. Afterward, he institutionalizes his philosophy with Latin treatises and finally exports it throughout the known world. Thümmig's *Institutiones* represents the second step of this project (Thümmig, 1725–1726/1982).

Wolff is therefore very careful in controlling the spread of his ideas at critical points. The terminological distinction between empirical and rational psychology (and the term psychology itself) appears neither in *German Metaphysics* (1720) nor in the *Anmerkungen* (Annotations) (1724). But the conceptual distinction already exists. According to the distinct points of view, he treats psychology in separate sections of the entire work. He titles the third chapter: "Of the soul in general, what we perceive of it." The fifth has the title: "Of the essence of the soul and of a spirit in general." In-between he places cosmology, as we have seen. Wolff uses periphrases in "popular" works to avoid terms that can evoke Scholastic philosophy.

Thümmig's goal with the *Institutiones* is to introduce the Latin terminology, reporting faithfully Wolff's definitions: "Psychology is the science of the soul or that part of Metaphysics which deals with it" (Thümmig, 1725–1726/1982, §.1, p. 115). Wolff "divides it into two parts, of which one is empirical (*empiricam*), the other is rational (*rationalem*)" (§.2, p. 115). Similar to Wolff's *German Metaphysics*, psychology is the most consistent part of Thümmig's treatise. But Thümmig changes the systematic order of the disciplines to reinforce deductive consistency. *Psychologia empirica* and *Psychologia rationalis* follow one another at the third place, after ontology and cosmology, and before natural theology.

Wolff adopts this new order in his Latin treatises. But the change does not involve a modification in any of the many new editions of the *German Metaphysics* for two reasons. First, he refuses to change the structure of his fundamental work to maintain the references with the other writings, which is why he entrusts a separate volume, the *Anmerkungen*, with possible adjustments and additions. Already at this stage, he is aware of drawing up a philosophical encyclopedia, in which the network of references is essential. The second reason is both pedagogical and methodological. The purpose of *German Metaphysics* is to treat the matter in a simple and clear way, making it understandable to everyone, even those who have no academic background and ignore Latin (Wolff, 1733/1973, §.16). Even if metaphysics is difficult to understand, following the method proposed by "rational thoughts," anyone can rise to the challenge (§.19).

That Wolff can't change the systematic order in his German fundamental work does not mean that he ignores the problem. He acknowledges Thümmig's reason for placing cosmology before psychology: the world, understood as the order of simultaneous and successive things connected to each other, is the conceptual context in which the mind itself exists. But if psychology is divided into two parts and empirical psychology holds its methodological autonomy in relation to other disciplines, from a didactic point of view, it is easier for the reader to deal with it before cosmology. After the abstractness of ontology, psychology confronts us with things we have daily experience of (Wolff, 1733/1973, §.79). This approach remains valid only for *German Metaphysics*, as Wolff hastens to specify (Wolff, 1751/1983b). In the more detailed Latin treatises, already announced in the Preface to the second edition, the order will be the same as in Thümmig's *Institutiones*.

12.4 Johann Peter Reusch (1691–1758)

Similar to Thümmig, who was the most faithful of Wolff's students, many followers publish compendia for schools and academic lessons, a habit that continues for decades (up to Kant and beyond) throughout Europe. Such an apparently less original, derivative production is important for the diffusion and refinement of terminology and concepts. Reusch, Wolff's former student in Halle and a professor in Jena, presents in his *Systema Metaphysicum* (1734) succinctly and coherently the doctrines of the philosophical "duumvirate" (Leibniz and Wolff). About empirical psychology, he affirms that no one "before Wolff had ever exposed more accurately what concerns the soul from the point of view of experience" (Reusch, 1735/1990, *Preface*, p. 4). For this reason, he grants much space to it, while, regarding rational psychology, he candidly admits that if he had not committed himself to expose Wolff's system exhaustively, he would have omitted some parts of it.[7]

12.5 Georg Bernhard Bilfinger (1693–1750)

Different is the case of Bilfinger, Wolff's disciple and popularizer from the first hour. In the same year as Thümmig (1725), he publishes a Wolffian textbook on metaphysics, *Dilicidationes philosophicae de Deo, anima humana, mundo* (Philosophical explanations of God, the human soul, the world), which secures him great popularity and prestigious assignments in Saint Petersburg and at the court of the Duke of Württemberg. In this work, he introduces psychology after cosmology, like Thümmig.[8] But interestingly, we find in Bilfinger no trace of the division between the two psychological disciplines. Even if he does not use the specific term, he means psychology as "experimental" in Wolff's specific meaning.

Bilfinger's text is enlightening because, far beyond the obsequious re-proposal or even trivialization of Wolff's text, it tries to create a synthesis between the two disciplines. He defines psychology as "the science of the human soul, since what we know of it through experience can be legitimately deduced and understood by a general concept" (Bilfinger, 1725/1982, §.233, p. 177). He brings together the a posteriori element determined by the analysis of consciousness with the a priori

[7] Within the crowded circle of Wolffian authors listed by Ludovici, the positions are diverse, moving from complete orthodoxy to more differentiated theories. Among the first, it is worth mentioning the Saxon Protestant Pastor Adam Heinrich Meissner (1711–1782), who publishes in 1737 the famous *Philosophisches Lexicon*. In the entry "Psychologia," he reports exhaustively definitions and concepts drawn from Wolff's German works, particularly the *Ausführliche Nachricht*. Johann Friedrich Stiebritz (1707–1772) publishes a Latin summary of metaphysics. Less orthodox is Samuel Christian Hollmann (1696–1787), active in Göttingen, a very successful author who influenced Johann Georg Heinrich Feder (1740–1821), helping to carry out Wolffian influences into the *Popularphilosophie* (popular philosophy).

[8] As Wolff observes in 1733/1973 (§.79).

aspect given by the foundation of the whole procedure in the soul's essence. Bilfinger reaffirms with the support of a large doctrinaire apparatus the originality of Wolff's contribution: ancient philosophers mix physics with pneumatology (§.232), while Wolff is the first one to deduce from a general concept of the soul all its faculties (§.234). The emphasis on the element of novelty shifts from empirical to rational psychology. Wolff seems to take the suggestion, and later on, in his *Psychologia rationalis* (1734), he expresses himself about this important accomplishment with the same words (Wolff, 1740/1972).

Furthermore, Bilfinger puts the spotlight on the definition of the soul as the "power to represent the universe from the position of its body in the world," which Wolff first provides in the rational psychology's part of *German Metaphysics* (Wolff, 1751/1983b, §.753, p. 468). So far, this happens not accidentally in the expository process of Wolff's metaphysics: topic and terminology intersect here with the body-mind problem. Bilfinger's definition of the soul implies, besides the modality of its union with the body, more determinations. For Bilfinger, in the soul, there is one power (*vis/Kraft*) and multiple faculties (*facultates/Wirkungen*); the power is related to inner data (*Bewusstsein/consciousness*), while the faculties are oriented toward the objects of our representations. Such a synthesis between Leibniz and Wolff we find again in Baumgarten (1779/1963, §.513).

Explaining in psychology the first five faculties, through the action of which the soul represents the world (sensation, imagination, attention, abstraction, and memory), Bilfinger wonders whether a discipline would be necessary to regulate them, such as Aristotle's logic in relation to the field of intellectual knowledge (Bilfinger, 1725/1982, §.268). He does not distinguish the lower cognitive faculties from the higher ones, but he advocates the introduction of a new discipline for the former, anticipating Baumgarten's *analogon rationis* (analogous to reason) and, with it, aesthetics.

Through Bilfinger's synthesis, we get the complete definition of soul, without jumping from one chapter to another of the *German Metaphysics* or from the 700 pages of the *Psychologia empirica* to the almost as many of the *Psychologia rationalis*: "the soul is the representative substance of the universe according to the position of its body" (Bilfinger, 1725/1982, §.288, p. 214).

To sum up: long before the publication of Wolff's Latin treatises, Bilfinger tries to deal with the duplicity in the consideration of the soul. Once again, we face the fact that within Wolff's school, the influence is not unequivocal from master to disciple, but it works also in reverse. All this terminological and conceptual elaboration and the dialectic between master and followers provide important material to the development of the concept of "I" as subjectivity. For this reason, psychology conceived as an investigation into the experience of the subject and aesthetics, as its further spin-off, set a turning point from *Aufklärung* (Enlightenment) to *Romantik* (romantism). The role of Wolff's first disciples in the elaboration of this process was so far underestimated. It is beyond doubt that they not only supported but even helped the master to develop and clarify his doctrines.

12.6 Friedrich Christian Baumeister (1709–1785)

Rector of the Görlitz gymnasium throughout his life, Baumeister meets at the University of Jena Heinrich Köhler (1685–1737),[9] to whom he owes his strong vocation for synthesis between Leibniz and Wolff. Knowing not only Wolff's texts but also those of Thümmig and Bilfinger, he attempts to contextualize them in the philosophical debate and to merge the two psychological disciplines together. Interestingly, he defines psychology with the same formula used by Wolff for rational psychology: "a science of things made possible by a human soul" (Baumeister, 1735/1978, §.8, p. 4; Wolff 1740/1972, §.1, p. 1). In his *Institutiones Metaphysicae* (Institutions of metaphysics) of 1738, he reunites psychology and natural theology under the title of "Pneumatology" and places them after cosmology, similar to Scharff (Baumeister, 1738/1988, pp. 331–334). According to Wolff, he divides psychology into empirical and rational "since what we know of the soul, we know it either through experience or by correct reasoning from experience data" (Baumeister, 1738/1988, §.479, p. 335). However, in his *Elementa Philosophiae Recentioris* (Elements of newest philosophy) (1747), a book reprinted for 40 years, he later reconciles the separation and treats them together in what he calls the "sweetest marriage" of experience and reason (Baumeister, 1747/2003, §.177, p. 237). In his *Philosophia Definitiva* (Philosophical definitions) (1733), he presents psychology as a single discipline, starting from a dual approach: "the complete notion of spirit includes not only simplicity or immateriality, but also intellect and will" (Baumeister, 1735/1978, p. 268). This passage represents paradigmatically that synthesis between Leibniz and Wolff operated under Köhler's influence.

12.7 Alexander Gottlieb Baumgarten (1714–1762)

Such a strong mediation characterizes the subsequent phase of the popularization of Wolff's psychology carried out by the second generation of students. Among them, Baumgarten is the dominant figure. His *Metaphysica* (1739) is, in fact, a commentary with notes on the teacher's lessons. But the direction taken by Baumgarten is original compared to that of his predecessors. Also for Baumgarten, as for Wolff's other disciples, psychology represents the most conspicuous part of metaphysics and systematically comes after cosmology and before natural theology. He divides it into empirical and rational; pneumatology, as a doctrine of nonhuman spirits, almost disappears from it.

---

[9] Köhler was Leibniz's secretary and copyist in the time when the latter was working on his texts on monads. He translates into German and publishes in 1720 Leibniz's *Principes de la Nature et de la Grâce* with a Preface by Wolff (Leibniz, 1720/2008; 1740/2010). In 1721, an anonymous Latin translation appears in the *Acta Eruditorum.* See Carboncini (2005).

Psychology is defined as the "science of the general predicates of the soul" (Baumgarten, 1779/1963, §.501, p. 173) and provides principles not only for logic, theology, and practical sciences, as it had been so far for all of Wolff's followers, but also for a new discipline: aesthetics (§.502). Baumgarten thinks out aesthetic during his years as a student in Halle. The term appears for the first time in his *Meditationes philosophicae de nonnullis ad poema pertinentibus* (Philosophical meditations on a few things concerning poetry), published in 1735 (Baumgarten, 1735/1993). The new discipline takes its steps there, where Leibniz's metaphysics is integrated by Wolff's empirical psychology. It starts from the acknowledgment that consciousness is the characteristic (*Merkmal*) of thought (Wolff, 1751/1983b, §.195). But the whole content of the soul does not correspond to consciousness: they do not coincide perfectly, as assumed by Descartes (Wolff, 1751/1983b, §.193; Bilfinger, 1725/1982, p. 182). There is a preconscious (to say it with Leibniz) or unconscious (according to the Wolffian terminology) area to which we cannot apply the concepts of logic. Darkness (lack of grasping and defining the differences) suspends consciousness (Wolff, 1751/1983b, §§.729–731). Wolff is not afraid of this dimension and does not conceal it; he just declares psychology not the right place to deal with it. Neither is logic (Wolff, 1740/1983c, §.31).

Baumgarten calls it *cognitio sensitiva* (sensible knowledge) and expands its relevance field. Every knowledge starts with experience. But the senses may perceive something that does not reach distinction and remains within the domain of the lower cognitive faculty. Baumgarten's *repraesentatio sensitiva* (sensible representation) originates from the *appetitus sensitivus* (sensual desire) (Wolff, 1738/1968, §.580); the sphere of the will follows, therefore, the cognitive dimension to a confused level. To keep both dimensions together is the role of the concept of perfectibility, linked to happiness, which pervades Wolff's philosophy. According to it, "the perfection of the soul consists in the conscious use of all the faculties of the soul, the higher as well as the lower ones" (Wolff, 1750/1970, §.15, p. 20).

Baumgarten is looking for a theoretical context to investigate sensibility, to consider that *fundus animae* (bottom of the soul), which psychology has discovered while investigating the lower cognitive faculties of the soul. Wolff merely analyzes the process of the single representation toward degrees of clarity and distinction; Baumgarten, who is interested in the totality of representations within the soul, resorts to monads (Lorini, 2014, p. 111). Officially rejected by Wolff, Leibniz's theory of monads plays nevertheless an important role in psychology, since the soul is a simple substance endowed with the strength to represent the world from the point of body position (Wolff, 1751/1983b, §§.599–753). The internal dialectic of the Leibniz-Wolffian philosophy, interpreted for a long time as a limit, has proven itself in recent studies to be one of the most interesting parts regarding the reception of both philosophers.

As for Leibniz, thoughts can be clear or obscure, distinct or confused. In clear but confused perceptions, however, it lurks something which we are not aware of, but which aspires to become object of our consciousness (to become apperceptions). The *petites perceptions* (small perceptions) push toward awareness and are, already in Leibniz, submitted to the sole judgment of taste (Leibniz, 1890/1978a,

§§.46–48; 1890/1978b, §.423). And while Baumgarten excludes the obscure perceptions from the process of knowledge, he makes the confused ones, spotted by psychology, the object of the new discipline of sensibility. They make up the deep ground of the soul, and psychology deals with them while investigating the lower faculties of the soul (Baumgarten, 1750/1961, §.80). The *fundus animae* with all its strength and impetus, with its vitalistic impulse, strives for consciousness and finds its theoretical field: "Aesthetics (theory of liberal arts, inferior epistemology, the art of thinking beautifully, the art of the analogy of reason) is the science of sensible knowledge" (§.1, p. 1).

12.8 Georg Friedrich Meier (1718–1777)

Georg Friedrich Meier, Baumgarten's student and successor in Halle, reaffirms the powerful connection between psychology and aesthetics, the former being the foundation of the latter (Meier, 1765/2007, §.477). The fine arts (*schöne Künste*) are therefore nothing but an instrument to improve the lower cognitive faculties of the soul. The conceptual range of the *fundus animae*, acknowledged but kept limited by Wolff, increases from Baumgarten to Meier to such an extent that Meier's definition of it in his *Metaphysik* is no longer expressed only in negative terms as the locus of unconscious knowledge, but also as the foundation of all knowledge. We are facing a dynamic model: we perceive some things clearly, others confusedly, others obscurely; but sometimes the same object presents itself to us differently—now clearly, now in a rather confused or obscure way (§.485).

With Meier, psychology becomes increasingly an anthropological discipline. It deals with the human soul, a thinking substance joined with a body. Pneumatology, defined as a "philosophical novel" (Meier, 1765/2007, §.472, p. 7), is drastically reduced to the doctrine of nonhuman spirits: nine paragraphs at the end of the book. As we can only know of finite spiritual substances, it is absurd to begin with what we do not know, as in the *Geister-Lehre* of the tradition of Protestant Aristotelianism. Only if we start from what the experience makes us know about our soul can we understand the characteristic elements common to all thinking substances. Four-fifths of the complete treatise deal, in Meier's words, with "experimental or empirical" psychology (*erfahrende oder empirische Psychologie*). However, the reduced dimension of rational psychology (*vernünftige Psychologie*) does not yet mean overcoming it. Meier recognizes its difficulty and abstractness but, like Baumgarten, cannot find another way to move from the particular, from what we have known about the soul in empirical psychology, to the universal, to the nature and essence of the soul, investigating the motives and modalities of its changes (§.474).

Meier also seems not to avoid that circularity that affects Wolff's psychology, according to which empirical psychology needs the validation of rational deduction at the same time that the latter needs the materials of experience (Lorini, 2014). This circularity reveals that psychology is a passing discipline, a breaking point of the so-called metaphysical dogmatism. In fact, in all theories we have reported so far, a question remains open, which Wolff probably tries to answer by changing the position of psychology over the years: can an empirical discipline take place in metaphysics? For Meier, the answer is affirmative: "Tearing out psychology from metaphysics and treating it as a particular discipline would create a great inconvenience" (Meier, 1765/2007, §.471, p. 6). Such a process would end up making it too superficial, for solid knowledge needs ontological and cosmological concepts. Those authors, who treat it as a branch of natural sciences, consider the soul in the same way as bodies.

If kept within metaphysics and handled with the strictness of science, psychology can be for Meier of utmost utility. Exact knowledge (*genaue Erkenntnis*) of our soul is the only way to get an adequate knowledge of God and, therefore, is essential both for natural and revealed theology (§.475). It also provides us with the principles of morality (§.476). Finally, psychology gives the foundations of logic, as it explains how our knowledge works (§.478). So far, the list of disciplines for which psychology is indispensable is not very different from Wolff's. But Meier was Baumgarten's friend and disciple and took over the breakthrough proposed by the master.

A further useful element of psychology concerns the foundation of fine arts and sciences (*schöne Künste und Wissenchaften*). These arts reflect their purpose in that beauty which depends on the perfection of the lower cognitive faculties of our soul and their regular use (Meier, 1765/2007, §.477 p. 15). While sharing Baumgarten's theories on sensible knowledge, Meier seems, however, to circumscribe more clearly the role of aesthetics as a particular discipline. Both understanding (*Theorie*) and exercise (*Ausübung*) of the fine arts live in the soul and in the action of its faculties. Beauty originates in sensible knowledge, but this is a too vast area that risks exerting dominant power over the soul. It requires a specific discipline to prevent this: aesthetics. It provides the rules for perfect sensitive knowledge and its presentation (Meier, 1765/2007, §.527).

12.9 Johann Georg Sulzer (1720–1779)

A Swiss theologian and scientist, Sulzer was a well-known gymnasium teacher in Berlin and a member of the local Academy of Sciences. Author of essays on aesthetics, including the famous *Allgemeine Theorie der schönen Künste* (General theory of fine arts) (1771–1774), he shares with Baumgarten and Meier the intuition of the implicit potentials of Wolff's psychology. Thus, we find in his *Kurzer Begriff aller Wissenschaften* (Concise notion of all sciences) (1745) not only a Wolffian definition of psychology and its division (Sulzer, 1786, §§.203–204) but also the claim of the importance of empirical psychology for a correct knowledge of the soul. Knowledge that requires sharpness and attention, because some changes happening in the soul are so obscure and sudden that it is difficult to notice them, hidden as they are in the "depth of the soul" (§.205). Sulzer recognizes Wolff's undeniable role in laying the foundations for a thorough knowledge of the soul, but he thinks this part of philosophy needs to be expanded and investigated.

Since the knowledge of the soul is the noblest part of the sciences, it is worthwhile that we dedicate ourselves to the development of empirical psychology and in particular to the "dark regions of the soul" neglected by Wolff (Sulzer, 1786, §.206). In such an expansion of perspectives, great attention should be devoted to the study of the diseases of the soul because of the alterations they cause in it. We find here the first opening of Wolff's empirical psychology to medical psychology (§.207). Sulzer does not underestimate the role of rational psychology, which he calls "explanatory" (*erklärende*). The "possibility" (the conditions of thinking) of the immortality of the soul, the improvement of intellect and will, and the principles of morality: all this has its foundation in rational psychology, and so the theory of the education of men (§.209). This last aspect, as we shall see, is very important.

12.10 Benedikt Stattler (1728–1797)

With the Bavarian theologian Stattler, we discover Wolff's influence in a different cultural environment: Catholicism in its more open and innovative fringes. Stattler was, in fact, a combative member of the Society of Jesus, who tries to reconcile rationalist philosophy with Catholic theology. For him, philosophy is the preparatory science for theology, and from this point of view, he composes a monumental work, *Philosophia methodo scientiis propria explanata* (Philosophy explained according to the method of sciences), of which *Psychologia* makes up one volume (1770). Stattler no longer distinguishes between empirical and rational psychology but treats the science of the soul as a philosophical science that mediates the passage to practical disciplines (Stattler, 1770/2011, §.1), including poetic and dramatic ones (§.2). The accent shifts to the truth of the psychic contents of the soul. This attention to psychological processes sets out an influential trend of Wolff's philosophy in the last part of the eighteenth and throughout the nineteenth century: pedagogy. As psychological explanation is the basis of both moral and theological education, psychology ends up becoming a didactic subject.

12.11 Johann Heinrich Campe (1746–1818)

From this point of view, Campe, a tutor at the home of the von Humboldt family, is an emblematic figure. He is the greatest pedagogue of the second half of the eighteenth century and the first author of children's literature in Germany. In his huge production, there is a work with many reprints, *Kleine Seelenlehre für Kinder* (Little doctrine of the soul for children) (1780), published also in French, in which he emphasizes the importance and usefulness of teaching children the basic notions of psychology. Campe's point of view is very interesting because it reproduces the dual function of psychology in a new context and almost 50 years after the publication of Wolff's Latin textbooks. On the one hand, psychology provides indispensable propaedeutic material for teaching religion and morals; on the other, it is an applied method. According to Campe, one of the fundamental principles of education is to promote a balanced development of all psychic faculties and not only of memory. Hence, the use of psychological conversations is the basis of his teaching activity (Campe, 1780, p. 2ff.).[10] Such ideas also influence the educational and scholastic systems of many German states (Jahnke, 1992, p. 84).

12.12 An Anonymous French Translator

Another important chapter in the influence of Wolff's psychology concerns France. We have already pointed out how the lexicographical and encyclopedic vocation underlies Wolffianism throughout and coincides with the systematic and utilitarian need of Wolff's philosophy. Where there is a lot of systematically ordered material, the temptation to make some popular extracts is irresistible. All the more so as in his country, the author has become a martyr for the *libertas philosophandi* (freedom of philosophizing). A cunning marketing operation, carrying in the title page the words "by Mr. Wolf," appears in Amsterdam: *Psychologie, ou traité sur l'âme* (Psychology, or treatise on the soul) (1745). In reality, it is an anonymous summary of the first part of Wolff's *Psychologia empirica*, concerning the *facultas cognoscendi* (cognitive faculty), adapted to the taste of the French public. Such an overview of psychology, called "experimental" to avoid any misunderstanding related to the Greek word "empirical" (Anonymous, 1745/1998, p. 32), evidently satisfies the interests of the author and the curiosity of the public. But despite all the references to Wolff's Latin manual and all praises, the persistent irony, which pervades the entire book, makes me doubt the sincerity of the author. Nonetheless, the work has been a powerful propaganda medium for Wolffian psychology.

12.13 Jean Deschamps (1709–1767)

A Protestant pastor of French origin and Wolff's disciple, Deschamps translates *German logic* into French and is the author of the *Cours abrégé de la philosophie wolffienne en forme de lettres* (Short course of Wolffian philosophy in the form of letters) (1743–1747), in which a volume is dedicated to both empirical and rational psychology. Deschamps, referring to the "imperfection" of the aforementioned *Traité*, proposes to extend psychology without neglecting its rational part, which,

---

[10] Among the numerous pedagogical texts, see also Anonymous (1787), which is a summary of Wolff's *Psychologia empirica*.

according to him, is the most interesting and in which lies the true novelty of the Wolffian approach (Deschamps, 1743–1747/1991). He chooses the epistolary form, more suited to a popular taste.

12.14 Johann Heinrich Samuel Formey (1711–1797)

The perpetual secretary of the Royal Prussian Academy of Sciences, Formey is definitely the most prolific Francophone popularizer of Wolff's philosophy, being convinced that the latter is, not only for its vastness and depth but also for its scientific rigor, a real encyclopedia (Formey, 1741–1753/1983). His numerous writings are translated into many languages, including English, Italian, and Russian. Having probably in mind Madame du Châtelet (1706–1749), Voltaire's (1694–1778) friend and enthusiastic Wolff's supporter,[11] he conceives an all-female philosophical exposition dedicated to women, entitled *La belle Wolfienne* (The Beautiful Wolffian Lady) (1741–1753). As for psychology, he introduces only the "experimental" part, considered the most innovative and useful. But that is not the only reason for his choice. By excluding rational psychology, Formey tries to free the moral sphere from the objections raised against Leibniz's pre-established harmony. Psychology is useful to morals, politics, and logic as it explains the reasons that determine the soul to act and to produce particular movements in the organs of the body. We do not need more. If you want to teach a person to write, you do not have to explain how the hand moves the pen. However, the principles explained in rational psychology are based on experience (Formey, 1741–1753/1983, pp. 6–8).

12.15 The Encyclopedias

The most important encyclopedias of the eighteenth century are notoriously the *Encyclopédie* of Diderot (1713–1784) and D'Alembert (1717–1783) and the so-called Yverdon "Protestant encyclopedia." Both contain many contributions written by Formey or by other authors inspired by Wolff's philosophy and sometimes translated word for word from his texts.[12]

In the anonymous entry "Psychologie" in Diderot's *Encyclopédie*, Wolff is mentioned as the inventor of psychology, a doctrine that provides principles not only for logic, morals, and natural theology but also for natural law. Moreover, it is emphasized that psychology replaces logic as introduction to philosophical knowledge

[11] See du Châtelet (1742/1988, pp. V–XI).

[12] See Carboncini (2007).

(Anonymous, 1765, p. 543). The article also contains Wolff's exact definition of psychology and its division into two parts.[13]

Very similar is the content of the homonymous entry in the Yverdon encyclopedia edited by Fortunato Bartolomeo De Felice (1723–1789). The author of the entry is the Swiss Reformed Pastor Gabriel Mingard (1729–1786), one of the main collaborators of the encyclopedia, a Freemason, and a strong supporter of an anthropological turnaround in knowledge.[14] With only a few more theological concerns, expressed in the introductory part, the content of the article is quite similar to that in the French encyclopedia, particularly in remarking how psychology is the "basis, the principle and the guide of every discipline" (Mingard, 1774, p. 512). Moreover, it is the relationship things have with us that makes them interesting. This anthropological vision of knowledge, based on self-consciousness, represents the great novelty of psychology, a science which, by its nature, "cannot and must not seek principles outside the object of its research" (p. 513). It is therefore necessary to start from the data of experience; and, for this reason, as Mingard writes, Wolff introduces as a first part the "psychologie empirique," which is the ground of rational psychology in the same way experimental physics is the ground of rational and systematic physics.

12.16 Emanuel Swedenborg (1688–1772)

In the second half of the eighteenth century, a Swedish philosopher, scientist, and theologian provokes a certain clamor for his controversial theses, especially regarding psychology, the theory of spirits, and theology, trespassing in mysticism, spiritism, and theosophy. Traveling to Germany in the first phase of his life (1730), Swedenborg comes into contact with Wolff's works, and in particular with the newly published *Psychologia empirica*, and writes a detailed commentary on some paragraphs, first published in 1923.

This circumstance suggests two interesting elements to conclude this exposition. Thinking of Kant's pamphlet *Dreams of a Spirit-Seer Illustrated by Dreams of Metaphysics* (1766), it is ironic to imagine that the critical comparison between Wolff and Swedenborg may have had much more substance than Kant could have possibly known. It is also possible to suppose that some reflections on the unconscious and the parallelism between material and spiritual world, later taken up by Gustav Jung (1875–1961), have also matured in Swedenborg through his reading of Wolff's empirical psychology.

---

[13] The question of the identity of the anonymous author of the article is not yet resolved. In my opinion, it is a compilation article and it is possible that the author used the *Traité*, as Rumore claims (Rumore, 2007, p. 180). But in my opinion, it is more likely that he also used Deschamps' *Cours abrégé* (Deschamps, 1743–1747/1991; see, e.g., pp. 1–7).

[14] See Donato (2016, pp. 105f.).

12.17 Conclusion

This excursus on the direct influences of Wolff's psychology through his disciples and followers must end here. When a philosophy achieves such a great and generalized success, it ends up pervading and influencing all cultural spheres. My goal was simply to show how the concept of an empirical science of the soul and its placement within the metaphysics generated debates within and around the Wolff school, and developed toward the birth of new autonomous disciplines such as psychology itself, aesthetics, anthropology, and pedagogy.

References

Anonymous. (1765). Philosophie. In D. Diderot (Ed.), *Encyclopédie, ou dictionnaire rai-sonné des sciences, des arts et des métiers par une société de gens de lettres* (Vol. XIII, p. 543). Faulche.

Anonymous. (1787). *Psychologia empirica, oder die empirische Psychologie, für die studierende Jugend in den fünf niedern Schulen*. Simon.

Anonymous. (1998). Psychologie ou traité sur l'âme. Contenant les Connoissances, que nous en donne l'Expérience par M. Wolf. In J. École (Ed.), *Christian Wolff: Gesammelte Werke* (Psychologie) (Vol. 46, Pt. III). Olms. (Original work published 1745).

Baumeister, C. F. (1978). Philosophia definitiva, h.e. definitiones ex systemate Wolfii. (New edition). In J. École (Ed.), *Christian Wolff: Gesammelte Werke* (Philosophia definitiva) (Vol. 7, Pt. III). Olms. (Original work published 1735).

Baumeister, C. F. (1988). Institutiones metaphysicae: Ontologiam, cosmologiam, psychologiam, theologiam denique naturalem complexae. In J. École (Ed.), *Christian Wolff: Gesammelte Werke* (Institutiones) (Vol. 25, Pt. III). Olms. (Original work published 1738).

Baumeister, C. F. (2003). Elementa philosophiae recentioris usibus iuventutis scholasticae accomodata. In J. École (Ed.), *Christian Wolff: Gesammelte Werke* (Elementa) (Vol. 83, Pt. III). Olms. (Original work published 1747).

Baumgarten, A. G. (1961). *Aesthetica*. Olms. (Original work published 1750).

Baumgarten, A. G. (1963). *Metaphysica* (7th ed.). Olms. (Original work published 1779).

Baumgarten, A. G. (1993). Meditationes philosophicae de nonnullis ad poema pertinentibus. In T. Gregory (Ed.), *Lessico Intellettuale Europeo* (Vol. 60). Olschki. (Original work published 1735).

Bilfinger, G. B. (1982). Dilucidationes philosophicae de Deo, anima humana, mundo, et generalibus rerum affectionibus. In J. École (Ed.), *Christian Wolff: Gesammelte Werke* (Dilucidationes) (Vol. 18, Pt. III). Olms. (Original work published 1725).

Campe, J. H. (1780). *Kleine Selenlehre für Kinder*. Bohn.

Carboncini, S. (2005). Nuovi aspetti del rapporto tra Christian Wolff e Leibniz. Il caso della Monadologie. In J. École (Ed.), *Christian Wollf: Gesammelte Werke* (Vol. 98, Pt. III, pp. 11–45). Olms.

Carboncini, S. (2007). Das Paradox der Aufklärung. Christian Wolff und die Encyclopédie. In J. École (Ed.), *Christian Wolff: Gesammelte Werke* (Vol. 101, Pt. III, pp. 73–94). Olms.

Couturat, L. (Ed.). (1961). *Opuscules et fragments inédits de Leibniz*. Olms. (Original work published 1903).

Deschamps, J. (1991). Cours abrégé de la philosophie wolffienne en forme de lettres. In J. École (Ed.), *Christian Wolff: Gesammelte Werke* (II.1. Cours abrégé) (Vol. 13, Pt. II). Olms. (Original work published 1743–1747).

Donato, C. (2016). Fortunato Bartolomeno De Felice e l'*Encyclopédie* d'Yverdon: Reti massoniche e cosmopolite nel *transfert* di conoscenza e cultura enciclopedica nel secolo dei Lumi. In S. Ferri (Ed.), *Fortunato Bartolomeo De Felice. Un intellettuale cosmopolita nell'Europa dei Lumi* (pp. 103–140). Franco Angeli.

du Châtelet, M. (1988). Institutions physique. Nouvelle édition. In J. École (Ed.), *Christian Wolff: Gesammelte Werke* (Vol. 28, Pt. III). Olms. (Original work published 1742).

Feuerhahn, W. (2002). Comment la psychologie empirique est-elle née? *Archives de Philosophie, 65*(1), 47–64.

Formey, J. H. S. (1983). La belle Wolfienne. In J. École (Ed.), *Christian Wolff: Gesammelte Werke* (Belle Wolfienne) (Vol. 16.1–16.2, Pt. III). Olms. (Original work published 1741–1753).

Frobes, J. N. (1998). Brevis ac dilucida systematis philosophiae Wolffiani delineatio. In J. École (Ed.), *Christian Wolff: Gesammelte Werke* (Dilucida delineatio) (Vol. 43, Pt. III). Olms. (Original work published 1734).

Galle, R. (2001). Entstehung der Psychologie. In A. A. Glaser (Ed.), *Die Wende der Aufklärung zur Romantik 1760–1820. Epoche im Überblick* (pp. 317–336). John Benjamin Publishing Company.

Jahnke, J. (1992). Psychologie für Kinder, Jugendliche und Ungelehrte. Zur Geschichte der Pädagogisierung von Psychologie. *Psychologie und Geschichte, 3*(3–4), 80–87.

Leibniz, G. W. (1978a). Meditationes de cognitione, veritate et ideis. In C. I. Gerhardt (Ed.), *G. W. Leibniz: Die philosophischen Schriften* (Meditationes) (Vol. IV, pp. 422–426). Olms. (Original work published 1890).

Leibniz, G. W. (1978b). Nouveaux Essais sur l'entendement par l'auteur du système de l'harmonie préétablie. In C. I. Gerhardt (Ed.), *G. W. Leibniz: Die philosophischen Schriften* (Nouveaux Essais) (Vol. V, pp. 39–509). Olms. (Original work published 1890).

Leibniz, G. W. (2008). Merckwürdige Schrifften Welche zwischen dem Herrn Baron von Leibnitz und dem Herrn D. Clarke über besondere Materien der Naturlichen Religion in Frantzös. und Englischer Sprache gewechselt, und mit einer Vorrede Herrn Christian Wolffens Nebst einer Antwort Herrn Ludwig Thümmigs auf die fünfte Englische Schrift, wegen ihrer Wichtigkeit in deutscher Sprache heraus gegeben worden von Heinrich Köhlern. In J. École (Ed.), *Christian Wolff: Gesammelte Werke* (Leibniz/Clarke Correspondance) (Vol. 113, Pt. III). Olms. (Original work published 1720).

Leibniz, G. W. (2010). Kleinere Philosophische Schriften mit einer Vorrede Herrn Christian Wolffs ehedem von dem Jenischen Philosophen Herrn Heinrich Köhler Deutsch übersetzet nun auf das neue übersehen von M. Caspar Jacob Hut. In J. École (Ed.), *Christian Wolff: Gesammelte Werke* (Leibniz Kleinere Schriften) (Vol. 114, Pt. III). Olms. (Original work published 1740).

Lorini, G. (2014). The origins of the transcendental subjectivity: On Baumgarten's psychology. *Philosophica, 44*, 107–126.

Luccio, R. (2014). *Dall'anima alla mente. Breve storia della psicologia*. Laterza.

Ludovici, C. G. (2001). "Wolff (Christian)" und "Wolffische Philosophie". Zwei Artikel aus Johann Heinrich Zedlers Universal-Lexicon aller Wissenschaften und Künste. In J. École (Ed.), *Christian Wolff: Gesammelte Werke* (Wolff und Wolffische philosophie) (Vol. 68, Pt. III). Olms. (Original work published 1748).

Meier, G. F. (2007). Metaphysik. Dritter Theil. Die Psychologie. In J. École (Ed.), *Christian Wolff: Gesammelte Werke* (Metaphysik) (Vol. Pt. III, 108.3). Olms. (Original work published 1765).

Meissner, H. A. (1970). *Philosphisches Lexicon: Darinnen die Erklärungen und Bechreibungen aus Herrn Christian Wolffens sämmtlichen teutschen Schrifften.* Stern. (Original work published 1737).

Mingard, G. (1774). Psychologie. In F. B. De Felice (Ed.), *Encyclopédie, ou Dictionnaire Universel Raisonné des Connoissances Humaines* (Vol. 35, pp. 511–513). De Felice.

Reusch, J. P. (1990). Systema Metaphysicum antiquiorum atque recentiorum item propria dogmata et hypotheses exhibens. In J. École (Ed.), *Christian Wolff: Gesammelte Werke* (Vol. 27, Pt. III). Olms. (Original work published 1735).

Riebow, G. H. (2002). Fernere Erläuterung der vernünfftigen Gedanken des Herrn Hof-Rath Wolfens von Gott, der Welt und der Seele des Menschen. In J. École (Ed.), *Christian Wolff: Gesammelte Werke* (Vol. 70, Pt. III). Olms. (Original work published 1726).

Rumore, P. (2007). Materiali per la ricostruzione della prima diffusione e ricezione tedesca della psicologia empirica di Wolff. In J. École (Ed.), *Christian Wolff: Gesammelte Werke* (Vol. 106, Pt. III, pp. 177–103). Olms.

Scharff, J. (2014). Pneumatica seu Pneumatologia, hoc est Scientia spirituum naturalis. In J. École (Ed.), *Christian Wollf: Gesammelte Werke* (Vol. 140, Pt. III). Olms. (Original work published 1629).

Stattler, B. (2011). Philosophia methodo scientiis propria explanata. Pars IV. Psychologia. In J. École (Ed.), *Christian Wollf: Gesammelte Werke* (Vol. 129.4, Pt. III). Olms. (Original work published 1770).

Sulzer, J. G. (1771–1774). *Allgemeine Theorie der schönen Künste: In einzeln, nach alphabetischer Ordnung der Kunstwörter auf einander folgenden, Artikeln abgehandelt* (Vol. 1–4). Weidmann.

Sulzer, J. G. (1786). *Kurzer Begriff aller Wissenschaften und anderer Teile der Gelehrsamkeit: Worin jeder nach seinem Inhalt, Nutzen und Vollkommenheit kürzlich beschrieben wird* (6th ed.). Langenheim.

Swedenborg, E. (1923). *Psychologica. Being notes and observations on Christian Wolff's Psychologia Empirica.* Swedenborg Scientific Association.

Thümmig, L. P. (1982). Institutiones philosophiae Wolfianae in usus academicos adornatae. In J. École (Ed.), *Christian Wolff: Gesammelte Werke* (Intitutiones) (Vol. 19, Pt. III). Olms. (Original work published 1725–1726).

Wolff, C. (1978). Vernünftige Gedancken von den Kräften des menschlichen Verstandes und Ihrem Gebrauche in Erkenntnis der Wahrheit. In J. École (Ed.)., *Christian Wolff: Gesammelte Werke* (German Logic) (Vol. 1, Pt. I) Olms. (Original Work printed 1754).

Wolff, C. (1968). Psychologia Empirica, methodo scientifica pertractata, qua ea, quae de anima humana indubia experientiae fide constant, continentur et ad solidam universae philosophiae practicae ac theologiae naturalis tractationen via sternitur (New edition emended from the prior, with privileges). In J. École (Ed.), *Christian Wolff: Gesammelte Werke* (Psychologia Empirica) (Vol. 5, Pt. II). Olms. (Original work published 1738).

Wolff, C. (1970). Philosophia moralis sive Ethica, methodo scientifica pertractata. In J. École (Ed.), *Christian Wolff: Gesammelte Werke* (Ethica) (Vol. 12, Pt. II). Olms. (Original work published 1750).

Wolff, C. (1972). Psychologia Rationalis, methodo scientifica pertractata, qua ea, quae de anima humana indubia experirientiae fide innotescunt, per essentiam et naturam animae explicatur, et ad intimiorem naturae ejusque autoris cognitionem profutura proponitur (New edition amended from prior, with privileges). In J. École (Ed.), *Christian Wolff: Gesammelte Werke* (Psychologia Rationalis) (Vol. 6, Pt. II). Olms. (Original work published 1740).

Wolff, C. (1973). Ausführliche Nachricht von seinen eigenen Schrifften, die er in deutscher Sprache heraus gegeben (New edition). In J. École (Ed.), *Christian Wolff: Gesammelte Werke* (Ausführliche Nachricht) (Vol. 9, Pt. I). Olms. (Original work published 1733).

Wolff, C. (1983a). Philosophia rationalis sive Logica, methodo scientifica pertractata et ad usum scientiarum atque vitae aptata. Praemittitur Discursus praeliminaris de philosophia in genere. In J. École (Ed.), *Christian Wolff: Gesammelte Werke* (Discursus praemlinaris) (Vol. 1.1, Pt. II, 3rd ed.). Olms. (Original work published 1740).

Wolff, C. (1983b). Vernünfftige Gedancken von Gott, der Welt und der Seele des Menschen, auch allen Dingen überhaupt. In C. A. Corr (Ed.), *Christian Wolff: Gesammelte Werke* (Deutsche Metaphysik) (Vol. 2, Pt. I, 11th ed.). Olms. (Original work published 1751).

Wolff, C. (1983c). Der Vernünfftigen Gedancken von Gott, der Welt und der Seele des Menschen, auch allen Dingen überhaupt, Anderer Theil, bestehend in ausführlichen Anmerckungen. In C. A. Corr (Ed.), *Christian Wolff: Gesammelte Werke* (Anmerkungen zur deutschen Metaphysik) (Vol. 3, Pt. I, 4th ed.). Olms. (Original work published 1740).

Chapter 13
Wolffians and the Emancipation of Aesthetic Faculties

Stefan Heßbrüggen-Walter

13.1 Introduction

In his *Psychologia rationalis* (Rational psychology), Wolff qualifies faculties as 'mere possibilities of change' (*nudae agendi possibilitates*), and the side note to this paragraph indicates that he understands this as a claim about 'the non-existence of faculties' (*inexistentia facultatum*) (Wolff, 1740/1972, §.81, pp. 59–61). Later followers of Wolff distanced themselves to some extent from this view.

Simultaneously, these thinkers began to develop philosophical theories of beauty, taste and the fine arts.[1] Baumgarten (1714–1762) maintains that beautiful thoughts, the subject matter of aesthetics, are the effect of faculties as the immediate cause (Baumgarten, 1750, §.27, p. 11). Aesthetic rules, i.e. rules of beautiful thinking, are grounded in the psychological constitution, the 'character', of a good 'aesthetician' (*bonus aestheticus*, Baumgarten, 1750, §.68, pp. 28ff). In a similar vein, Baumgarten's student Meier (1718–1777) states that aesthetic rules must be based on the psychology of the inferior faculties (Meier, 1748, §.3, p. 6), because beautiful thoughts are caused by inferior faculties (Meier, 1748, §.4, p. 7). Sulzer (1720–1779) holds that in order to understand art, we must understand how sensations originate in the soul (Sulzer, 1771, p. 20). This is because the goal of art is to provoke sensations (Sulzer, 1771, p. 311).

None of these claims make sense, if these faculties do not exist. If there are no faculties, faculties cannot be an immediate cause of representations. They cannot be emended to create a good aesthetic character. If art is meant to provoke sensations, sensations and their concomitant faculty must in some sense be real entities.

[1] Buchenau (2018a, p. 416) claims that it was Baumgarten's aesthetics that motivated his psychological views.

S. Heßbrüggen-Walter (✉)
School of Philosophy and Cultural Studies, HSE University, Moscow, Russia

S. d. F. Araujo et al. (eds.), *The Force of an Idea*, Studies in History and Philosophy of Science 50, https://doi.org/10.1007/978-3-030-74435-9_13

As we will see in detail, one option for resolving this situation is to redefine the notion of a mental faculty itself and give it a *realistic* interpretation. This is the path Baumgarten and Meier have chosen. Gottsched's (1700–1766) and Sulzer's *realism* about faculties is slightly more oblique. Both authors are in favour of abandoning pre-established harmony in favour of *physical influence*, i.e. direct intersubstantial causation; they affirm that substances can cause changes in each other without divine 'ideal influence'. But then it is obvious that there must be at least one real and not merely fictitious difference between faculties, namely, whether or not they are active or passive faculties.

However, a *realist* stance about faculties faces its own challenges. In order to identify representations as being one of a kind and therefore as products of one and the same faculty is not easy. Baumgarten, the *realist*, at the same time acknowledges the existence of eight different faculties that in one way or another relate to sense perceptions.[2] Meier and Sulzer do not dispute realism about faculties as such. But they try to unify the various faculties introduced into the debate as various modes of directing our attention.

13.2 Faculties in Wolff

As is well-known, Wolff believed that the power to represent the world (*vis repraesentativa*) is the essence of the soul (Wolff, 1740/1972, §.66). But in taking this stance, Wolff faces the challenge to explain how one and the same power can be responsible for the apparent variety of representations we find in us, e.g. sense perceptions, memory, beliefs or inferences (Heßbrüggen-Walter, 2004, p. 80).

In *Psychologia rationalis*, we find two complementary strategies to resolve this problem of how one power can be responsible for multiple qualitatively different realisations. The first contrasts faculties as expressing the possibility of a change with a power (*vis*) as a 'continuous striving to act' (*continuus agendi conatus*) (Wolff, 1740/1972, §.54, p. 35). Only a force can *cause* a change, while a faculty circumscribes the changes that are possible in principle for the force to actuate. In Wolffian parlance, only a force can be a sufficient reason for why this rather than that change becomes actual. The concept of a faculty captures the different kinds of change that are possible for a given substance (Wolff, 1740/1972, §.55).

The second strategy is based on the identification of the power to represent as the essence of the soul. The essence contains the ground of all other properties a substance has or may possibly have. Since the power to represent is the essence of the soul, it follows that all changes in the soul must in some way be explicable through this essence. Such an explanation must appeal to certain patterns and regularities in the succession of states of the soul. The observation that the soul is capable of

[2] Since we are here concerned with the possibility of a philosophy of sensible cognition, we will discuss only the so-called 'inferior' faculties and leave aside the 'higher' faculties of understanding, reason, etc.

different kinds of change is then reducible to the thesis that the soul is subject to laws and that these laws constrain the kinds of changes that are possible for a soul (Wolff, 1740/1972, §.81). In the end, both strategies converge: the concept of a faculty of the soul can be used as a valid explanation, if there is a law that determines what changes are possible for the soul in a given situation. In this sense, Wolff claims, laws in philosophical psychology are modelled after laws of motion, as he develops them in his *Cosmology* (Wolff, 1740/1972, §.84, Heßbrüggen-Walter, 2004, p. 82).

Wolff formulates only two laws for inferior faculties, the law of sensation and the law of the imagination.[3] The law of sensation associates changes in the soul with changes in a sense organ. Wolff is careful to formulate it in such a manner that it is neutral with regard to the question how such an association comes to pass:

> If change is brought about by some sensible object in some sense organ, there coexists some sensation in the mind that can be explained by this [change in the sense organ] in a comprehensible manner (Wolff, 1740/1972, §.85).[4]

The law imposes a condition on when to count a representation as a sensation: there must be a concomitant change in a sense organ, and this change must be connected to an interaction with a sensible object. The fact that there is such a change can then count as an explanation of why the representation in question qualifies as a sensation. Such laws then describe or predict what will happen, if a given condition is fulfilled. It is worth noting that Wolff is careful to formulate the law of sensation in such a manner that it does not presuppose any stance in how exactly the change in the sense organ and the concomitant change in our soul hang together. This attitude is intentional: within rational psychology, all claims must be neutral with regard to the different systems of mind-body interaction.[5]

The law of imagination concerns the relation between different states of the soul: if we produce the representation of one object, the imagination produces the perception of another one. 'If we perceive things once and if the perception of one [thing] is produced, the imagination produces the perception of another [thing]' (Wolff, 1738/1968, §.76).[6]

There must be a perception of one object. If there is one such perception, the soul brings about the representation of another object. So in both cases, Wolff argues that only the power to represent the world is a real entity, while faculties are *mere possibilities*. They serve to explain patterns that we can observe in the sequence of representations brought about by the power to represent. These patterns can be expressed as laws that determine a condition under which it is possible to explain the existence of a representation.

---

[3] So if faculties are indeed conceptually related to laws, there are for Wolff only two inferior faculties in the strict sense.

[4] In all quotations, internal references have been removed. All translations by the author, unless indicated otherwise.

[5] See Wolff (1740/1972, §.549) and Falk Wunderlich's contribution in this volume.

[6] Emended by comparing to Wolff (1738/1968, §.104)

13.3 Inferior Faculties in Baumgarten and Meier

As already mentioned, Baumgarten believes that beautiful thoughts must be caused by faculties. This is incompatible with the understanding of faculties developed by Wolff who denied a causal role to faculties. Instead, we find in Baumgarten a conception that is strikingly similar to that of Wolff's antagonist Christian August Crusius (1715–1775): faculties are real properties that can be distinguished not only on a conceptual level but in reality as well.[7] At the same time, Baumgarten agrees with Wolff that such faculties are governed by laws. However, it is not really clear whether these laws are meant to be descriptive or normative.

According to Baumgarten, all faculties are real, because they are, in his terminology, powers. All powers have actual existence and are more than 'mere possibilities'. Powers differ in whether they are necessary or sufficient grounds for determinations. Those that are sufficient grounds are at the same time substantial (Baumgarten, 1757/2013, §.198, p. 137).

At the same time, Baumgarten agrees with Wolff that we cannot think about powers or faculties without thinking about laws. Laws are propositions or representations of the agreement between a ground of a property and the property itself: 'A proposition that expresses a determination in conformity with a ground is a norm (rule, law) and indeed a norm in the wider sense is the representation of a determination in conformity with a ground' (Baumgarten, 1757/2013, §.83, p. 115).

If it is possible that a determination is not in such conformity with its ground, we must therefore take the law to express what shall be the case rather than what is the case. Only if a determination is in fact in agreement with its ground, the corresponding law can be understood descriptively.

If a determination has a ground in this sense, e.g. when accidents inhere in a substance, this is brought about by a power in a broad sense of the word. If the ground of the determination is sufficient for the determination, this ground is a power in the strict sense:

> If accidents inhere in a substance, then there is some ground of inherence, or power, in the broader sense (efficacy, energy, activity), and a sufficient ground. This [latter] is power in the stricter sense (and sometimes called simply power for the sake of brevity) (Baumgarten, 1757/2013, §.197, p. 137).

In sum, if a determination is in agreement with its ground, this means that the determination inheres based on a law. The ground of this inherence is called power in a broader sense, if it is a necessary ground. It is called power in a strict sense, if it is a sufficient ground.[8]

---

[7] See Crusius (1745/1964, §.79, pp. 135ff) and Heßbrüggen-Walter (2004, pp. 93ff). The influence of Thomasians on Baumgarten prompts Schwaiger (2011, p. 27) to qualify him as a 'pietist proponent of the Enlightenment' (*pietistischer Aufklärer*).

[8] Baumgarten does not address the question how faculties of the soul as powers in the broad sense could make a causal contribution as necessary grounds of representations, if the power to represent

We are now in a position to understand Baumgarten's conception of psychological laws. For the present context, the 'law of sensation' is the most important: 'The law of sensation is: Just as the states of the world and my states follow one another, so too should the representations of their present states mutually follow one another' (Baumgarten, 1757/2013, §.541, p. 206).

This law does not merely *describe* that representations of states of the world follow those states themselves. It rather *prescribes* how representations *should* follow.[9] But if faculties were reducible to lawful regularities, as Wolff had maintained, there would be no leeway for any deviation from these laws. So a prescriptive understanding of laws of faculties is only possible if these faculties are understood as real causes and if these causes do not always bring about their effects.

So on Wolff's understanding of a faculty, it would be impossible to understand how a philosophical discipline could be useful in emending it. This means that the idea of a discipline of aesthetics that emends inferior faculties is only coherent if inferior faculties are conceived as real causes and if their laws are prescriptive rather than descriptive. Baumgarten's formulation of the law of sensibility is therefore the psychological foundation for the very possibility of philosophical aesthetics as a theory of beautiful thought.

Second, we can note that the law of sensation agrees with Baumgarten's general definition of a law: determinations ('representations') must conform with their ground ('states of the world', perceived in outer sense, and 'my states', perceived in inner sense). This grounding relation is not causal. When we compare the law of sensation to its counterpart in Wolff, we see that Baumgarten does not mention objects at all, but refers only to states of the world or the self that is correlated with representations of such states. And a state of the world or of the self plays no role in explaining the corresponding representation.

Why this is the case becomes obvious when we look at Baumgarten's distinction between active and passive faculties. Baumgarten does acknowledge that the distinction exists (Baumgarten, 1757/2013, §.216, p. 140). But he denies that when one substance acts on another, the resulting relation of 'influence' is 'real'. Rather, the receptivity of the 'passive' substance is only 'ideal' (Baumgarten, 1757/2013, §.448, p. 185). This means that the change in the passive substance depends in no way on the active power of the active substance:

> For, in physical influence, the suffering of the substance that really suffers does not have a sufficient ground in its own powers. In pre-established harmony, the suffering of the suffering substance has a sufficient ground (1) in its own powers and (2) in the substance ideally influencing [it]. Hence, in pre-established harmony, the influencing substance is equally as fecund as in physical influence, while the suffering substance is however more fecund than in physical influence (Baumgarten, 1757/2013, §.459, p. 188).

---

the world is a power in the strict sense and therefore in itself a sufficient ground of representations.

[9] This interpretation relies on the fact that the Latin original uses the subjunctive in the consequent. The law of imagination will use the indicative.

The concept of 'suffering' a change is for Baumgarten in no way linked to an existing active substance exerting a 'real' influence on the 'suffering' substance. What takes place within the suffering substance is explicable through its own power alone. The additional influence of an acting substance is 'ideal', not 'real'. Any such 'ideal' influence must ultimately be traced back to the creator of pre-established harmony, i.e. God.

To sum up, Baumgarten's understanding of faculties differs in many important respects from Wolff's. For Baumgarten, faculties are not mere possibilities, but real properties of substances. They differ from powers in that they are only necessary rather than fully sufficient conditions of a change. Wolff and Baumgarten agree that powers and faculties are subject to laws. But at least in the case of sensibility, these laws are according to Baumgarten prescriptive rather than descriptive. Baumgarten explicitly endorses pre-established harmony and does not believe that psychology must be neutral with regard to the choice of a system of interaction between the mind and body. This leads him to an understanding of passive faculties that allows only for an ideal influence of other substances, while the real change as such is effected only by the suffering substance itself.

Still, Baumgarten's commitment to faculties as real properties is not unproblematic. Everyone entertaining the idea that mental powers are real must explain what constitutes a specific kind of representations that is qualitatively different from other kinds and therefore caused by a specific power.[10] For it is not parsimonious to assume that each and every individual representation is grounded by a specific power that grounds only this individual representation.

This problem is particularly pressing, because Baumgarten also deviates from Wolff's material analysis of inferior faculties, introducing eight of them. The faculty of sensing itself represents either the state of the soul in internal sense or the state of the body, i.e. of sense organs, in outer sense (Baumgarten, 1757/2013, §.535, p. 205). The imagination represents absent things, e.g. past states of my soul (Baumgarten, 1757/2013, §.558, p. 211). Perspicaciousness or 'acute wit' is the faculty to perceive correspondences between things (Baumgarten, 1757/2013, §.573, p. 215). Memory perceives the identity between a representation I reproduce and a representation I had had earlier (Baumgarten, 1757/2013, §.579, p. 217). The power of invention separates and combines images (Baumgarten, 1757/2013, §.589, p. 219).[11] Foresight is the ability to be aware of the future state of the world (Baumgarten, 1757/2013, §.595, p. 221). Taste is judgement of sensible objects and perceives their perfections or imperfections (Baumgarten, 1757/2013, §§.606–608, pp. 223–224). Anticipation consists in the identification of a foreseen perception as a representation that I will have in the future (Baumgarten, 1757/2013, §.610, p. 225). All these faculties reappear in the *Aesthetica* as part of what Baumgarten calls 'elegant ingenuity' (*ingenium venustum*), a *superfaculty* that consists in a

[10] Heßbrüggen-Walter (2004, p. 109) raises this point for Crusius. It also applies to Baumgarten.

[11] It is thus purely sensual and differs in this respect from the poetic faculty in Gottsched. See Mirbach (2014, p. 123).

coordinated effort of all eight faculties to bring about a 'beautiful cognition' (Baumgarten, 1750, §.29, p. 12). Thus, a critical reconstruction of Baumgarten's aesthetics would have to answer the question why it is exactly these faculties that deserve to be distinguished as real properties of the soul and whether it is feasible to identify criteria for the distinction of classes of representations that are sufficiently different to justify the distinction of these faculties.

Another area in need of clarification is the connection between aesthetics, inferior faculties and their laws. First, it should be noted that with the exception of the law of sensibility quoted above, none of the other laws of inferior faculties is explicitly formulated as a prescriptive law. Hence, whether these laws can serve to elucidate the aesthetic functions of the respective faculties is not clear. The textual evidence in Baumgarten himself is inconclusive.

Baumgarten's student Georg Friedrich Meier also subscribes to the idea that faculties are real properties of substances (Meier, 1755, §.158, pp. 258f). And he agrees with Baumgarten's general analysis of a law of a faculty as expressing the agreement between a ground of a property and the property itself (Meier, 1755, §.80, pp. 140f). However, he differs from his teacher in that for him all laws governing the use of our inferior faculties must be understood as prescriptive or ideal. This is possible, because according to Meier, our inferior faculties are subject to voluntary control (*Herrschaft*, Meier, 1748, §.279, p. 43):

> This use [of the sensual faculty of cognition] is either a natural or an arbitrary use of this faculty. [...] The arbitrary use of the sensual faculty of cognition depends on our freedom. We can exercise this use or we can refrain from it, we can practice it one way or another, because these exercises depend on our choice (Meier, 1748, §.277, pp. 38ff).

This somewhat surprising thesis becomes clearer when we address the role of attention in Meier's psychology. According to Meier, attention and its counterpart, abstraction, modify the clarity of representations. If I direct my attention to a representation, it becomes clearer (Meier, 1757, §.506, p. 56). Abstraction serves to make a representation more obscure. So depending on what we attend to representations of an object or trait of an object become either clearer or more obscure. Meier believes that all inferior faculties introduced by Baumgarten can be explained as the result of interactions between these two interdependent and fundamental capabilities (Meier, 1748, §.283, p. 48). He also maintains that attention and abstraction are the fundamental modes of exercising our force of representation (Meier, 1757, §.524, p. 83). In this way, Meier systematises the various inferior faculties of cognition which Baumgarten had introduced as necessary factors in sensual cognition.

Meier introduces sensation as the faculty of representation of something that is present to us, either representations of changes in the soul or in a sense organ (Meier, 1748, §.330, pp. 148f). It is equivalent to attention, insofar as it is directed towards such changes (Meier, 1748, §.331, p. 150; 1755, §.530, p. 90). The notion of imagination is limited to what Kant would have called 'reproductive imagination'. It consists in attention directed at our past states (Meier, 1755, §.555, p. 130). More precisely, this attention is turned towards a past representation that had been obscured by abstraction (Meier, 1748, §.373, p. 261). Imagination is thus the precondition of the persistence of representations, because without imagination, there would be no memory, so that all representations would momentarily vanish, since we could not retain them (Meier, 1748, §.371, p. 257).

Memory consists in the attention for the sameness of an imagination I now have with what I have represented in the past (Meier, 1755, §.578, p. 171). While memory does depend on imagination, it is at the same time closely related to wit, because we compare imaginations we now have with past sensations and look for similarities (Meier, 1748, §.438, p. 436). Wit is explained as attention for correspondences in our representations (Meier, 1755, §.567, p. 153). Meier analyses it as a composite faculty depending on the senses and imagination (Meier, 1748, §.401, p. 330). The poetic faculty which plays a role similar to Kant's 'productive imagination' is based on four faculties, the senses, imagination, memory and wit and thus relies on a complex interplay of attention and abstraction (Meier, 1748, §.457, p. 186; 1755, §.588, p. 187). Two of the faculties constituting the inferior cognitive faculty deal with future representations. The faculty to anticipate is triggered by a representation that determines the imagination to call up a past representation (Meier, 1755, §.599, p. 204). It is the basis for foresight (*praesagitio*), the faculty for conjectures or the expectation of similar cases (Meier, 1748, §.482, p. 239; 1755, §.609, p. 222). Moreover, we have a sensual faculty of judgement that consists in attention for the perfections and imperfections of things (Meier, 1755, §.617, p. 235). Judgement that is concerned with the beautiful or ugly properties of things is taste (Meier, 1748, §.468, p. 506).

Baumgarten and Meier divert from Wolff's psychology and instead follow his main adversary Crusius, as far as their understanding of the concept of a faculty is concerned. Faculties are no longer just explanatory devices; instead, they are a part of reality, since they ground properties of substances. The representative force of the soul and its various other faculties are nevertheless different: the representative force is a sufficient ground for a state of the soul, while faculties are only necessary grounds. Baumgarten seems to assume that a faculty is proven to be real if we are able to give a law that governs its sphere of activity. Whether or not such laws are prescriptive or descriptive is ultimately not to decide. His student Meier is unequivocal: laws of faculties are prescriptive. This means that we control our inferior faculties.

Meier seems to assume that the multifarious faculties that make up our inferior faculty of cognition are in need of systematisation. At the same time, he must explain how exactly we exert control over our sense perceptions and related representations. Both aims are served by his analysis of the notion of attention. We can direct our attention, so all faculties that make up the inferior faculty can be expressed as various modes of its use of attention and, at times, its counterpart, abstraction.

13.4 Inferior Faculties in Gottsched and Sulzer

Gottsched's general understanding of the notion of a faculty is closer to Wolff's than Baumgarten's and Meier's. According to Gottsched, we must distinguish power (*Kraft*) and faculty (*Macht*): a power strives to achieve a change, while the notion of a faculty captures only a possibility to act (Gottsched, 1762/1983, §.300, p. 253). When we apply this distinction to the soul, this means that one power to act can bring about different kinds of change. These different kinds of change can be explained through different possibilities to act, i.e. faculties (Gottsched, 1762/1983, §.1016, p. 563). Ultimately, however, every such explanation must appeal to the force of representing the world, because it is this force that effects every change in the soul (Gottsched, 1762/1983, §.1019, p. 564). In contrast, faculties as such cannot effect change, because they are 'insufficiently determined', i.e. mere possibilities in the Wolffian sense (Gottsched, 1762/1983, §.391, p. 253).

However, as Eric Watkins has shown, Gottsched defends at least 'unofficially' one distinction of faculties that serves not merely explanatory purposes, namely, the distinction between 'really' active and 'really' passive faculties (Gottsched, 1762/1983, §.300, p. 253; Watkins, 1995, pp. 302f). And while Baumgarten had formulated this distinction in a way that lets 'passive' change still originate in the 'suffering' substance, in a system of 'real' influence, the substance suffering a change is no longer itself a sufficient ground of all its changes. Passive changes in such a substance depend at least in part on the effect of the causally active substance. But then the capability to suffer such changes which we could call 'receptivity' is distinct from any faculty for instigating change. In other words, Baumgarten's conclusion that there could be one power for representing the world in the soul without any further distinction between active and passive faculties is incompatible with a system of 'real' influence favoured by Gottsched.

Instead of the rich model of inferior faculties suggested by Baumgarten and Meier, Gottsched favours a less complicated solution and distinguishes only four different inferior faculties. The definitions he uses are by and large quite similar to Baumgarten's and Meier's. Besides the senses as such, we need wit, the ability to easily perceive similarities between things. Wit presupposes acuity, the ability to perceive many different aspects of a thing which is opposed to the dullness of sense (Gottsched 1742/1973, §.11, p. 152). Imagination recalls former sensations evoked by some present sensation, if there is any similarity between both (Gottsched 1742/1973, §.14, p. 54).[12]

So while Baumgarten and Meier divert from Wolff's psychology in their very understanding of the concept of a faculty, Gottsched is more reticent in this regard and closer to Wolff's understanding of faculties as mere possibilities. He disagrees with Baumgarten's endorsement of pre-established harmony. And it is here that, even

[12] According to Gottsched, imagination is the only faculty that has its own law. But he qualifies it as merely empirical (Gottsched, 1762/1983, §.892, p. 519). See Mirbach (2014, p. 118).

though he nowhere acknowledges the fact explicitly, he is implicitly committed to a real distinction between faculties, namely, between active and passive faculties.

Sulzer addresses the distinction between active and passive faculties in his 1763 memorandum for the Berlin Academy which was first published in French in 1770 and in German translation in his collected philosophical writings in 1773.[13] We must distinguish two faculties of the soul, a 'faculty of representation' and a 'faculty of sensation'. Both are fundamental:

> Although the effects of our soul seem to be so multifarious, they still are ultimately based on the application of two faculties which are the sources of all its other determinations and changes. One is the faculty to represent something or to cognise the qualities of things; the other [sc. is] the faculty to sense and to be touched in an agreeable or disagreeable manner (Sulzer, 1773, p. 225).

At the same time, Sulzer is attracted to Meier's strategy to explain faculties as different uses of attention. In sensation, we look at a totality of representations without distinguishing its parts from each other. In representation, we distinguish some of the parts of a totality, so that we can refer to them verbally and thereby conceptualise them as traits (*Merkmale*) (Sulzer, 1773, pp. 226ff). We could then assume that it is attention rather than representation and sensation that must count as a fundamental faculty which can produce both kinds of states, representations and sensations, depending on whether it is directed towards a totality of representations or at some of its parts.

Still, the difference between both faculties is fundamental.[14] Sensations are based on impacts by objects, while representations are spontaneous acts of the soul. This becomes obvious in Sulzer's way to present the differences between the states evoked by the faculty of representation and the faculty of sensation. When we try to represent a simple concept in complete distinctness, all other forces of the soul are dormant, we forget about ourselves (Sulzer, 1773, p. 228).[15] This is true not only for our mental self-awareness but for our bodily self as well: when we represent the world, there is nothing happening in our body. In sensing, on the contrary, we are always aware of our bodily state as well: 'During reflection nothing happens in the body that could awaken the idea of ourselves; everything is completely quiet and calm; contrary to this, the state of sensation is always associated with some sensual sensation' (Sulzer, 1773, p. 232).

[13] The translated German title of the text refers to the two main faculties as the faculty of representation (*Vermögen etwas vorzustellen*) and the faculty of sensation (*Vermögen etwas zu empfinden*). As far as I can see, it has not yet been noted that the original French title refers to the faculty of representation as *apperception*. The implications of the theory presented in Sulzer (1773) for his understanding of apperception have therefore not been noted either.

[14] This is why Heinz (2011, p. 92) refers to Sulzer's position as a 'doctrine of two faculties' (*Zweivermögenslehre*).

[15] Buchenau (2018b, p. 41) reads the passage as being limited to philosophical contemplation. That would however lead to the somewhat counterintuitive consequence that non-philosophers are never engaged in the representation of objects.

While we reflect, we become an 'abstract being' that seems to have no connection to the world (Sulzer, 1773, p. 229). The state of sensation is directly opposed. In sensation, our attention is directed towards ourselves rather than towards the object that was the occasion of the sensation (Sulzer, 1773, p. 233). This means that somewhat paradoxically, the state of sensation has nothing in common with any object, even though it is caused by an object (Sulzer, 1773, pp. 229f).

Still, we may concede that this alone is no conclusive evidence that Sulzer wants to base sensation on interaction with a body.[16] But we must pay attention to a second essential property of sensations: they cannot be controlled:[17]

> It is therefore certain that a human being is not in control with regard to the first motions of its soul. There is not the least degree of freedom to sense or not to sense. All we can do is to inhibit the effect of sensation, i.e. to put a stronger sensation against it. From this I conclude that sensations and their immediate consequences are involuntary acts of the soul. (Sulzer, 1773, p. 242)

This however means that Sulzer cannot share Baumgarten's and Meier's guiding intuition that the inferior faculty is subject to voluntary control. We sense what we sense and there is nothing we can do about it. The very idea of prescriptive laws of sensing or of activities of the inferior faculties in general is then incoherent.

We should note in passing that between the extremes of sensation and representation, we are capable of a third state, contemplation (*Betrachtung*). Superficially, it seems as if we were able to direct our attention both towards an object and towards my own state. Sulzer believes that this may be explained as an oscillation between the more fundamental states of reflection and sensation (Sulzer, 1773, p. 236).

Sulzer's views on particular inferior faculties can be found in his *Allgemeine Theorie der schönen Künste* (General theory of the fine arts) (1771). He discusses only three such faculties, in this again closer to Gottsched than to Baumgarten and Meier: imagination, the power of invention and wit. Imagination is the faculty to represent objects of sense when they are not present (Sulzer, 1771, p. 291). The power of invention creates representations of objects that we have never sensed (Sulzer, 1771, p. 259). Wit is the ability to understand and sense relations between objects (Sulzer, 1771, p. 1273). Taste is the ability to sense the beautiful (Sulzer, 1771, p. 461). There are also minor differences between his views on the nexus between sensation and reflection in 1763 and 1771. Now sensation and reflection are no longer taken to be diametrically opposed. Rather, whether or not a given state qualifies as sensation or reflection is a matter of degree (Sulzer, 1771, pp. 310f). In terms of his approach in 1771, we could say that now most states are states of contemplation unifying sensation and reflection.

In sum, Sulzer's position is under tension. On the one hand, he contends that we become distinctly aware of objects only in reflection, i.e. in a state in which all sensory activity has stopped. In contrast, sensation is a state that makes us aware only of ourselves. On the other hand, the state of reflection is in no way connected to

[16] See Heinz (2011, p. 91).

[17] Dumouchel (2018, p. 32) shows how this thesis fits into Sulzer's overall psychology of sensations.

sensing objects through our bodies. And it is only in sensation that any causal interaction with the object of sensation takes place. Still, it is obvious that, whatever the merits of this solution may be, it depends crucially on the distinction between the active faculty of reflection and the passive faculty of sensation.

13.5 Conclusion

Wolff's psychology is no adequate foundation for aesthetic theories, if such theories must be based on the real existence of faculties. Faculties are for Wolff mere possibilities. Only the substantial 'force of representation' is real. The sequence of representations in the soul can be equally well explained appealing to laws of representations governing the force of representation. Therefore, faculties have no actual existence as properties of substances.

None of the authors we have discussed shares this view. We can distinguish two argumentative strategies against Wolff's 'nominalism' and in favour of a 'realistic' conception of faculties. Baumgarten and Meier redefine the concept of a faculty. For them, faculties are 'grounds of properties' and have as such real existence. Both also believe as well that the representative force is bound to laws, but at least Meier is explicit that such laws are normative and must be understood as norms or ideals rather than in analogy to laws of nature, as Wolff had maintained. Gottsched and Sulzer think about faculties in terms of their role in the interaction with other substances, i.e. objects of cognition. In Gottsched's case, his alignment with those who believe that intersubstantial causality is real commits him implicitly to a real distinction between active and passive faculties. Sulzer does not discuss the interaction of cognitive faculties and objects of cognition in causal terms. But he explicitly acknowledges that there is not just one representative power in the soul, but rather two fundamental faculties. Both remain silent with regard to possible laws for such faculties.

However, Baumgarten's notion that all faculties are real properties invites suspicions, since he assumes no less than eight different inferior faculties. Meier resolves this problem in that he shows how each of these faculties can also be explained as a distinct mode of directing our attention towards certain aspects of reality and abstracting from others, an analysis that is not completely alien to Sulzer either.

If this analysis is convincing, it raises some interesting questions for the interpretation of eighteenth-century German aesthetics. If aesthetics does presuppose the real existence of mental faculties, we may begin to wonder which of the faculties mentioned, e.g. by Baumgarten and Meier, are really required for an adequate characterisation of this domain, or whether a more parsimonious model of the aesthetic mind along the lines of Sulzer and Gottsched is the superior solution. We also may have to rethink the role of aesthetic rules and their connection to psychological laws, in particular if we believe that such laws formulate ideal or normative presuppositions of aesthetic thought.

References

Baumgarten, A. G. (1750). *Aesthetica*. Kleyb.

Baumgarten, A. G. (2013). *Metaphysics: A critical translation with Kants elucidations, selected notes and related materials*. Bloomsbury. (Original work published 1757).

Buchenau, S. (2018a). Wolffs Rezeption in der Ästhetik. In R. Theis & A. Aichele (Eds.), *Handbuch Christian Wolff* (pp. 405–425). Springer VS.

Buchenau, S. (2018b). Sulzers "Physik der Seele" zwischen Medizin und Philosophie. In É. Décultot, P. Kampa, & J. Kittelmann (Eds.), *Johann Georg Sulzer—Aufklärung im Umbruch* (pp. 36–50). De Gruyter.

Crusius, C. A. (1964). *Entwurf der nothwendigen Vernunft-Wahrheiten, wiefern sie den zufälligen entgegen gesetzet werden*. Olms. (Original work published 1745).

Daniel Dumouchel, D. (2018). "Tiefen der Seele". Veränderte Zustände und psychologische Paradoxe. In É. Décultot, P. Kampa, & J. Kittelmann (Eds.), *Johann Georg Sulzer—Aufklärung im Umbruch* (pp. 14–35). De Gruyter.

Gottsched, J. C. (1973). *Versuch einer critischen Dichtkunst: Erster allgemeiner Theil*. de Gruyter. (Original work published 1742).

Gottsched, J. C. (1983). *Erste Gründe der gesammten Weltweisheit: Theoretischer Teil*. de Gruyter. (Original work published 1762).

Heinz, M. (2011). Sulzer und die Anfänge der Dreivermögenslehre bei Kant. In F. Grunert & G. Stiening (Eds.), *Johann Georg Sulzer (1720–1779): Aufklärung zwischen Christian Wolff und David Hume* (pp. 83–100). Akademie-Verlag.

Heßbrüggen-Walter, S. (2004). *Die Seele und ihre Vermögen: Kants Metaphysik des Mentalen in der 'Kritik der reinen Vernunft'*. Mentis.

Meier, G. F. (1748). *Georg Friedrich Meiers öffentlichen Lehrers der Weltweisheit zu Halle Anfangsgründe aller schönen Wissenschaften* (Vol. 1). Hemmerde.

Meier, G. F. (1755). *Georg Friedrich Meiers, öffentlichen ordentlichen Lehrers der Weltweisheit zu Halle, und der königlichen Academie der Wissenschaften in Berlin Mitgliedes, Metaphysik* (Vol. 1). Gebauer.

Meier, G. F. (1757). *Georg Friedrich Meiers, öffentlichen ordentlichen Lehrers der Weltweisheit zu Halle, und der königlichen Academie der Wissenschaften in Berlin Mitgliedes, Metaphysik* (Vol. 3). Gebauer.

Mirbach, D. (2014). Gottsched und die Entstehung der Ästhetik. In E. Achermann (Ed.), *Johann Christoph Gottsched (1700–1766): Philosophie, Poetik und Wissenschaft*. Akademie-Verlag.

Schwaiger, C. (2011). *Alexander Gottlieb Baumgarten—Ein intellektuelles Porträt: Studien zur Metaphysik und Ethik von Kants Leitautor*. Frommann-Holzboog.

Sulzer, J. G. (1770). Observations sur les divers états oú làme se trouve en exercant sesfacultés primitives, celle d'appercevoir et celle de sentir. In *Histoire de l'Académie Royale des Sciences et Belles Lettres, Annee 1763* (pp. 407–420). Haude et Spener.

Sulzer, J. G. (1771). *Allgemeine Theorie der Schönen Künste: In einzeln, nach alphabetischer Ordnung der Kunstwörter auf einander folgenden, Artikeln abgehandelt*. Weidemanns Erben und Reich.

Sulzer, J. G. (1773). Anmerckungen über den verschiedenen Zustand, worinn sich die Seele bey Ausübung ihrer Hauptvermögen, nämlich des Vermögens, sich etwas vorzustellen und des Vermögens zu empfinden, befindet. In J. G. Sulzer (Ed.), *Johann George Sulzers vermischte Philosophische Schriften: Aus den Jahrbüchern der Akademie der Wissenschaften zu Berlin gesammelt* (Vol. 1, pp. 225–243). Weidmann und Reich.

Watkins, E. (1995). The development of physical influx in early eighteenth-century Germany: Gottsched, Knutzen, and Crusius. *The Review of Metaphysics, 49*(2), 295–339.

Wolff, C. (1968). Psychologia Empirica, methodo scientifica pertractata, qua ea, quae de anima humana indubia experientiae fide constant, continentur et ad solidam universae philosophiae practicae ac theologiae naturalis tractationem via sternitur (New edition emended from the prior, with privileges). In J. École (Ed.), *Christian Wolff: Gesammelte Werke* (Psychologia Empirica) (Vol. 5, Pt. II). Olms. (Original work published 1738).

Wolff, C. (1972). Psychologia rationalis methodo scientifica pertractata, qua ea, quae de anima humana indubia experientiae fide innotescunt, per essentiam et naturam animae explicantur, et ad intimiorem naturae ejusque autoris cognitionem profutura proponuntur (New edition emended from the prior, with privileges). In J. École (Ed.), *Christian Wolff: Gesammelte Werke* (Psychologia rationalis) (Vol. 6, Pt. II). Olms. (Original work published 1740).

Chapter 14
Wolff and the Beginnings of Experimental Psychology in the Eighteenth Century

Andreas Rydberg

14.1 Introduction

Although the term *psychology* has been in use at least since the sixteenth century, the distinction between empirical and rational psychology has typically been attributed to Christian Wolff (1679–1754) (Leventhal, 2019, p. 113; Luccio, 2013; Vidal, 2011, pp. 89–97; Rudolph & Goubet, 2004; Brozek, 1999; Lapointe, 1972). Scholars have, furthermore, highlighted Wolff's references to psychometrics, sometimes situating these in the immediate context of the time, sometimes inscribing them in the slightly longer history of measurement and transcendental philosophy (Feuerhahn, 2007; Hatfield, 2006; Sturm, 2006; Métraux, 1983; Ramul, 1960). What has caught less scholarly attention, however, is that Wolff also initiated what became a rather disparate discourse on psychological experiments and experimental psychology.

In this chapter, I analyze this discourse as well as some of the attempts to translate it into specific psychological experiments or even into an experimental psychology. Rather than depicting a homogenous development, I draw attention to three rather different discourses. First, there was the experimental-philosophical discourse, representing an attempt to replace the experimental intervention, which had proven so successful in natural philosophy, with behavioral intervention. Second, there was the iatromechanical discourse, whereby the body was seen as a complex machine circulating subtle bodily fluids. Since within this framework mental phenomena were thought to correspond with movements in the nerves, the challenge here became how to surgically measure the nerves, nerve membranes, and nerve liquids. Third, there was the metaphysical and ethical discourse on perfection, according to which the human being is morally obliged to perfect the self as much as possible. Here, the pursuit of perfection boiled down to daily observations and

A. Rydberg (✉)
Uppsala University, Uppsala, Sweden
e-mail: andreas.rydberg@idehist.uu.se

S. d. F. Araujo et al. (eds.), *The Force of an Idea*, Studies in History and Philosophy of Science 50, https://doi.org/10.1007/978-3-030-74435-9_14

experiments on the self, a notion of experiment that differed widely from both behavioral and surgical intervention.

By highlighting these three discourses, I wish to draw attention to the heterogeneity and complexity of the discussion of experimentation and measurement in the wake of Wolff's psychology. Rather than a unified concept of psychological experiment or experimental psychology, there was a plethora of disparate ways of—and reasons for—measuring and experimenting on the soul. Finally, I also wish to warn against anachronistic readings that might inscribe these discourses within a disciplinary history of the emergence of scientific psychology. Such readings tend to prioritize and legitimize the present at the expense of the understanding of the past.

14.2 Experiments, Psychological Experiments, and Experimental Psychology

For Aristotle (384–322 BC), experience or *empeiría* was the result of many perceptions stored in the memory (Aristotle, 1995c, II.19). Although deriving from perceptions, an experience here typically took the form of a universal and evidently true statement about how things are in the world (Dear, 1995, pp. 22–23). In the scholastic world, it was this meaning of experience that was transferred into the Latin *experientia*, *experimentum*, and *historia* (Park, 2011; Pomata, 2011; Dear, 2006). Whereas *experientia* and *historia* maintained the meaning of evidently true facts, *experimentum* gradually acquired a new meaning (Daston, 2011). In Francis Bacon's (1561–1626) *Novum organum* (*The New Organon*), experiments thus typically referred to specific situations designed to force nature to reveal otherwise concealed truths (Bacon, 1620/2000). In the second half of the seventeenth century, this notion of experiment provided the basis for Robert Boyle's (1627–1691) famous experiments with the air pump, and a century later it had become fairly established in English, German, and French (Daston, 2011, p. 86; Shapin & Schaffer, 1985).

Wolff was deeply influenced by the Aristotelian-scholastic tradition at the same time as he was an equally eager proponent of the new experimental philosophy (École, 1979; Mühlpfordt, 1992). In his autobiography, he thus describes how he initially struggled when he arrived at the university in Halle, and how his breakthrough first came when he took over the chair in experimental philosophy from the physician Friedrich Hoffmann (1660–1742) in 1709 (Wolff, 1841/1980a, p. 146). Having procured the right instruments, Wolff started a successful career as a teacher and practitioner of the new experimental philosophy as well as a prolific writer of voluminous treatises on the subject (Mühlpfordt, 1992). In his publications, Wolff leaned heavily on the British tradition, with numerous references to prominent British natural philosophers. That Wolff drew on the fairly established Baconian notion of experiment also becomes clear from his own definitions. In *Psychologia empirica* (*Empirical Psychology*), he thus wrote that "observation is experience, that is centered around the facts of nature, obtained without our intervention.

Experiment is experience, that is centered around the facts of nature, which cannot be obtained without our intervention" (Wolff, 1732, §.456, p. 357).[1] The idea of intervention also went into another distinction that Wolff made in *Philosophia rationalis sive logica* (*Latin Logic*) between *common historical knowledge* (*cognitio historica communis*) and *secret historical knowledge* (*cognitio historica arcana*):

> While some facts of nature are hidden (§20), others are so apparent (§1) that they require only attention and, of course, some acumen. The hidden facts must be brought to light by skilled investigators, and even then they are not known unless reason gives its assistance to the senses. As a result we distinguish between common and secret historical knowledge. (Wolff, 1728, §.21, p. 10; 1963, p. 12)[2]

The concepts of experience and secret historical knowledge both reflect the view that there are facts of nature that can only be revealed through intervention, preferably through the use of technical devices such as air pumps, telescopes, microscopes, and the like. An experiment required intervention, and as long as this condition was met, the soul, in principle, could be an object of experimentation. In fact, in *Psychologia empirica*, Wolff stressed that experiments can indeed be used in all parts of philosophy:

> The art of experimenting consists of truths derived from experiments. Thus far almost only the physicists have used this art. Nonetheless there is room for it in all forms of philosophy, even in natural theology. Because of this I remember that I have often urged that there is also an experimental natural theology; experimental philosophy in every aspect can indeed be extended to all parts of philosophy. (Wolff, 1732, §.459, p. 358)

That Wolff was open to the use of experiments is in line with the criteria of intervention in the course of nature. When it comes to the particular case of psychology, Wolff compared it with experimental physics in *Philosophia rationalis sive logica*. "It is clear that empirical psychology corresponds to experimental physics, and thus pertains to experimental philosophy" (Wolff, 1728, §.111, p. 51; 1963, p. 56). In *Psychologia empirica*, he repeated this claim, now adding that empirical psychology uses experiments in the same way as physics:

> We have mentioned elsewhere (*Disc. Praelim.* # 111 not.) that empirical psychology is similar to experimental physics. For it is also the case that experimental physics supplies principles for dogmatic physics. … Empirical psychology is similar to experimental physics; for we use experiments—either directly or by deducing something from them—to examine the tenets of dogmatic physics. (Wolff, 1732, §§.4–5, pp. 3–4; 1980b, pp. 231–232)

However tempting it may be to take these passages as evidence for psychological experiments or even an experimental psychology, there are in fact no descriptions of or references to specific psychological experiments. Even if Wolff did not exclude such experiments, a more moderate interpretation would be that experience derived

---

[1] For similar distinctions and definitions, see also Wolff, 1713, §.13, pp. 66–67; 1720b, §.325, pp. 159–160; 1728, §.747, p. 539.

[2] All translations are my own unless modern English translations are cited, alone or in addition to the original sources (which is the case here). In cases where modern English translations have been used, these have been checked against the originals.

from experiments could also be used to draw conclusions within the field of empirical psychology. If this is how we should understand Wolff's references to the use of experiments in psychology, it explains the otherwise enigmatic absence of any discussion of the more specific nature of psychological experiments.

If Wolff neither confirmed nor denied the existence of specific psychological experiments, others would. Clearly, the most fascinating and intriguing example is the Wolffian philosopher Gottlieb Friedrich Hagen (1710–1769). Hagen became acquainted with Wolff on an educational journey in 1731 (Albrecht, 2010). In 1734, he published a shorter work titled *Programma de mensurandis viribus intellectus* (*Program for the Measure of the Intellectual Powers*), as well as the much more extensive treatise *Meditationes philosophicae de methodo mathematica* (*Philosophical Meditations on the Mathematical Method*) (Hagen, 1734a, b). As the titles suggest, the works revolved around the possibilities and limitations of mathematics. In *Meditationes philosophicae*, Hagen distinguished between *observatio* and *experimentum* on the basis of whether experiences involved intervention or not (Hagen, 1734a, §.6, §.28, pp. 114, 127). With both feet on firm Baconian ground, he then did what Wolff had not done: he discussed in detail the difference between different kinds of experiments. Although the discussion included political and moral experiments, Hagen was particularly interested in the difference between physical and psychological experiments. Whereas the former were a means of studying physical bodies, the latter focused on simple things or souls:

> Another class of experiments is the one that in simple things expounds transformations produced through a voluntary act and is called psychological. Such an experiment is the case if someone wants to frighten a person and depicts various evils that will soon threaten him and observes the effects of conversations and whether he is inclined to fear or not. (Hagen, 1734a, §.37, p. 132)

Two things are especially worth noticing. First, in contrast to Wolff, with Hagen it is no longer just a question of using experiments in psychology, but of conducting specific psychological experiments. Second, these experiments were designed around intervention followed by observations of behavioral change. In the shorter treatise *De mensurandis viribus*, Hagen described a number of similar experiments. Attention could thus be measured by assessing the formation and contemplation of ideas over a period of time, or the faculty of judgment by assessing the process of forming judgments. The most concrete and elaborate experiment, however, concerned the measurement of the natural powers of the intellect:

> Swiftness of operation and the number of performed acts characterize the measurement of the natural powers of the intellect, which will be illustrated with immediate clarity through an example. The magnitude of the natural powers that two boys have for learning history could easily be determined. Thus carry out this examination: any unknown fact should be narrated to both boys, with all the circumstances that are to be attended to in history, such as proper names, dates, natural sequence, etc. And when they are asked, has anything of the previously told information been learned? If not, retell the story for them both until everything has been committed to memory. Repeat this until the whole fact has been thoroughly known, and it will be possible to determine who possesses the largest intellectual powers and who has the highest ratio of attention. The measure of the intellectual powers is given here by means of the speed of activity and the number of individual acts accomplished: the

> smaller the number of repetitions, the greater the speed. This experiment has to be repeated several times before the measure of the two boys can be determined. (Hagen, 1734b, pp. 4–5)

It is hard not to be struck by the astonishingly modern character of Hagen's experiment. In fact, the kind of experiment depicted here is not supposed to exist until the second half of the nineteenth century. Although we cannot know for sure that Hagen actually conducted this experiment, it is clearly the case that he constructed it as practically feasible, and that the epistemic structure of the experiment revolved not only around intervention but also around replication of the whole experiment in order to secure evidence.

Hagen's experiments are indeed fascinating and intriguing. On the one hand, Hagen clearly went beyond Wolff in arguing and describing in detail particular psychological experiments. These experiments were meticulously thought out and designed. On the other hand, despite the potentially revolutionizing nature of his discovery (at least in the eyes of a modern reader), no one seemed to pick it up or even care about it. In fact, Hagen himself would spend the following years translating and publishing Wolff's writings, after which he would disappear from the academic radar (Albrecht, 2010). And with this, his psychological experiments would fall into oblivion seemingly without influencing psychology in any particular way. Instead, the task of introducing an experimental psychology would be shouldered by a new generation of anthropologically inclined physicians.

14.3 Anthropological Medicine and the Measurement of Nerves and Nerve Liquids

The eighteenth-century discussion of psychological experiments and measurability drew on distinctions that went back to ancient philosophy. Aristotle thus divided quantities into numerable pluralities and measurable magnitudes (Aristotle, 1995b, V.13). He further argued that although qualities are neither numerable nor measurable, some of them nevertheless permit a "more or less" (Aristotle, 1995a, VIII). This was, for instance, obvious with affective qualities such as various emotional states. The distinction between different quantities and qualities was further elaborated by Euclid (fl. ca. 300 BC), and in the Middle Ages, a further distinction was established between extensive and intensive magnitudes (Michell, 2006, pp. 416–417; Sturm, 2006, p. 360; Crombie, 1994; Sylla, 1971). While length and volume were categorized as having extensive magnitude, temperature or affectional states such as hatred were now considered to have intensive magnitude. The distinctions between quantities and qualities and multitudes and magnitudes may seem abstract, but in fact corresponded to an increasing interest in quantification and measurability. As a result, a number of magnitudes such as temperature, velocity, and density were turned into real measures in the medieval and early modern periods (Michell, 2006, p. 417; Murdoch, 1975, p. 287). With these successful attempts also followed questions of how to measure mental magnitudes. After all, are not most mental phenomena in fact intensive magnitudes? In the early modern age, two different strategies were developed for how to turn such mental, intensive magnitudes into real measures. The first strategy is the one that we have just seen in Wolff and Hagen, that is, to maintain that these are in fact mental and to find out ways of measuring them. While Wolff had little to offer in terms of practical solutions, Hagen addressed the issue by translating the powers of the intellect into the capacity to remember and retell a story. The second strategy was rather different and instead sought to solve the problem by reducing the mental to the physical. Drawing on the new mechanical philosophy, the so-called iatromechanical theory charted the body as a hydraulic machine circulating bodily fluids (Clericuzio, 2016; Distelzweig, 2016; Müller, 1991). The iatromechanical model gained ground and popularity through authorities such as William Harvey (1578–1657) and René Descartes (1596–1650) at the same time as it struggled to explain processes such as healing and generation. By way of solving these problems, philosophers and physicians turned to chemistry to develop a new vitalist-mechanical model in which the nerves circulated a fine liquid referred to as life spirit (Shackelford, 2016; Chang, 2011). One of the main proponents of this model was Hoffman, whose chair in experimental philosophy Wolff took over in 1709. In the 1690s, Hoffmann had developed and presented his own brand of iatromechanical medicine in *Fundamenta medicinae* (*Foundations of Medicine*) and other writings. Whereas the heart circulates blood and the lymphatic fluid, the nerves circulate life spirits (*spiritus animales*). "This very fine fluid, too fine to be seen, which is contained in the nerves, is nothing but very fine matter, endowed with a limited mechanical power suitable for bringing about ideational and ordered motions in the body. Hence it can properly be called spirit" (Hoffmann, 1695, p. 31; 1971, p. 22). Hoffmann further argued that the living spirits were the causes behind mental phenomena such as sensation and perception. "The animal spirits not only move the body but also provide perception, and no sensation can take place without motion" (Hoffmann, 1695, pp. 35–36; 1971, p. 25).

Hoffmann's theory of how mental phenomena corresponded to the movement of the life spirits was picked up by Johann Gottlob Krüger (1715–1759). Krüger had studied Wolff's philosophy in the 1730s before becoming a student of Hoffmann. The study of medicine for Hoffmann led to a doctoral degree in 1742 and eventually to a professorship in philosophy and medicine at the University of Halle (Sturm, 2010; Nowitzki, 2003, p. 34).[3] In the 1740s, Krüger published a programmatic work on a new scientific medicine, *Grundriß eines neuen Lehrgebäudes der Artzneygelahrheit* (*Outline of a New Foundation for the Medical Science*, 1745), as well as a three-volume work on the new natural philosophy titled *Naturlehre* (*Natural Philosophy*, 1740–1749). In the second volume, the *Physiology*, Krüger declared that most people take the brain to be a machine fueled by a fine nerve liquid (*Nervensaft*), which is sometimes also referred to as *life spirits* (*Lebensgeister*) (Krüger, 1748, §§.294–295, pp. 542–545). Concerning the more exact nature of this

[3] For additional studies, see also Borchers (2011), van Hoorn (2006), and Zelle (2001).

liquid, Krüger emphasized that philosophers had come up with numerous theories, but that most of these tended to be based on abstract deductions rather than experience and experiments, on metaphysical speculation rather than science. To this, he then added that he could accept a physiological but never a metaphysical explanation of the connection between body and soul (Krüger, 1748, §.296, p. 548). As for his own theory, he drew on the Italian physician Giorgio Baglivi's (1668–1707) work *De fibra mortice* (*On Nerve Movement*, 1700). Baglivi conducted experiments on living animals, for instance, by trepanning the cerebral membrane of living dogs and observing the effects (Krüger, 1748, §§.306–309, pp. 568–575). Krüger was highly impressed by these experiments, which he thought proved the role of the nerve membrane in perception. Drawing on Baglivi, he argued that perceptions are transmitted through vibrations in the nerve membrane and that the strength of the vibration corresponds to the intensity of the perception. Krüger formulated this proportional relation between perception and vibration in the nerve membrane into a law of perception:

> It is that one perception = S, the second = s, the impact that one body causes = V, the second = v. The nerve-tension in the first case = T, in the second = t; so is S:s = VT:vt (§.315.). When now T = t, then S:s = V:v. This means that the vivid part of the perceptions is related to the powers of the bodily impact from which it is produced, when the nerve tensions are equally strong. It is further that V = v, then S:s = T:t. That is why the sensations behave like the tensions of the nerve membranes when the effects of the bodies that touch the limbs of the senses are the same. (Krüger, 1748, §.316, p. 589)

As Krüger saw it, the way to scientifically study the soul and its processes was by studying the brain and the movements and vibrations of the nerves. Drawing on mechanical philosophy, and particularly on iatromechanical medicine and physiology, Krüger held that the way to unravel the body-soul mystery was, contrary to Hagen's manipulations of behavior, by conducting surgical experiments in the tradition of Baglivi. That these were not framed as specific psychological experiments did not hinder them from revealing truths about the soul.

In 1751, Krüger published the *Diät oder Lebensordnung* (*Diet or Order of Life*), in which he argued that medicine must also be able to provide an overall theory of the healthy life (Krüger, 1751). The same year, he left Halle to become a professor of medicine and philosophy at the University of Helmstedt (Sturm, 2010). Krüger had by now become a leader of a small group of physicians and philosophers, and in his inaugural speech at Helmstedt, he outlined a medical education wherein physicians should take the whole human being into account (Krüger, 1752). Both writings marked a gradual shift from physiology to a more holistic medical anthropology. The most radical expression of this change is the *Versuch einer Experimental-Seelenlehre* (*Attempt at an Experimental Psychology*) (Krüger, 1756). In contrast to the *Naturlehre*, which operated wholly within the domain of physiology, the *Experimental-Seelenlehre* aimed "to show the philosophers, who are no physicians, the good that medicine might do for them within psychology" (Krüger, 1756, preface). Echoing the critique of metaphysical speculation, Krüger stated in the preface that his ambition was to "describe the human soul as it is and not as it should be" (Krüger, 1756, preface). When it came to the question of what an experimental psychology would actually be like, Krüger opened by rhetorically anticipating a rather obvious objection:

> One would maybe take it for a mere joke if I said that I have taken on the task of showing how one can know the soul through experiments. Experiments, one would say, can only be conducted on bodies. Would one really bring the soul under the air pump, observing its gestalt through the magnifying glass, measuring its powers? This thought is so obvious that I think it would occur to most people who might face these pages. (Krüger, 1756, §.1, p. 1)

The remark shows both that Krüger himself was well aware of the difficulties of making the soul an object of experimental study and that he expected his readers to be so too. And yet, despite the apparent logic of this objection, he considered it to be an error deriving from too narrow a concept of experiment. "It is nevertheless mistaken, and the mistake is that one assumes that no other instruments can be used to conduct experiments on the soul than those that are used in the instrument chamber of the natural philosopher" (Krüger, 1756, pp. 1–2). Krüger's solution was to broaden the concept of experiment so that new forms of experiments on the body would reveal truths of the soul:

> I thus presuppose about what is founded in experience not only that mutations of the body can be known through mutations of the soul but also that the mutations of the soul can be known from the mutations of the body. From this I conclude that one can produce changes in the soul, that would not have occurred in the normal course of nature, by producing extraordinary changes in the body, and this means nothing less than that it is possible to conduct experiments on the soul. (Krüger, 1756, §.7, p. 18)

Krüger's broader concept seems to pave the way to the kind of physiological experiments conducted by Baglivi and others. However, well aware that such experiments on the soul would practically mean to trepan the skull and put needles in the brain, he dismissed them in what is probably one of the first ethical discourses of its kind:

> But as I suggested this, I met an objection that I must admit is of the greatest importance, and that must be so since it reaches me from the voice of humanity itself. What? It exclaims: Cruel barbarian and beast, you who deny nature and excel even the Mexicans in inhumanity. You want to open rational human skulls in order to discover the seat of their reason, you want to cut their brains into pieces to experience where the memory has its seat, you want to pierce the flesh and the periosteum so that they can tell you if they feel anything in these parts, you want to tear the heart out of their bodies and pierce it with needles and ask them if they feel anything. (Krüger, 1756, §.7, pp. 18–19)

The passage accentuates the difference between Hagen's and Krüger's experiments. Whereas the former resorted to harmless manipulation of behavior, the latter used the knives and nails of the surgeon. The dismissal of physiological experiments on normal human beings left Krüger with the three alternatives of experiments on criminals, experiments on animals, and lastly, what he referred to as examination of medical cases. Of these, the first was immediately rejected on the same grounds as experiments on any other human being, and this despite the potential gains and benefits. "Yet I confess the truth, and admit, that my heart would be far too sensitive to conduct such experiments; so I am also equally assured that, through such experiments, if undertaken with appropriate wisdom, psychology would benefit greatly through medicine" (Krüger, 1756, p. 20). The second option, to conduct experiments on animals, remained a possibility at the same time as it was questionable how valuable such experiments would be. After all, not everyone agreed that animals even had souls. In the end, Krüger was thus left with the third option, to draw conclusions from medical cases:

> The sisterly connection, which exists between medical science and philosophy, will provide us with a means to get an experimental psychology without staining our hands with blood. Thus, except that one can conduct many experiments on animals, the observations of the medical science, made in all times, provide circumstances where the soul, through extraordinary changes of the body, is placed in such extraordinary and rare states that they quite reasonably can be regarded as experiments conducted on the soul. (Krüger, 1756, §.7, p. 20)

Carsten Zelle has argued that "Krüger abandons 'experiment' in favour of 'observation' and thus upgrades descriptions of medical cases" (Zelle, 2001, pp. 100–101). Although one reading would indeed be that Krüger abandons "real" experiments in favor of medical observations, one must not forget that he explicitly emphasized that experimental psychology should be based on a broader concept of experiment. This broader concept of experiment must, in turn, be situated in the larger epistemic context of early modern medical experience. Although medical experience boiled down to actual clinical observations, just as the Aristotelian concept of *empeiría* fell back on actual perceptions, clinical physicians typically diagnosed patients by consulting medical examples or cases collected and systematically organized in large catalogues referred to as *observationes*, *curationes*, or *historiae* (Leventhal, 2019; Pomata, 2010, 2005; Gadebusch Bondio & Ricklin, 2008). With the help of these catalogues, physicians were able to determine diseases and prescribe cures. "In physics experience can best be sought from knowledge of mechanics, chemistry, and anatomy; in medical practice experience derives most abundantly from the observations of diseases, and from more accurate histories and cures" (Hoffmann, 1695, p. 2; 1971, p. 5). As a physician and student of Hoffmann, Krüger was of course well acquainted with the role of medical experience. Against this background, the idea of psychological experiments was to find extraordinary cases that would reveal truths of the soul. Extraordinary cases in which, for instance, skull injuries corresponded to pathological behaviors could thus be used to contribute important truths of the soul. In these cases, the extraordinary carried the burden of proof that was placed on devices such as air pumps in normal experiments. The great advantage here was that since the experiment had, so to speak, already happened, no real-time intervention was necessary.

The context of anthropological medicine depicted above further complicates the picture of psychological experiments and experimental psychology. To Wolff's discourse on the use of experiments in psychology and Hagen's attempt to construct such experiments around behavioral intervention must now be added, first, the physiological experiments in the tradition of Baglivi, and second, Krüger's later attempt to launch an experimental psychology wherein extraordinary medical cases carried the burden of proof. The latter was epistemically different from the former in the sense that it did not require any real-time intervention. In fact, since the experiments had, so to speak, already been conducted by nature, one only needed to locate and interpret them.

Another somewhat related attempt was made by the Wolffian philosopher Christian Albrecht Körber (d. 1747?). While little is known about Körber's life, he did produce an extensive work titled *Versuch einer Ausmessung Menschlicher Seelen und aller Dinge überhaupt* (*Attempt at the Measurement of Human Souls and All Things Whatsoever*) in 1746. In contrast to both Hagen and Krüger, Körber's work was a partly impenetrable hodgepodge of Wolffian metaphysics, mathematical formulas, and references to nerves and nerve liquids.

Drawing on the established apparatus of measurement, Körber distinguished between countable quantities and graded qualities (Körber, 1746, p. 6). To the latter belonged simple, finite, and indivisible things—that is, souls. He further argued that souls have the power to change or act, that the ideas and concepts springing from them are temporal and spatial, and that they move with velocity and direction. The consequence of all this was not only that the acts of the soul can be measured but also that it is possible to discover the laws of movement that the soul obeys. Körber's *Versuch* was essentially an attempt to convert Wolff's metaphysics, and particularly his psychology, into mathematical formulas. As such the work is full of formulas, of which some are easier to follow than others. In one of the more intelligible examples, Körber argues that sensation of sound can be measured using the formula MC. M here signifies the mass of air that produces the impact on the ear, whereas C stands for the speed of sound. As sound hits the ear, ideas of a certain clarity are produced given the ratio of M to C and the level of consciousness of the receivers. From this it follows, Körber argues, that the difference between two persons' abilities to perceive can be measured by comparing the clarity of their conceptions of one and the same sound (Körber, 1746, pp. 178–189). Throughout the *Versuch*, Körber added new formulas to the ones already presented, thereby increasing the complexity of his text to levels challenging for even the most patient reader. This was, for instance, the case with a formula that was supposed to describe the relation between perception (*Empfindung*) and imagination (*Einbildung*) (Fig. 14.1).

While it is questionable how much sense formulas such as these made for an early modern reader (or a modern one), Körber sometimes used colorfully described situations to illustrate and substantiate his points. This was the case with attention, where Körber describes what happens as the character Sempronius enters a classroom. First, he chats with his classmates and perceives what they say with a certain clarity. But as the teacher notices and reprimands him with a strong voice, the words of the classmates scatter, and Sempronius immediately becomes attentive to what the teacher has to say:

> The strong shouting causes in the ideal ears of Sempronius an impulse with greater speed (174.§.), which surpasses the impetus of the previous speech of the professor in his ears. And since he loses his remaining thoughts about it, the acts of distinction become weaker, and thus lose their velocity (153.155.§.); but the acts of distinction of the subsequent speech of the professor exceed the previous ones and influence their directions through an impulse. He now attends more to the present speech of the professor than he did before, regardless that the speech before was felt just as well as the present and that he was aware at the time

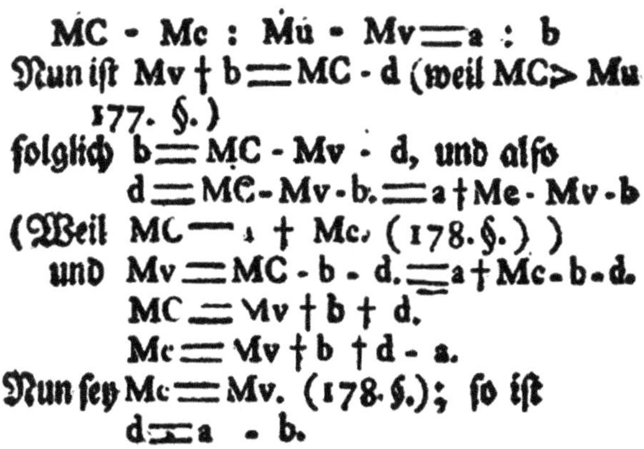
MC - Mc : Mu - Mv = a : b
Nun ist Mv + b = MC - d (weil MC > Mu. 177. §.)
folglich b = MC - Mv - d, und also
d = MC - Mv - b. = a + Mc - Mv - b
(Weil MC = a + Mc, (178. §.))
und Mv = MC - b - d. = a + Mc - b - d.
MC = Mv + b + d.
Mc = Mv + b + d - a.
Nun sey Mc = Mv. (178. §.); so ist
d = a - b.

Fig. 14.1 Formula describing the relation between perception and imagination (Körber, 1746, p. 224)

> that the professor spoke, at the same time as he attended to his classmates, and talked with them about other things; only that the thoughts before did not have the same grade of clarity, even though he knew what the professor was saying. How can it be that he attends more to the present speech than the previous? (Körber, 1746, §.185, pp. 235–236)

Overall, the whole course of events corresponds to the replacement of some sounds with others of a greater mass and velocity. As these impact the ear, they produce corresponding nerve signals which in turn increase the grade of clarity of perceptions. In addition, Sempronius not only shifts his attention but actually increases it, with the result that he now perceives the voice of the professor with even greater clarity. The explanation is that the professor's reprimanding voice gives rise to a sense of fear that corresponds to accelerated velocity in the nerves. Fueled by fear, the level of attention rises, and it will now require a stronger negative impact to stop it. Sempronius is now strongly motivated and focuses on what the professor has to say. It is also at this point that the measure of the soul comes in. By keeping relevant variables constant, it would be possible to compare, measure, and thereby determine who has the most developed faculty of attention.

Körber's attempt to convert Wolff's metaphysics of graded qualities into mathematical formulas borrowed from a number of discourses. On the one hand, it expanded the discourse on graded qualities, temporality, spatiality, velocity, and direction that was already present in Wolff's writings. To further concretize his argument, Körber sometimes used illustrative cases such as the one above. This is something that we also find in Wolff, but that was perhaps used more frequently by Hagen, for whom concrete cases were at the core of psychological experiments. On the other hand, Körber borrowed the discourse on nerves and nerve liquids from physiology, a discourse that Wolff was typically critical toward. At the same time as Körber embraced this discourse, his metaphysical and theoretical hypotheses were of exactly the kind that Krüger dismissed as metaphysical speculation without basis in experimental facts.[4] Taken together, Körber's work again illustrates the heterogeneity of the approaches and attempts to measure and experiment on the soul in early- and mid-eighteenth-century Germany.

14.4 The Metaphysics of Perfection and Introspective Experiments on the Self

Throughout Krüger's work runs a critique of what he referred to as metaphysical speculation. The critique not only aimed at a straw man, but targeted Gottfried Wilhelm Leibniz's (1646–1716) monadology as well as rationalist metaphysics more broadly. But what were these theories actually about, and how did they tap into the discussion of measurement and experimentation on the soul? In this third section, I chart the Leibniz-Wolffian metaphysics of perfection as a third major discourse in which measurability and experimentation played a rather special role.

At the core of the metaphysics of perfection was God as the most perfect being. Created in the image of God, the human being is, according to Leibniz, obliged to refine and perfect itself as much as possible and in every possible respect (Leibniz, 1694–1698?/1988, pp. 83–84; 1686/1989a, p. 35; 1714/1989c, pp. 218–219). On a more particular technical level, the perfection of the mind corresponded to the formation of as clear and distinct ideas as possible (Leibniz, 1686/1989a, p. 35; 1684/1989b, pp. 23–24). To lead a fulfilling human life was here essentially a matter of increasing the overall level of perfection by cultivating, refining, and increasing the levels of clarity and distinctness through the practice of philosophy and science. The point I wish to make here is that the idea of magnitude was at the very heart of the metaphysics of perfection, and that this in turn paved the way for discussions of degrees, measurability, and, as we will see, a rather peculiar notion of observations and experiments on the self.

If we now turn to Wolff, it is well-known that he drew heavily on Leibniz, especially when it came to the metaphysics of perfection. In *Vernünfftige Gedancken von Gott* (*German Metaphysics*), Wolff argues that things as well as our knowledge of things can be more or less perfect depending on, in somewhat simplified terms, the level of complexity (Wolff, 1720b). Humans are thus more perfect than plants in the sense that they involve more parts and more complex connections between these parts. In a similar vein, knowledge of plants and humans can be more or less perfect depending on the number of parts it involves and the understanding of the relations between them. In *Vernünfftige Gedancken von Gott*, Wolff over and over again emphasizes that there are grades of perfection and, more specifically, grades of clarity and distinctness. In the parts of the work that treat the soul, he further elaborates

[4] Despite this difference and the fact that I have not been able to establish any referentiality between the works, Körber's choice of topics and cases sometimes makes one wonder if he did not draw directly on Krüger's physiology.

this topic by ascribing duration and velocity to thoughts (Wolff, 1720b, §§.736–737, pp. 406–407). If we now turn to *Psychologia empirica*, we find that it is in this metaphysical context of grades, and more particularly of grades of pleasure and disgust, that Wolff emphasizes "that a mathematical knowledge of the human soul and thus a psychometry is possible" (Wolff, 1732, §.522, p. 404). Since psychometrics is treated elsewhere in this anthology, I will only briefly comment on the historical context here. Rather than inscribing the few and fragmentary remarks on psychometrics in a long history of measurement, or arguing that Wolff founded a discipline of psychometrics, I suggest that the immediate context was that of the metaphysical discourse of grades and perfection. Situated in this context, we see that it was not actual measurement that was important. What mattered was instead to maintain the view of a perfect mathematically ordered universe, a universe where, as Wolff put it in the *Vernünfftige Gedancken von der Menschen Thun und Lassen* (*German Ethics*), "nature obliges us to do what makes us and our state more perfect and to avoid that which makes us and our state imperfect" (Wolff, 1720a, §.19, p. 15).

The metaphysics of perfection was further elaborated first by Alexander Gottlieb Baumgarten (1714–1762) and then by Georg Friedrich Meier (1718–1777), both of whom drew heavily on Wolff. Baumgarten belonged to a new generation of nonorthodox Wolffian philosophers. Today, he has become famous for his aesthetics, but at the time, he also wrote extensively on ethics. In *Ethica philosophica* (*Philosophical Ethics*), Baumgarten followed Wolff when formulating the moral imperative "Perfect yourself, that is, perfect yourself in the natural state as much as possible, in the natural state. ... Do what is good, omit what is bad as much as possible: do in everything what optimizes yourself" (Baumgarten, 1740, §.10, p. 4). To the moral imperative of self-perfection corresponded a broad spectrum of spiritual exercises wherein practitioners were urged to examine their own cognition as clearly and distinctly as possible and with respect to both quantity and quality. The practitioners should, furthermore, attend to and reflect on the cognitive states separately as well as compare them (Baumgarten, 1740, §.157, p. 66). In this context, he also encouraged his readers to systematically reflect on the past, present, and future, and stressed that "DIARIES, which are tools to support your memory, should not be scorned, or the day-to-day notes of things pertaining to you" (Baumgarten, 1740, §.160, p. 67). As Baumgarten addressed the specific duties toward the soul further into the treatise, he emphasized the importance of empirical, rational, and mathematical knowledge of the soul. "Know your soul, empirically, rationally, and mathematically as much as you can, and do not only know it so much, but cure and amend it" (Baumgarten, 1740, §.202, p. 89). This was far from exceptional, as Baumgarten time and again urged his readers to "know, experience, and measure" the various faculties of the soul (Baumgarten, 1740, §§.201–249, pp. 88–123).

Although Baumgarten did not explicitly refer to these operations as experiments, Meier would do exactly this. Having studied with Baumgarten in the late 1730s, Meier became a loyal friend, admirer, and popularizer of Baumgarten's philosophy (Schenk, 1994). In the introduction to the third volume of his metaphysics, which concerned psychology, Meier argued that psychology is crucial for practical philosophy, and more particularly for the duties toward the self. Since moral action requires knowledge of the soul, "a great part of the duties toward ourselves cannot be discovered, acknowledged, or exercised without psychology" (Meier, 1757, §.476, p. 14). Similar points were made in Volume 2 of *Philosophische Sittenlehre* (*Philosophical Ethics*), wherein Meier devoted more than 300 pages to self-knowledge. To enhance self-knowledge, one should examine the self on a daily basis, preferably in the morning when the mind is still unoccupied (Meier, 1762/2007, §.407, p. 391). Meier also suggested that one should observe and conduct experiments on the self. "We must conduct experiments and inquiries of ourselves and our powers, and we must collect as many experiences, observations, and inquiries that could lead to self-knowledge, as possible" (Meier, 1762/2007, §.422, p. 434). Concerning the more specific nature of these, Meier thought of them neither in terms of behavioral intervention nor as physiological operations. Instead, he launched yet another category of introspective observations and experiments on the self:

> If I, for instance, want to know whether I'm able to overcome myself, is there any better way to convince myself than by trying? Than by trying and seeing whether I'm able to keep my passions in check? And so on. It would certainly be an extraordinarily useful endeavor to examine more closely the art of conducting experiments on the self, and on the whole how to obtain experiences of oneself. (Meier, 1762/2007, §.422, p. 434)

Meier, it seems, here suggests yet another strand of experiments on the soul. In sharp contrast to behavioral intervention, surgical trepanning of the brain, or the consultation of medical cases, Meier's experiments revolved around introspective examinations of the self. In this regard, they fell back on a long tradition of introspective spiritual exercises represented within philosophy but perhaps even more so within the Christian devotional tradition. In the immensely popular and widespread *Exercitia spiritualia* (*Spiritual Exercises*), Ignatius of Loyola (1491–1556) thus encouraged his readers to organize their lives around such daily examinations of the self. Later on, such Christian spiritual exercises fueled and shaped the philosophy and epistemology of Descartes, Leibniz, and in the longer run, Wolff, Baumgarten, and Meier (Hatfield, 1986; Hunter, 2001; Rydberg, 2017). Being thus well established, they also attracted criticism. Among these critics was Krüger, who remarked that philosophers have traditionally examined the soul through the internal rather than the external senses (Krüger, 1756, §.5, p. 13). As Krüger saw it, this method had little to do with experiments, the latter requiring that "we put the thing in a condition that it would otherwise not be in, and that we thereby force nature to reveal things that it intended to keep hidden from our eyes" (Krüger, 1756, §.6, p. 15). Meier was certainly also familiar with the distinction between observations and experiments. In fact, one reason why he did not explicitly address it in the context of experiments on the self may have been that he simply did not see any particular problem in adding an element of intervention. In any case, the discourse further illustrates the extraordinary heterogeneity and complexity of the eighteenth-century discussion of experiments on the soul.

14.5 Conclusion

How should we understand Wolff's psychology, and particularly the attempts to measure and experiment on the soul? What kind of project was this? In this final section, I will suggest two contexts. The first, somewhat narrower, context is that of early modern experimental philosophy, and particularly of the nature and limits of experimentation. Measurement and experimentation on the soul here accentuated questions of what defined an experiment as such. The second context is larger and revolves around the recent reorientation within the history of philosophy and science from abstract theories to concrete practices and spiritual exercises. Situated in this context, the early modern interest in the soul becomes a very different project from the one traditionally brought forth by epistemologically oriented historians of philosophy and historians of psychology.

To start with the first context, I have argued that there was no such thing as a unified project of psychological experiments or experimental psychology in early- and mid-eighteenth-century Germany. Instead, there was a heterogenous plethora of attempts to discuss and perhaps even practically measure and conduct experiments on the soul. Although these often overlapped, the discourse on experimentation was particularly disparate. On the one hand, for most experimental philosophers, it was rather clear what an experiment was. An experiment was an experience involving intervention in the course of nature. While most agreed regarding the necessity of intervention, there was little consensus as to what, more specifically, counted as an intervention. Whereas Wolff clearly thought that experiments could be used in all scientific disciplines including psychology, there is no indication that he ever acknowledged the existence of specific psychological experiments. Instead, such specific experiments were introduced first by Hagen, who solved the problem of intervention by measuring the effects of various forms of behavioral interventions. In sharp contrast to this solution, the early Krüger instead argued for a physiological strand of experiments on the soul. In place of Hagen's "light" behavioral interventions, Krüger introduced sharp instruments and the trepanning techniques of the surgeon. As Krüger shifted focus in the 1750s from physiology to a more holistic anthropology, he abandoned surgical intervention in favor of the consultation of medical cases. With this shift, intervention became epistemically detached from the experimental situation as a temporally and spatially situated event. The experiments, it seemed, had already been conducted by nature and now it remained to discover and interpret them. Yet another approach to measurement and experimentation was developed through the Leibniz-Wolffian metaphysics of perfection. In this project, intensive magnitudes or graded qualities were crucial, but, paradoxically, not actual measurement. Following this discourse to Baumgarten and Meier, yet another conception of experiments is revealed in the form of introspective observations, examinations, and experiments on the self. Rather than being a child of the new experimental philosophy, this strand of experimentation drew on a long tradition of Christian spiritual exercises.

Taken together, there is nothing in the discourses on the measurement and experimentation on the soul that suggests a unified project of eighteenth-century experimental psychology. The only attempt at such an endeavor was made by Krüger, who advocated a rather particular concept of experimentation on the soul, and who in the end fell into a rather traditional Wolffian account of the soul. Instead, there was a broad interest in exploring the limits of experimental philosophy. What exactly was required for something to be called an experiment? And how could the soul be made an object of experimentation? There was not one answer to these questions, but rather a number of traditions employing different epistemic techniques to provide a number of disparate hypotheses and attempts.

The context of experimental philosophy invites another much larger question. What kind of project was natural and experimental philosophy? What was the purpose of engaging in this project? While the traditional answer has been that it was a project of knowledge, of searching and obtaining knowledge for its own sake, recent historians of philosophy and science have started to question this picture. Drawing partly on the French historian of philosophy Pierre Hadot's reinterpretation of ancient philosophy as a way of life, a number of scholars have argued for a similar approach to early modern philosophy and science (Hadot, 1995). Sorana Corneanu has thus argued that Bacon's call for a new scientific method in fact took shape in the larger context of what she calls the early modern *cultura animi* tradition (Corneanu, 2011). Through close reading of otherwise often neglected sources and passages, she shows that Bacon launched his account of science and scientific method as a cure of a soul perceived as diseased and perturbed by the passions. The new method was here presented as a systematic regimen, the purpose of which was to temper, control, and cure the mind. Similar readings have been made of the philosophy and mathematics of Descartes and others (Jones, 2006; Gaukroger, 2001; Hatfield, 1986). Mathematics here became a way of cultivating and refining the mind and its faculties, cognitively as well as morally. If we turn to the German context, Ian Hunter has in a similar way argued that Leibniz's metaphysics of perfection should be understood in terms of spiritual exercise, and more specifically as a self-transformative contemplative practice whereby practitioners perfect their own selves by contemplating the perfection of the universe (Hunter, 2001, pp. 102–107).

While Corneanu has highlighted the context of *cultura animi*, other scholars have approached the same problematic in terms of therapy or moral psychology. Ted Brennan thus argues that the Stoic account of the soul must be understood not as a psychology in the modern sense of the word but as a moral psychology in which the perfect human psyche belongs to the perfect ethical agent, "violations of ethical standards always reflect lapses in psychological hygiene, and our obligations are set for us by the actual practice of psychologically perfect agents" (Brennan, 2003, p. 259). Although I don't think that all early modern philosophy and science, or all aspects of it, can be understood in this way, I'm convinced that the two related concepts of *cultura animi* and moral psychology provide important contexts for understanding the Wolffian philosophy and psychology. Following Hunter's reading of Leibniz, the Wolffian version of a regimen of the mind here becomes the metaphysics of perfection, according to which the human being is obliged to perfect the self cognitively and morally. Wolff's often complicated discussions of the interdependence between different disciplines reflected the view that each and every discipline contributed to this vision by cultivating clear and distinct concepts within its own particular domain. The crucial role of psychology here was to refine the knowledge of the mind and its workings, something that was seen as necessary for all other knowledge and especially for practical philosophy, where virtuous action was the direct result of knowledge. In the *Vernünfftige Gedancken von der Menschen Thun und Lassen*, Wolff thus emphasized, "Since the intellect must judge what is good and evil, and what is best among the good, so does the will become more perfect or improved as one brings people to a more vivid knowledge of the good. Thus, one cannot improve the will in any other way than through the intellect" (Wolff, 1720a, §.373, pp. 238–239). Similar points regarding the relation between knowledge and virtuous action were made by Meier in the immediate context of psychology. "Without psychology one can take no certain and reliable step in practical philosophy, since it contains the authentic and true foundations of all human rights and obligations" (Meier, 1757, §.476, p. 13). The modern understanding of scientific disciplines as distinct, meaningful, and valuable quite separately from other scientific disciplines would have been unintelligible for Wolff and the Wolffians as well as for most early modern intellectuals. That the project of psychology was simply not meaningful if separated from practical philosophy and the questions of how to lead a virtuous life needs to be taken seriously. To ignore this by anachronistically inscribing these discourses in a long history starting with embryotic and flawed theories and ending with modern scientific psychology is to prioritize the legitimization of the present at the expense of the understanding of the past. Such prioritization of the present risks closing us off, making us blind and unable to learn from the genuine otherness of historically specific contexts.

References

Albrecht, M. (2010). Hagen, Gottlieb Friedrich (1710–69). In H. Klemme & M. Kuehn (Eds.), *Dictionary of eighteenth-century German philosophers* (Vol. 2). Continuum.

Aristotle. (1995a). Categories. In J. Barnes (Ed.), *The complete works of Aristotle: The revised Oxford translation*. (One-volume digital edition). Princeton University Press.

Aristotle. (1995b). Metaphysics. In J. Barnes (Ed.), *The complete works of Aristotle: The revised Oxford translation*. (One-volume digital edition). Princeton University Press.

Aristotle. (1995c). Posterior analytics. In J. Barnes (Ed.), *The complete works of Aristotle: The revised Oxford translation*. (One-volume digital edition). Princeton University Press.

Bacon, F. (2000). *The new organon* (L. Jardine & M. Silverthorne, Trans.). Cambridge University Press. (Original work published 1620).

Baumgarten, A. G. (1740). *Ethica philosophica*. Carl Herrmann Hemmerde.

Borchers, S. (2011). *Die Erzeugung des "ganzen Menschen": Zur Entstehung von Anthropologie und Ästhetik an der Universität Halle im 18. Jahrhundert*. De Gruyter.

Brennan, T. (2003). Stoic moral psychology. In B. Inwood (Ed.), *The Cambridge companion to the Stoics* (pp. 257–294). Cambridge University Press.

Brozek, J. M. (1999). From "psichiologia" to "psychologia": A graphically documented archival study across three centuries. *Journal of the History of the Behavioral Sciences, 35*(2), 177–180.

Chang, K. (2011). Alchemy as studies of life and matter: Reconsidering the place of vitalism in early modern chymistry. *Isis, 102*, 322–329.

Clericuzio, A. (2016). Mechanism and chemical medicine in seventeenth-century England: Boyle's investigation of ferments and fermentation. In P. M. Distelzweig, B. Goldberg, & E. Ragland (Eds.), *Early modern medicine and natural philosophy*. Springer.

Corneanu, S. (2011). *Regimens of the mind: Boyle, Locke, and the early modern cultura animi tradition*. University of Chicago Press.

Crombie, A. C. (1994). *Styles of scientific thinking in the European tradition: The history of argument and explanation especially in the mathematical and biomedical sciences and arts*. Duckworth.

Daston, L. (2011). The empire of observation, 1600–1800. In L. Daston & E. Lunbeck (Eds.), *Histories of scientific observation*. University of Chicago Press.

Dear, P. (1995). *Discipline and experience: The mathematical way in the scientific revolution*. University of Chicago Press.

Dear, P. (2006). The meanings of experience. In K. Park & L. Daston (Eds.), *The Cambridge history of science* (Early modern science) (Vol. 3). Cambridge University Press.

Distelzweig, P. M. (2016). "Mechanics" and mechanism in William Harvey's anatomy: Varieties and limits. In P. M. Distelzweig, B. Goldberg, & E. Ragland (Eds.), *Early modern medicine and natural philosophy*. Springer.

École, J. (1979). De la notion de philosophie expérimentale chez Wolff. *Les Études Philosophiques, 4*, 397–406.

Feuerhahn, W. (2007). Die Wolffsche "Denkschulung" in der psychometrie. In J. Stolzenberg & O.-P. Rudolph (Eds.), *Christian Wolff und die europäische Aufklärung: Akten des 1. Internationalen Christian-Wolff-Kongresses, Halle (Saale), 4.-8. April 2004, Teil 5: Wolff und seine Schule, Wirkungen Wolffs, Wolff in Halle—Vertreibung und Rückkehr*. Olms.

Gadebusch Bondio, M., & Ricklin, T. (2008). Einleitung. In *Exempla medicorum: Die Ärzte und ihre Beispiele, 14.-18. Jahrhundert*. Sismel.

Gaukroger, S. (2001). *Francis Bacon and the transformation of early-modern philosophy*. Cambridge University Press.

Hadot, P. (1995). *Philosophy as a way of life: Spiritual exercises from Socrates to Foucault*. (A. I. Davidson, Ed.; M. Chase, Trans.). Blackwell.

Hagen, G. F. (1734a). *Meditationes philosophicae de methodo mathematica*.

Hagen, G. F. (1734b). *Programma de mensurandis viribus intellectus*.

Hatfield, G. (1986). The senses and the fleshless eye: The meditations as cognitive exercises. In A. O. Rorty (Ed.), *Essays on Descartes' meditations*. University of California Press.

Hatfield, G. (2006). Empirical, rational and transcendental psychology: Psychology as science and as philosophy. In P. Guyer (Ed.), *The Cambridge companion to Kant and modern philosophy*. Cambridge University Press.

Hoffmann, F. (1695). *Fundamenta medicinae ex principiis naturae mechanicis in usum philiatrorum succincte proposita*.

Hoffmann, F. (1971). *Fundamenta medicinae*. (L. S. King, Trans.). Neale Watson Academic.

Hunter, I. (2001). *Rival enlightenments: Civil and metaphysical philosophy in early modern Germany*. Cambridge University Press.

Jones, M. L. (2006). *The good life in the scientific revolution: Descartes, Pascal, Leibniz, and the cultivation of virtue*. University of Chicago Press.

Körber, C. A. (1746). *Versuch einer Ausmessung menschlicher Seelen und aller einfachen endlichen Dinge überhaupt, wie solche der innern Beschaffenheit derselben gemäß ins Werck zu richten ist, wenn man ihre Kräffte, Vermögen un Würckungen recht will kennen lehrnen: In dreien Theilen*.

Krüger, J. G. (1748). *Naturlehre zweiter theil, welcher die physiologie, oder lehre von dem Leben und der Gesundheit der menschen in sich fasset* (2nd ed.). Carl Herrmann Hemmerde.

Krüger, J. G. (1751). *Diät oder Lebensordnung*. Carl Herrmann Hemmerde.
Krüger, J. G. (1752). *Zuschrift an seine Zuhörer von der Ordnung in welcher man die Artzneigelahrheit erlernen müsse*. Carl Herrmann Hemmerde.
Krüger, J. G. (1756). *Versuch einer Experimental-Seelenlehre*. Carl Herrmann Hemmerde.
Lapointe, F. H. (1972). Who originated the term "psychology". *Journal of the History of the Behavioral Sciences, 8*(3), 328–335.
Leibniz, G. W. (1988). Felicity. In P. Riley (Ed.), *Leibniz: Political writings* (2nd ed.). Cambridge University Press. (Original work published 1694–1698?).
Leibniz, G. W. (1989a). Discourse on metaphysics. In R. Ariew & D. Garber (Trans.), *Philosophical essays*. Hackett. (Original work published 1686).
Leibniz, G. W. (1989b). Meditations on knowledge, truth, and ideas. In R. Ariew & D. Garber (Trans.), *Philosophical essays*. Hackett. (Original work published 1684).
Leibniz, G. W. (1989c). The principles of philosophy, or monadology. In R. Ariew & D. Garber (Trans.), *Philosophical essays*. Hackett. (Original work published 1714).
Leventhal, R. S. (2019). *Making the case: Narrative psychological case histories and the invention of individuality in Germany, 1750–1800*. De Gruyter.
Luccio, R. (2013). Psychologia—The birth of a new scientific context. *Review of Psychology, 20*(1–2), 5–14.
Meier, G. F. (1757). *Metaphysik. Dritter Theil*. Gebauer.
Meier, G. F. (2007). *Philosophische Sittenlehre. Anderer Theil*. Zweyte verbesserte Auflage. Olms. (Original work published 1762).
Métraux, A. (1983). An essay on the early beginnings of psychometrics. In G. Eckardt & L. Sprung (Eds.), *Advances in the historiography of psychology*. VEB Deutscher Verlag der Wissenschaften.
Michell, J. (2006). Psychophysics, intensive magnitudes, and the psychometricians' fallacy. *Studies in the History and Philosophy of Biological and Biomedical Sciences, 17*, 414–432.
Mühlpfordt, G. (1992). Die organischen Naturwissenschaften in Wolffs empiriorationalistischer Enzyklopädistik. In S. Carboncini & L. C. Madonna (Eds.), *Nuovi studi sul pensiero di Christian Wolff*. Olms.
Müller, I. W. (1991). *Iatromechanische Theorie und ärztliche Praxis im Vergleich zur galenistischen Medizin: Friedrich Hoffmann, Pieter van Foreest, Jan van Heurne*. Franz Steiner.
Murdoch, J. E. (1975). From social to intellectual factors: An aspect of the unitary character of late medieval learning. In J. E. Murdoch (Ed.), *The cultural context of medieval learning: Proceedings of the first International Colloquium on Philosophy, Science, and Theology in the Middle Ages, September 1973*. Reidel.
Nowitzki, H.-P. (2003). *Der wohltemperierte Mensch: Aufklärungsanthropologien im Widerstreit*. De Gruyter.
Park, K. (2011). Observation in the margins, 500–1500. In L. Daston & E. Lunbeck (Eds.), *Histories of scientific observation*. University of Chicago Press.
Pomata, G. (2005). Praxis historialis: The uses of historia in early modern medicine. In G. Pomata & N. G. Siraisi (Eds.), *Historia: Empiricism and erudition in early modern Europe*. MIT Press.
Pomata, G. (2010). Sharing cases: The observationes in early modern medicine. *Early Science and Medicine, 15*(3), 193–236.
Pomata, G. (2011). Observation rising: Birth of an epistemic genre, 1500–1650. In L. Daston & E. Lunbeck (Eds.), *Histories of scientific observation*. University of Chicago Press.
Ramul, K. (1960). The problem of measurement in the psychology of the eighteenth century. *American Psychologist, 15*, 256–265.
Rudolph, O.-P., & Goubet, J.-F. (2004). Einleitung: Die Psychologie Christian Wolffs: Systematische und historische Untersuchungen. In O.-P. Rudolph & J.-F. Goubet (Eds.), *Die Psychologie Christian Wolffs: Systematische und historische Untersuchungen*. Niemeyer.
Rydberg, A. (2017). *Inner experience: An analysis of scientific experience in early modern Germany*. Uppsala Universitet.

Schenk, G. (1994). *Leben und Werk des halleschen Aufklärers Georg Friedrich Meier*. Hallescher Verlag.

Shackelford, J. (2016). Transplantation and corpuscular identity in Paracelsian vital philosophy. In P. M. Distelzweig, B. Goldberg, & E. Ragland (Eds.), *Early modern medicine and natural philosophy*. Springer.

Shapin, S., & Schaffer, S. (1985). *Leviathan and the air-pump: Hobbes, Boyle, and the experimental life*. Princeton University Press.

Sturm, T. (2006). Is there a problem with mathematical psychology in the eighteenth century? A fresh look at Kant's old argument. *Journal of the History of the Behavioral Sciences, 42*(4), 353–377.

Sturm, T. (2010). Krüger, Johann Gottlob (1715–59). In H. Klemme & M. Kuehn (Eds.), *Dictionary of eighteenth-century German philosophers* (Vol. 2). Continuum.

Sylla, E. (1971). Medieval quantifications of qualities: The "Merton School". *Archive for the History of Exact Sciences, 8*(1), 9–39.

van Hoorn, T. (2006). *Entwurf einer Psychophysiologie des Menschen: Johann Gottlob Krügers Grundriss eines neuen Lehrgebäudes der Artzneygelahrtheit (1745)*. Wehrhahn.

Vidal, F. (2011). *The sciences of the soul: The early modern origins of psychology*. (S. Brown, Trans.). University of Chicago Press.

Wolff, C. (1713). *Vernünfftige Gedancken von den Kräften des menschlichen Verstandes*. Renger.

Wolff, C. (1720a). *Vernünfftige Gedancken von der Menschen Thun und Lassen*. Renger.

Wolff, C. (1720b). *Vernünfftige Gedancken von Gott, der Welt und der Seele des Menschen, auch allen Dingen überhaupt*. Renger.

Wolff, C. (1728). *Philosophia rationalis sive logica, methodo scientifica pertractata*. Renger.

Wolff, C. (1732). *Psychologia empirica methodo scientifica pertractata*. Renger.

Wolff, C. (1963). *Preliminary discourse on philosophy in general*. (R. J. Blackwell, Trans.). Bobbs-Merril.

Wolff, C. (1980a). Christian Wolffs eigene Lebensbeschreibung herausgegeben mit einer Abhandlung über Wolff von Heinrich Wuttke. In H. W. Arndt (Ed.), *Christian Wolff Biographie*. Olms. (Original work published 1841).

Wolff, C. (1980b). Christian Wolff's prolegomena to empirical and rational psychology: Translation and commentary (R. J. Richards, Trans.). *Proceedings of the American Philosophical Society, 124*(3), 227–239.

Zelle, C. (2001). Experiment, experience and observation in eighteenth-century anthropology and psychology—The examples of Krüger's Experimentalseelenlehre and Moritz' Erfahrungsseelenkunde. *Orbis Litterarum, 56*, 93–105.

Chapter 15
The Science of the Soul and the Unyielding Architectonic: Kant Versus Wolff on the Foundations of Psychology

Michael Bennett McNulty

15.1 Introduction

Christian Wolff earned pride of place in the early history of psychology for his novel work on the foundations and methodology of the fledgling science.[1] Inspired by the renowned successes in the science of bodies of the prior century achieved by luminaries such as Galileo (1564–1642), Descartes (1596–1650), and Newton (1643–1727), Wolff sought to appropriate the same sorts of methods to study the human soul. He famously innovated the distinction between empirical and rational psychology and argued that their respective observational and demonstrative methods were equally essential to achieving well-grounded knowledge of the human mind (Wolff, 1738/1968, §§.1, 434; 1740/1972, §.1). According to Wolff (1740/1972, §.112), we ought first to collect empirical data about the soul, garnered through observation as well as experimentation, and subsequently to formulate hypotheses and explanations through rigorous, precise reasoning. Such theories thereby developed are then to be tested with respect to the empirical data. Ideally, we ought additionally to mathematize our theories of the mind and to formulate quantitative explanations of occurrences in the soul, both to precisify our theories and to enhance our certainty in them (Wolff, 1738/1968, §§.522, 616; 1740/1983a, §§.27–28). Finally, Wolff developed a comprehensive theory of knowledge and metaphysics within which these various pieces fit together, a framework that proved fruitful if not critical for the early development of psychology as well as physiological studies of thinking and perception.[2]

[1] For helpful overviews of the relevant history, see Hatfield (1995) and Vidal (2011).

[2] Recently, Hatfield (2018) interrogates this standard prioritization of the Wolffian tradition in the history of psychology, emphasizing the importance of Descartes' physiological-mechanical study of the nervous system.

M. B. McNulty (✉)
University of Minnesota, Minneapolis, MN, USA
e-mail: mcnu0074@umn.edu

S. d. F. Araujo et al. (eds.), *The Force of an Idea*, Studies in History and Philosophy of Science 50, https://doi.org/10.1007/978-3-030-74435-9_15

Immanuel Kant (1724–1804) was likewise a well-known admirer of the advances in natural science that preceded him. For example, Kant modeled his approach to metaphysics on the "secure path of science," which refers to the approach to natural science inaugurated by the scientific revolution along with earlier advancements in the nonnatural sciences of logic and mathematics (1787/1911a, pp. 7–13; see also Kant, 1784/1911c, pp. 279–280, 327). Furthermore, he sought even to provide what he saw as a necessary metaphysical grounding for the new science of bodies (Kant, 1787/1911a, pp. 126–127; 1786/1911b, pp. 471–477; 1784/1911c, pp. 294–326).

In contrast to Wolff, however, Kant was sharply critical of the prospects for the advancement of the inchoate science of the soul. Kant rejected Wolffian psychology root and branch. Not only did Kant famously challenge the fruitfulness of the metaphysical doctrine of rational psychology in the Paralogisms chapter of the *Critique of Pure Reason*, but in other texts—most prominently, the preface of the *Metaphysical Foundations of Natural Science*—he sharply condemned the empirical side of the investigation of the soul. To wit, Kant denied both the utility and prudence of observing mental occurrences and refused empirical psychology the engine of the natural scientific advancements of the prior century: mathematics (Kant, 1786/1911b, p. 471; 1798/1917, pp. 121, 132–133, 161; 1942, p. 237; 1970a, p. 679).

Although the contrast between Wolff's and Kant's ultimate, divergent evaluations of psychology is clear enough, the basis for Kant's departure from Wolff is not simple to discern. In the following, I demonstrate that Kant's divergences from the Wolffian tradition that bring about his negative evaluation of the natural science of the soul lie relatively deep within his metaphysics and theory of knowledge. In particular, Kant breaks subtly, but consequentially, from Wolff regarding the basic sorts of cognition, the essence of natural science, the relation between the empirical and the rational, and the nature and application of the mathematical method. It is these methodological and conceptual modifications that undergird Kant's departure from Wolff and rejection of a fruitful empirical psychology.

Beyond clarifying the precise relations between their conceptualizations of psychology, consideration of these differences also epitomizes an essential distinction between Kant's and Wolff's respective philosophical approaches, which I take to be the primary upshot of my chapter. Particularly, diagnosing this schism contrasts the rigor of Kant's Critical philosophical system with the pliability of Wolff's. Wolff's philosophical system is dynamic, allowing for the combination of various sorts of cognition, methods, and doctrines, whereas Kant's is substantially more regimented. For Kant, everything has its particular, ordained niche in the architectonic; from this perspective, Wolff's dynamism appears as a jumbled hodgepodge. Indeed, in Kant's eyes, Wolff succumbs to tantalizing, but dangerous, lures, like using the mathematical method beyond its strict limits and surreptitiously utilizing empirical cognition in a priori contexts. These clashing philosophical models ultimately prompt their varying evaluations of the prospects for a positive science of psychology. For Wolff, the consolidation of observation, experimentation, deduction, mathematics, and quantitative measurements in psychology supports optimism for the science, whereas Kant's rigorously structured system allows for none of this confused interplay. To put it succinctly, Wolff's science of the soul finds no home in the unyielding architectonic of Kant.

In part 1 (Sect. 15.2), I provide an overview of the major tenets of Wolffian psychology to which Kant responds. In particular, I explain the distinction between empirical and rational psychology as well as the interrelations between the two (Sect. 15.2.1); Wolff's views on the mathematical method, which finds application in rational psychology (Sect. 15.2.2); and his account of psychometrics, the mathematics of mental phenomena (Sect. 15.2.3). In part 2 (Sect. 15.3), I detail in turn Kant's negative reactions to each of these tenets of Wolffian psychology, which both casts into relief his fundamental divergences from Wolff; highlight heretofore underappreciated dimensions of Kant's views on psychology, natural science, and mathematics; and demonstrate the austerity of his Critical system.

15.2 Wolff on Psychology

In the following, I detail three noteworthy aspects of Wolff's conception of psychology: first, his distinction between empirical and rational psychology; second, his views on the application of the mathematical method to psychology (and philosophy, more generally); and, third, his conception of psychometrics.[3]

15.2.1 Empirical and Rational Psychology

For Wolff, psychology is the doctrine of the human soul, within which distinguishable subsidiary doctrines—its empirical and rational species—consider the soul in disparate ways. The distinction between empirical and rational psychology is first explicitly coined in the *Preliminary Discourse* (Wolff, 1740/1983a) and fleshed out in (perhaps arduous) detail in the paired *Empirical Psychology* (Wolff, 1738/1968) and *Rational Psychology* (Wolff, 1740/1972), although it is also implicit in the earlier *German Metaphysics* (Wolff, 1751/1983b).

Rational and empirical psychology differ from one another in terms of both the sorts of knowledge contained in each and their respective methods of achieving the said knowledge. Whereas empirical psychology is based on *experiences* of the soul and concerns psychological occurrences, rational psychology is based on *demonstration* and seeks explanations for the occurrences in the soul. Thus, in the *Empirical Psychology*, Wolff (1738/1968) explains that "*Empirical psychology* is the science that establishes principles through experience, whence reason is given

[3] For overviews of Wolff's account of psychology, see Richards (1980), Hatfield (1995, pp. 197–200), Dyck (2014, pp. 19–42), Kim (1994, pp. 35–52), Hinske (1999), Sturm (2009, pp. 56–68), Vidal (2011, pp. 89–95), Goubet (2018), and Rumore (2018).

for what occurs in the human soul" (§.1, p. 1),[4] where such "reasons" are to be provided in rational psychology. Empirical psychology is paradigmatically introspective; as Wolff puts it, "we come to know the subjects dealt with in empirical psychology by attending to those occurrences in our souls of which we are conscious" (§.2, p. 2).[5] We fundamentally rely on *apperception* in empirical psychology: whereas perception is the soul's awareness of things, apperception is its consciousness of its own perceptions (Wolff, 1738/1968, §.25).[6] Thus, according to Wolff, empirical psychology is concerned not with the nature of the soul; its goal is rather simply to describe those events that can be apperceived of the soul in daily experience. One needs no special training to make these observations. So, Wolff (1751/1983b) reports, he "wants to invoke nothing more than what everyone who pays attention to themselves can cognize" (§.191, pp. 106–107).[7] That said, empirical psychology is no mere list, so to say, of immediate observations of the soul; it also involves the postulation of faculties of the mind, whose effects are witnessed in experience, and their principles (throughout Wolff, 1738/1968; 1751/1983b, §§.191–539). Thus, the body of the *Empirical Psychology* is a catalogue of the chief mental faculties—cognition and appetition—and their various capacities: sensation, imagination, reason, will, and so forth (see Dyck, 2014; Rumore, 2018). All in all, empirical psychology provides both a survey of the characteristic capacities of the soul and their principles as well as a detailed account of mental occurrences.

As mentioned above, rational psychology provides the "reasons" for these phenomena catalogued in empirical psychology. Wolff (1740/1972) defines rational psychology as "the science of whatever is possible through the human soul" (§.1, p. 1).[8] More helpfully, Wolff (1740/1983a) elsewhere explains that "In rational psychology we derive a priori from a unique concept of the human soul all of those features observed a posteriori that pertain to it, as well as those deduced from these observations, insofar as this is proper to philosophy" (§.112, p. 51). Rational psychology is paradigmatically concerned with the nature, or essence, of the soul, from which the observed phenomena and faculties discussed in empirical psychology may be inferred and thereby explained. In particular, according to Wolff (1740/1972, §.7), in rational psychology, we seek *demonstrations* from the nature of the soul for occurrences therein. In such demonstrations, "only definitions, indubitable experiences, axioms, and propositions already demonstrated are assumed as principles of demonstration" (§.3, p. 2).[9] Thus, empirical and rational psychology are fundamentally distinct in terms of methodology. Empirical psychology involves introspection and the cataloguing of occurrences and capacities, whereas rational psychology requires a demonstrative, theoretical approach.

[4] Translation from Richards (1980, p. 230).

[5] Translation from Richards (1980, p. 231).

[6] See also Richards (1980, p. 228) and Rumore (2018, pp. 182–184).

[7] Also noted by Dyck (2014, p. 28).

[8] Translation from Richards (1980, p. 234).

[9] Translation from Richards (1980, pp. 234–235).

Absolutely critical to Wolff's conception of both philosophy in general and psychology is his commitment to the interplay and reciprocity between reason and experience. As Richards (1980) nicely puts it, "empirical psychology shows *that* something is true of the soul, and rational psychology explains *why* it is true" (p. 238n.). The occurrences and faculties of the mind discovered through our everyday experiences of the soul are explained via deductions from indubitable grounds—the essence of the soul—in rational psychology. Thus, as Wolff (1751/1983b) describes, the observations of empirical psychology serve as the "touchstone" (*Probier-Stein*) for rational psychology (§.727, p. 453). Moreover, empirical psychology also plays another important role with respect to its rational counterpart. Sometimes in the course of rational explanation of psychological occurrences, it is necessary to pose hypotheses. Yet, at all times, rational hypotheses take experience as their touchstone; empirical psychology constitutes not only the starting point for rational psychology but also the source of (dis)confirmation of rational hypotheses (Wolff, 1738/1968, §.5).[10] To detail this, consider Wolff's account of philosophical hypotheses from the *Preliminary Discourse*. There he describes philosophical hypotheses as describing grounds for experienced phenomena that as yet can only be assumed and not proven (Wolff, 1740/1983a, §.126). From such hypotheses, however, we can extract other consequences that are, themselves, testable with respect to experience via observation or experimentation. When experience conforms to the hypothesis, our confidence in it is increased, whereas when observations contradict it, it must be false (§.127). Such philosophical hypotheses are utilized in rational psychology: potential explanations are proposed and subsequently tested with respect to experience. Thus, in psychology as well as philosophy more generally, the rational and the empirical are thoroughly intermixed. As Wolff (1738/1968) himself puts it, in philosophy, we aspire to the "marriage of reason and experience" (§.497, p. 379).[11] Through conjoining the observations and demonstrations of, respectively, empirical and rational psychology, Wolff claims that the psychologist achieves *mixed* cognition (§.434).[12] A cognition is a posteriori when it derives from experience, a priori when from demonstration, and *mixed* when it derives from a combination of both (§§.434–438, 490–492, 495–496).[13] In isolation, empirical and rational psychology would be respectively a posteriori and a priori. However, as mentioned, the two work in concert—empirical psychology provides the principles and confirmation for rational psychology—and thereby produce mixed cognition.[14]

[10] Wolff (1738/1968, §.6) even generally claims that historical cognitions confirm philosophical ones.

[11] See also Dyck (2014, pp. 19–42) and Rumore (2018, pp. 180–182).

[12] See also Wolff (1751/1983b, §.382), Vidal (2011, pp. 94–95), Dyck (2014, pp. 39–40), and Rumore (2018, p. 180).

[13] See also Wolff (1751/1983b, §.372) and Kim (1994, pp. 23–26).

[14] Indeed, only pure mathematics is a priori in the strict sense, for Wolff (1751/1983b, §.372).

As noted above, Wolff analogizes the methods of physics and psychology. This analogy not only discloses his optimism for the science of psychology—may psychology develop as physics did in the preceding century—but additionally illuminates the reciprocity between empirical and rational psychology. Wolff differentiates experimental (empirical) physics from dogmatic (rational) physics. Dogmatic physics, like rational psychology, is structured demonstratively and aims at the explanation of occurrences catalogued in its empirical counterpart (Wolff, 1740/1983a, §§.59, 108). Experimental physics, like empirical psychology, provides the *principles* for explanation in dogmatic physics (§§.107, 109). The explicit mirroring of psychology and physics described in the following passage elucidates both Wolff's general conception of the interplay between reason and experience and the connection between empirical and rational psychology:

> [T]he psychologist imitates the astronomer, who derives theory from observations and corroborates theory through observation, and who, by the aid of theory, is led to observations which he otherwise might not make. And thus the demonstrations of rational psychology suggest what ought to be considered in empirical psychology. And wherever empirical psychology is established and rational psychology cultivated, we are enriched by many principles which otherwise would have to be secured with great difficulty. Thus the best thing is for one constantly to join the study of rational psychology with that of empirical psychology, even though we have considered it wise to treat them separately. (Wolff, 1738/1968, §.5, pp. 4–5)[15]

In both physics and psychology, progress is made through leveraging the reciprocal relationship between theory and observation, or reason and experience. We make observations, use them to develop theories, and confirm those theories in experience. In physics, Newton deduced his theory of gravitation from the phenomena, like Kepler's laws, which were justified by ample observations. Then further consequences were elicited from the theory, which were subsequently tested and verified by experience. Likewise, according to Wolff, we marry reason and experience in psychology by developing descriptions through observation, creating theories that account for said descriptions based on the nature of the soul, and verifying these theories by returning to experience, namely, by testing their empirical consequences. Such is the holy bond of reason and experience in psychology.[16]

15.2.2 The Mathematical Method

Rational psychology, like dogmatic physics and all other sciences, proceeds demonstratively, according to Wolff. This is an aspect of his commitment to the wide-ranging utility of the so-called mathematical method, modeled on Euclidean-style,

[15] Translation from Richards (1980, pp. 232–233).

[16] Baumgarten and Meier follow Wolff in many regards, including distinguishing these two sorts of psychology (Baumgarten, 1739/2013, §.503; Meier, 1755–1759, 3:§.474). Baumgarten, however, is apparently not as sanguine as Wolff about the confirmatory role of experience vis-à-vis rational principles (Fugate & Hymers, 2013, p. 22; Mensch, 2019, p. 196).

demonstrative geometry. By means of the mathematical method, we aim to achieve certain knowledge via indefeasible inferences from an indubitable basis.[17]

Wolff details the mathematical method in his "Short Instruction of the Mathematical Method or Way of Teaching," which serves as introduction to his *Foundations of All Mathematical Sciences* (originally published in 1710). There Wolff explains that the mathematical method is an ordering of cognitions that begins with definitions, proceeds to axioms that follow therefrom, and then results in theorems (Wolff, 1750/1973, §.1). Axioms (*Grundsätze*), for Wolff, are immediately inferred from definitions (§.29), whereas theorems (*Lehrsätze*) are inferred mediately via the comparison of definitions (§.37). He goes on to discuss proofs of theorems (§§.42–45), concluding that the methods of proof in mathematics are none other than the rules of logic or reason, that is, the rules of syllogistic inference (§.45). Altogether, the mathematical method achieves indubitable grounding for given cognitions by deducing them as theorems from certain axioms via valid, syllogistic inferences. At the end of the "Short Instruction," he concludes that the mathematical method "is general" and "should be used in all sciences" (§.51, p. 30), a claim repeated throughout Wolff's philosophical corpus. For instance, in Chap. 4 of the *Preliminary Discourse*, wherein Wolff describes "the philosophical method" in detail, he states that the mathematical and philosophical methods are the same (Wolff, 1740/1983a, §.139). More generally, Wolff (1751/1983b, §§.361–371) defines a science as a system of cognitions interconnected in irrefutable inferences from noncontradictory grounds, meaning that any science must make use of the mathematical method of demonstration.[18]

In various cases, including in the above quote describing the method of rational psychology, Wolff asserts that "indubitable experiences" can play a role in the mathematical method. Experiences, according to Wolff (1750/1973, §§.33–34), concern particular things, insofar as they are based on sensation and one can only sense particular things. Nonetheless, experiences can serve for the inference to general conclusions insofar as they can be certain (Wolff, 1750/1973, §.35; 1751/1983b, §§.330, 390). The inclusion of experiences on the list of legitimate, indubitable bases for demonstration may lift an eyebrow, but it is a distinctive bit of the Wolffian picture of methodology, and one that fits with the abovementioned interplay between empirical and rational doctrines and the consequent production of mixed cognitions. The aspiration in rational psychology is to explain all our experiences of the soul through such sure inferences from indubitable foundations. In this process, the truths of empirical psychology can serve in a few ways. First, insofar as experiences of the soul may be indubitable, they may be premises in certain inferences. Second, the experiences may serve as conclusions of such inferences: that is, we may seek to explain a particular experience of the soul. Finally, as discussed above, hypotheses of rational psychology must be tested with respect to such indubitable experiences.

[17] For discussions of Wolff's mathematical method from a variety of perspectives, see Frängsmyr (1975), Dunlop (2013), and Frketich (2019).

[18] For more on using the mathematical method in natural science, see Wolff's (1754/1965, §.25) example of mathematically demonstrating that air has elastic force and the commentaries on it in van den Berg (2014, pp. 31–32) and Frketich (2019, pp. 340–342).

15.2.3 Mathematical Cognitions and Psychometrics

Not only does psychology adopt a mathematical method insofar as it aims at a demonstrative system of cognitions, but moreover, Wolff claims that there is a mathematics of psychology. That is, he defended the existence of a quantitative doctrine of the soul, or *psychometrics*.[19]

As Wolff describes in the *Preliminary Discourse*, there are fundamentally three different sorts of cognition (*cognitio*) considered with respect to *content*: historical, philosophical, and mathematical. Historical cognitions concern occurrences in the world: "The *cognition* of that which is and occurs, be it in the material world or immaterial world, we name *historical cognition*" (Wolff, 1740/1983a, §.3, p. 2). He provides a few examples, such as the knowledge that the sun rises in the morning and sets in the evening or that trees bud in the spring. Philosophical cognition, in contrast, is "of the ground of that which is or occurs" (§.6, p. 3). That is, philosophical cognition is paradigmatically knowledge of the explanation of that that is cognized historically. Wolff's example of philosophical cognition is that of knowing how the motion of water in a riverbed (an occurrence) depends on the slope of the land and the pressure expressed by the upper parts of the water on the lower (the grounds). To achieve philosophical cognition, one must not only know these grounds of an occurrence, rather one must also be "capable of understandably explaining" or "proving" that it *is* the ground (§.9, p. 4). Wolff contrasts historical and philosophical cognition by observing that it is one thing to know *that* the river flows in its riverbed (historical cognition) while it is quite another to know that it occurs *because* of the slope of the ground of the pressure of the upper parts of the water (philosophical cognition). Mathematical cognition, finally, is "cognition of the quantity of things" (§.14, p. 6). So, Wolff explains, one has a mathematical cognition of the sun when one knows the midday heat of the sun on a given day as a ratio or proportion with a given standard, say, the midday heat of the sun at the summer or winter solstice. For another example, one has a mathematical cognition of the river when one knows its flow velocity.

Crucially, the mathematical method and mathematical cognitions are not essentially connected, for Wolff. The mathematical method concerns the form of cognitions in a system, whereas the preceding tripartite division of cognitions has to do with their content. Indeed, the axiomatic-deductive procedure of the mathematical method described above can be applied to all different sorts of cognitions: historical, philosophical, and mathematical. That is, each sort of cognition can serve as premises and conclusions in a deductive system. As discussed above, even historical truths derived from experience function as premises in deductive inferences. That is, historical cognitions can be used as a basis for philosophical cognitions, in particular, when one uses a fact as a ground for other facts (Wolff, 1740/1983a, §.10). So, Wolff illustrates, one can historically cognize via experience—say, through

[19] For more on the history of psychometrics in the eighteenth century, see Ramul (1960) and Feuerhahn (2004, 2010).

experiments with an air pump—that air has weight and is elastic. Even without knowing the ground for this phenomenon, one can use this fact to explain other phenomena, such as how water is sprayed from artificial fountains, like Heron's. Furthermore, Wolff provides a variety of examples wherein the sorts of cognitions intermingle. For instance, when "astronomers derive mathematical cognition of the motions of the heavenly bodies from observations," they utilize historical cognition to achieve mathematical cognition (§.18, p. 9). Likewise, when one computes the heat of the midday sun from the density of light rays and their angle of incidence (as mentioned above), one begins with the philosophical cognition—that the density and angle of incidence of the light rays ground the heat of the sun—to achieve mathematical cognition of the calculated heat. Through the combination of mathematical and philosophical cognition as in this example, the highest possible certainty is achieved (§§.27–28). In particular, Wolff claims that the best possible evidence for the connection between a ground and a consequence (philosophical cognition) is the demonstration that the quantity of the consequence corresponds to the force of the cause (mathematical cognition).[20]

The domain of mathematical knowledge is quite extensive: as Feuerhahn (2010, pp. 70–73) notes, Wolff is committed to the universal applicability of mathematics to finite things. According to Wolff, all finite things, in virtue of being finite, can be increased or decreased, and everything that can be increased or decreased has a quantity. Thus, all finite things, as such, have a determinate quantity, and mathematical cognition thereof is in principle possible (Wolff, 1740/1983a, §.13). This thesis applies equally to immaterial things, including souls. Wolff states that, therefore, a doctrine of mathematical cognitions of the soul, *psychometrics*, is possible (Wolff, 1738/1968, §§.522, 616).[21] For Wolff, all psychological states possess a particular degree—such as of intensity—that can be measured. Wolff provides many examples, asserting that desire and aversion (Wolff, 1738/1968, §.522), attention (Wolff, 1740/1983a, §§.13–14; 1751/1983b, §.270), memory (Wolff, 1740/1972; §.242), pleasure (Wolff, 1740/1972; §.616), and cognition of freedom (Wolff, 1739/1979; 1:§§.607–608) possess measurable degrees. Although the *German Metaphysics* predates his explicit coining of the empirical/rational psychology distinction as well as that of psychometrics as the mathematical science of the soul, therein Wolff claims that there are degrees—quantities—of various psychic phenomena, such as clarity and distinctness of ideas (Wolff, 1751/1983b, §§.208–211), cognition (§.279), the power of imagination (§.260), attention (§.270), reason (§.370), pleasure (§§.409–410), desire, and aversion (§.438). Wolff also asserts that individuals' powers of memory have particular degrees, which can be measured by the number of things that they can remember (§§.260–264). Hence, although Wolff does not

[20] See also Feuerhahn (2004, p. 299) and Kim (1994, pp. 26–27).

[21] Baumgarten and Meier echo both Wolff's endorsement of psychometrics and his justification of its possibility (Baumgarten, 1739/2013, §§.249, 743, and 747; Meier, 1755–1759, Vol. 1, §.191, Vol. 3, §§.739, 752). Baumgarten even goes a bit further than Wolff by developing a comprehensive account of the first principles of the mathematics of intensive—non-extended—quantities (Baumgarten, 1739/2013, §§.165–190; see Fugate & Hymers, 2013, p. 22).

expound extensively on psychometrics, he repeatedly mentions the mathematics of mental phenomena, states that such phenomena have degrees, suggests methods of measuring such degrees in particular cases, and proffers a few quantitative relations involving psychological occurrences.[22]

15.3 Kant's Developments and Their Implications

Kant adopts much of the Wolffian framework, including the distinction between empirical and rational psychology as well as the tripartite division of cognitions into historical, philosophical, and mathematical. However, the sharing of terminology belies Kant's estrangement from the Wolffian tradition and the way it results in his rejection of empirical psychology as a science. As I argue in this section, although Kant adopts the nomenclature of his forerunners, he nonetheless overhauls the conceptual apparatus he inherited from the Wolffian tradition. And it is precisely Kant's modifications along with the rigor of the resulting conceptual system that fundamentally bring about his pessimism about the natural science of psychology.

In this section, I describe three intertwined departures of Kant's from the Wolffian tradition. First, Kant reconceptualizes the tripartite division of cognitions as one having to do with the *source* of cognitions, not their content. This seemingly insignificant alteration has substantial implications. Chief among them will be the isolation of rational psychology from empirical psychology. Second, Kant also rejiggers the relationship between the two sorts of psychology by conceiving of the rational part as the *foundation* for the empirical. These two commitments result in Kant's theses that rational psychology is both the primary doctrine—coming "first" conceptually—and takes no reciprocal influence from empirical psychology. Third, Kant proffers a new understanding of the mathematical method and claims that mathematical cognitions are precisely those that are produced by this method. In virtue of this commitment, Kant maintains that mathematical cognitions are not achievable in all domains of knowledge, particularly not in psychology.

15.3.1 The Genetic Conception of Cognition

Kant emulates Wolff by distinguishing between the three species of cognition: historical, philosophical, and mathematical. In the Architectonic chapter of the *Critique of Pure Reason*, he demarcates historical from rational cognition and subsequently two types of rational cognition—philosophical and mathematical (Kant, 1787/1911a,

[22] Later researchers, especially the so-called rational physicians [*Vernünftige Ärzte*] of Halle (such as Johann Gottlob Krüger and Johann August Unzer), more profitably developed this mathematical science of the mind (see Ramul, 1960; Nowitzki, 2003, pp. 33–162; Sturm, 2009, pp. 53–126; Feuerhahn, 2010).

pp. 540–541).[23] However, the mere use of the same labels disguises a *fundamental* reconceptualization of the tripartite distinction, the consequences of which cascade through Kant's philosophy and partially ground his negative evaluation of psychology as a natural science. The alteration appears prima facie benign: Kant replaces a distinction based on the contents of cognition with one regarding their source. Historical cognition, for Kant, is "*cognitio ex datis*"—that is, cognition *from* data—while rational cognition is "*cognitio ex principiis*"—that is, cognition *from* principles (p. 540). Throughout the passage on the sorts of cognition from the Architectonic, Kant consistently refers to the distinct sources of cognition. For example, he states that historical cognition is "given" and does not "arise **from** reason," whereas rational cognitions must be "drawn out of universal sources of reason." Further, philosophical cognitions are rational cognitions "*from* concepts," whereas mathematical cognitions are those "*from* the construction of concepts" (p. 541, my italic emphases; see also p. 469).

This genetic conception of the three sorts of cognition has the consequence of relatively isolating the sorts from one another. After all, a cognition can have one and only one source. So, when one learns something by rote, even a mathematical theorem or philosophical system, one possesses historical cognition. To achieve philosophical or mathematical cognition, one must achieve that cognition through the appropriate means: philosophical or mathematical reasoning. The strict siloing of cognitions based on their respective sources is most clear in the Discipline of Pure Reason in its Dogmatic Use, wherein Kant's overarching aim is to isolate the methods of philosophy and mathematics from one another: "mathematics and philosophy are two entirely different things [...] thus [...] the procedure of the one can never be imitated by that of the other" (Kant, 1787/1911a, p. 477).[24]

This isolation of the sources of cognition makes impossible the intermixing of cognition that Wolff envisaged. Briefly, according to Wolff, the different contents of cognitions can overlap, allowing for the possibility of hybrid cognitions—e.g., one can have a cognition of the *quantity* of a *ground*, meaning that said cognition is mathematical *and* philosophical. And, indeed, Wolff maintains that such combined philosophical-mathematical cognition constitutes the most certain sort of knowledge. Kant denies this possibility, because he reconceptualizes the distinction as one being in terms of sources, of which a cognition can only have one. Kant's revision to the conceptualization of the three types of cognition and their consequent isolation have major consequences on his theoretical philosophy, philosophy of science, and conceptions of psychology, anthropology, and practical philosophy. The first implication—that empirical and rational psychology are separated from one another—is covered in the next section.

---

[23] For a more comprehensive treatment of Kant's views on historical cognition in relation to Wolff's, see Albrecht (1982).

[24] Translation from Kant (1781/1998, p. 637).

15.3.2 A New Relation Between Empirical and Rational Psychology

As Corey Dyck (2014, pp. 70–103) has colorfully put it, whereas Wolff married reason and experience, Kant seeks a divorce. Although Kant notably adopted the Wolffian framework of empirical and rational doctrines of nature, he broke decisively with Wolff by sequestering the two sorts of doctrines from one another. Indeed, Dyck has well shown that one of the aims of Kant's Paralogisms was to block empirical information about the soul from serving as a legitimate source for rational psychology, *pace* the Wolffian tradition. I maintain that this divergence from the Wolffian approach to psychology is supported by the aforementioned genetic conception of the sorts of cognition. Furthermore, I contend that Kant thought that *rational* doctrines are supposed to ground the possibility of their empirical counterparts, in contrast to Wolff, for whom *empirical* doctrines of nature—like physics and psychology—are primary.

In the Architectonic of Pure Reason of the *Critique of Pure Reason*, Kant situates the rational and empirical doctrines of psychology and physics in his all-encompassing system of knowledge. The rational manifestations of these sciences are species of metaphysics, in the narrow sense, which is the doctrine of speculative philosophical cognitions (Kant, 1787/1911a, pp. 543–544). According to Kant, a species of metaphysics, rational physiology (or the immanent doctrine of nature), can be differentiated into two subdisciplines, "rational physics" and "rational psychology," which consider respectively outer, corporeal nature and inner, thinking nature (pp. 546–547). As noted, Kant also pairs the immanent doctrines of nature with empirical counterparts: empirical physics and empirical psychology. Although they stand in relation with their metaphysical, rational counterparts, these sciences lay elsewhere in the architectonic, namely, outside of metaphysics in applied—empirical—philosophy (p. 548).[25] Thus, Kant, like Wolff, divides the doctrines of inner and outer sense—respectively psychology and physics—into rational and empirical counterparts.[26]

However, Kant refashions Wolff's distinction between empirical and rational doctrines in two ways. First, Kant categorically rejects the influence of empirical information or principles upon rational doctrines. Here is where the seemingly benign, genetic reconceptualization of the tripartite division of cognition described above (§.2.1) rears its head. Whereas Wolff leaves open the possibility of empirical sources for metaphysical doctrines like psychology, for Kant, metaphysics definitionally consists of philosophical cognitions, conceived of as those that have the

[25] Kant (1786/1911b, p. 468) uses "empirical" and "applied" interchangeably in the context of doctrines of nature.

[26] Precisely this way of dividing up the doctrines of nature is repeated throughout Kant's lectures on metaphysics (see Kant, 1970a, pp. 656, 670; b, pp. 221–223; c, pp. 364–365; 1983, pp. 875–876).

right source: conceptual analysis and transcendental deduction.[27] The cognitions of empirical psychology are historical, for Kant; insofar as they are derived from experience, such cognitions have the wrong genesis and hence have no place in rational psychology (Kant, 1784/1911c, pp. 265, 369).[28] Thus, in student notes from his metaphysics lectures, Kant is attributed the view that "This [sc. rational psychology] should be derived merely a priori, entirely independent from empirical principles" (Kant, 1983, p. 903). Elsewhere in lecture notes one finds the view that empirical psychology is only included in metaphysics as a "stranger and guest" (p. 876) and that "empirical psychology and empirical physics do not belong in metaphysics at all" (Kant, 1970b, p. 223).[29]

Secondly, Kant modifies Wolff's distinction between empirical and rational psychology insofar as, for Kant, rational doctrines of nature *ground* their empirical counterparts. Whereas Wolff thought of empirical doctrines as primary—providing principles and confirmation to rational theories—Kant reversed the order of dependence. Kant's commitment in this regard is most clear in his discussions of the doctrine of body from the *Metaphysical Foundations of Natural Science*: "natural science must derive the legitimacy of this title only from the pure part—namely, that which contains the *a priori* principles of all other explanations—and why only in virtue of this pure part is natural science to be proper science" (Kant, 1786/1911b, pp. 468–469).[30] The pure (rational) part of a natural science, like physics or psychology, contains both the principles of the science's "propriety"—that is, those that ground its containing a priori laws, achieving apodictically certain cognitions, and allowing for the application of mathematics (pp. 468, 470)—as well as the principles of all the explanations of the total science, including the empirical part (pp. 470–471, 472, 473). So rational physics includes all that may be cognized a priori of the concept <matter>, including various laws concerning its categorial determinations—its quantity, quality, relation, and modality—as well of principles for the mathematical construction of features of matter, namely, its motion and the communication thereof (pp. 473–477, 490–493, 544–547). Furthermore, rational physics develops the explanatory framework for physics in general; according to Kant, "everything real in the objects of outer sense, which is not merely a determination of space (place, extension, and figure), must be viewed as moving force" (p. 523).[31] The same would go for psychology, were it to be a proper natural science. Were psychology to constitute a proper natural science, rational psychology would

[27] These are described as the fundamental methods of philosophical cognition in the Discipline (see Kant, 1787/1911a, pp. 470–472, 474, 480–481).

[28] In this regard, Kant follows Crusius, who also constructs a firewall between empirical and rational doctrines. Crusius criticizes Wolff's account of psychology on precisely these grounds of illegitimately violating the segregation of the two sorts of doctrines (1745, §§.4–5, 424; see also Dyck, 2014, p. 52).

[29] See also Kant (1970b, p. 221; c, pp. 366–367).

[30] Translation from Kant (1786/2002, p. 184). See also Kant (1787/1911a, p. 40n.; 1784/1911c, p. 295).

[31] Translation from Kant (1786/2002, p. 233).

need to contain the a priori principles and laws concerning thinking things, the model for explanations in empirical psychology, as well as principles making possible the application of mathematics to the mind. In sum, Kant propounds a distinct conception of the relation between rational and empirical psychology from Wolff, insofar as, for Kant, rational psychology precedes and ought to provide the general, a priori framework for the total science.

Kant's reconceptualized distinction between empirical and rational doctrines of nature thus go some way to clarifying the basis for Kant's pessimism about the natural science of psychology. Whereas Wolff can take as a basis observation of mental events accessible to every individual, Kant blocks off this source of information for rational psychology. Coupled with his strenuous objections to the independent development of rational psychology through a priori ratiocination in the Paralogisms of the *Critique of Pure Reason*, the science of psychology is in rough shape from the outset, for Kant.

15.3.3 Construction as the Method of Mathematics

Although Kant concurs with Wolff that the method of mathematics allows us to achieve certain knowledge, he fundamentally and famously reconceives the method of mathematics. For Kant, the mathematical method is not the method of demonstration but rather an especial method of reasoning that requires the irreducible products of an essentially distinct faculty of cognition: intuition. This conception substantially restricts the applicability of the mathematical method. Kant's revision to the conceptualization of the mathematical method couples with the aforementioned isolation of the three types of cognition to constitute a more decisive, comprehensive break with Wolff. That is, since Kant reconceptualizes mathematical cognition to be not of a particular *content*, but to have a particular *genesis*—that is, to be a product of his newfangled mathematical construction—the derivative domain of mathematical cognition is substantially restricted in comparison to Wolff.

Kant agrees with Wolff that we aspire to apodictic certainty in natural science and that the mathematical method yields it (Kant, 1786/1911b, pp. 470–472).[32] However, analogously with with earlier apparent agreements between the two, in this case, the commitment means something entirely different for Kant because he espouses a distinct conception of the mathematical method.[33] As noted above, according to Kant (1787/1911a), mathematics distinctively makes use of mathematical construction, where "to **construct** a concept means to exhibit *a priori* the

[32] Kant disagrees with Wolff, however, regarding whether historical cognitions can be certain—Kant rejects this (Kant, 1787/1911a, p. 478; 1786/1911b, p. 468; 1966a, p. 229).

[33] The broad literature on Kant's conception of the mathematical method in contrast to Wolff's includes Hinske (1998, pp. 108–111), Shabel (1998), Sutherland (2010), Dunlop (2014), Heis (2014), Gava (2018), and Frketich (2019).

intuition corresponding to it" (p. 469).[34] For Kant, mathematics uniquely makes use of individual, intuitive representations in service of a priori conclusions. One constructs a mathematical concept, like <circle>, by drawing a particular circle in intuition. The intuitive representation possesses information that goes beyond the mere content of the associated concept, and aspects of this intuitive content can be inferred to hold universally of everything falling under the concept. In this way, by leveraging the information provided by the independent, irreducible faculty of intuition, mathematical construction is capable of a unique level of certainty, "intuitive certainty," which philosophical cognitions are unable to attain (Kant, 1787/1911a, p. 481). As Kant himself illustratively puts it, "intuitive principles," that is, those grounded on mathematical construction, are "self-evident, which the philosophical principles, for all their certainty, can never pretend to be, any synthetic principle of pure and transcendental reason is infinitely less obvious (as is stubbornly said) than the proposition that **Two times two is four**" (p. 481).[35]

The domain of mathematical construction is likewise limited by this procedure of mathematical construction. Not all concepts are capable of exhibition in intuition. Only insofar as it refers *merely* to the pure forms of intuition—space and time—can a concept be mathematically constructible (Kant, 1787/1911a, p. 475). Because mathematical construction functions to ground synthetic a priori knowledge about the formal features of appearances, mathematical cognitions must refer only to the pure forms of intuition—space, time, and their combination. The concepts of metaphysics refer to the content of appearances or their existence and hence do not admit of construction. Another way to put this point, for Kant (1787/1911a), mathematical concepts must be definable, which is "to exhibit originally the exhaustive concept of a thing within its boundaries" (p. 477).[36] Philosophical and empirical concepts are, however, not *definable* (they are only *explicable*).

Thus, in virtue of Kant's idiosyncratic conception of the mathematical method, according to which it requires a priori exhibition of a concept in intuition, the application of the method is considerably limited in comparison with Wolff. Whereas Wolff argued that the mathematical method ought to be used in philosophy, for Kant, this commitment misunderstands the strict distinction between the methods. Indeed, according to student notes to his lectures on logic, Kant repeatedly emphasizes that Wolff's appropriation of the mathematical method was a mistake and baldly states that "Wolff expounded philosophy in accordance with [the mathematical] method, which cannot be done" (Kant, 1966b, p. 783).[37] Moreover, as is clear from the passage and above considerations, since Kant adopts a genetic conception of the sorts of cognitions, the divergence between philosophical and mathematical methods entails that they essentially produce different kinds of cognitions—respectively, philosophical and mathematical.

---

[34] Translation from Kant (1781/1998, p. 630).

[35] Translation from Kant (1781/1998, p. 640). See also Kant (1784/1911c, p. 327; 1966a, p. 229; b, p. 747; c, pp. 857–858).

[36] Translation from Kant (1781/1998, p. 637).

[37] See also Kant (1966a, p. 272).

Furthermore, the bar for applying mathematics in natural sciences—like physics or psychology—is even higher. As mentioned above, throughout the preface of the *Metaphysical Foundations of Natural Science*, Kant emphasizes that we must make use of the mathematical method to achieve certainty in a natural science. Nonetheless, he also reiterates that we do not get the requisite application of the mathematical method for free, as it were. Particularly, Kant (1786/1911b, pp. 472, 479) criticizes those natural scientists—the "mathematicians"—that avoid metaphysics and simply assume their objects to be amenable to mathematical treatment. Such assumptions of the mathematization of the objects of natural science necessarily smuggle in metaphysical assumptions. For Kant, we require a sort of metaphysical validation for the application of mathematics to a natural science in the manner necessary for the derivation of a priori laws and the achievement of apodictically certain cognitions.

Kant agrees with Wolff, to an extent, that all finite things have magnitudes. By the arguments of the Axioms of Intuition and the Anticipations of Perception of the *Critique of Pure Reason*, all intuitions and sensations are, respectively, extensive and intensive magnitudes (Kant, 1787/1911a, pp. 148–158). But crucially, while *knowledge of magnitudes* is mathematical cognition for Wolff, it is not for Kant (at least, in the sense necessary for proper natural science). To achieve mathematical cognition, for Kant, one needs to make use of the method that he describes: mathematical construction. But to construct mathematically the objects of a natural science, one requires a priori metaphysical validation, lest one simply (and illegitimately) *assume* it—à la the metaphysics-averse and maligned "mathematicians." Thus, in a punch line, Kant agrees with Wolff that all things possess quantities, but the acquisition of *mathematical cognition* of the objects of natural science requires more.[38]

15.4 Conclusion

A detailed examination of Kant's views on the science of psychology against the contrasting backdrop of the Wolffian conception reveals a great deal about his break from the then-traditional accounts of cognition, natural science, and psychology. In particular, my considerations establish that Kant's disagreement with Wolff regarding the science of psychology has its roots deep within his philosophical system. Kant's overhauled conception of the sorts of cognition and of the mathematical method, coupled with the strictness with which he applies these notions—segregating distinct cognitions, methods, and sorts of doctrines from one another—undergirds a substantial part of his objections to and skepticism of psychology. To emphasize again the central thesis of the chapter, Kant's rejection of a positive natural science of psychology has to do fundamentally with the rigorous application of these overhauled conceptual categories.

[38] Thus, Feuerhahn's (2004, p. 299n.7) reading—according to which the Wolffians accept, whereas Kant rejects, the universal applicability of mathematics—is too simplistic. See also Dunlop (2014).

Particularly, Kant's revisions to the Wolffian conceptions of the three forms of cognition, the distinction between empirical and rational doctrines, and the mathematical method leave empirical psychology secluded in the architectonic. In virtue of Kant's commitments, empirical psychology can bear only historical cognitions, is incapable of the mathematical cognitions, and cannot be a source for rational psychology. Given the deficiency of its foundations, empirical psychology can thus be only a historical doctrine of nature, a collection of historical cognitions, lacking the explanatory systematization and mathematization of a genuine science of nature, like physics.

Acknowledgments I thank the members of a reading group on Wolff's *German Metaphysics*—namely, Katerina Mihaylova, Hirohito Mita, and Emanuel Stobbe—at Martin-Luther-Universität Halle-Wittenberg for our fruitful discussions as well as Michael Walschots and John Walsh for their feedback on an earlier draft. I also appreciate the thorough comments of the Research Group in Classical German Philosophy at KU Leuven, especially those of Henny Blomme, Karin de Boer, Stephen Howard, and Kwangchul Kim.

This work was supported by a Research Fellowship from the Alexander von Humboldt Foundation.

References

Albrecht, M. (1982). Kants Kritik der historischen Erkenntnis—Ein Bekenntnis zu Wolff? *Studia Leibnitiana, 14*(1), 1–24.

Baumgarten, A. (2013). *Metaphysics: A critical translation with Kant's elucidations, selected notes, and related materials* (C. Fugate, & J. Hymers, Trans.). Bloomsbury. (Original work published 1739).

Crusius, C. A. (1745). *Entwurf der nothwendigen Vernunft-Wahrheiten wiefern sie den zufälligen entgegen gesetzt werden*. Gleditsch.

Dunlop, K. (2013). Mathematical method and Newtonian science in the philosophy of Christian Wolff. *Studies in History and Philosophy of Science, 44*(3), 457–469.

Dunlop, K. (2014). Arbitrary combination and the use of signs in mathematics: Kant's 1763 prize essay and its Wolffian background. *Canadian Journal of Philosophy, 44*(5–6), 658–685.

Dyck, C. (2014). *Kant and rational psychology*. Oxford University Press.

Feuerhahn, W. (2004). Die Wolffsche psychometrie. In O.-P. Rudolph & J.-F. Goubet (Eds.), *Die Wolffsche Psychologie: Systematischer Ort, Konstitution und Wirkungsgeschichte* (pp. 227–236). Max Niemeyer.

Feuerhahn, W. (2010). Die Wolffsche 'Denkschulung' in der Psychometrie. In J. Stolzenburg & O.-P. Rudolph (Eds.), *Wolffiana II: Christian Wolff und die europaïsche Aufklärung* (pp. 69–85). Olms.

Frängsmyr, T. (1975). Christian Wolff's mathematical method and its impact on the eighteenth century. *Journal of the History of Ideas, 36*(4), 653–668.

Frketich, E. (2019). Wolff and Kant on the mathematical method. *Kant-Studien, 110*(3), 333–356.

Fugate, C., & Hymers, J. (2013). Introduction to the translation. In C. Fugate & J. Hymers (Eds.), *Metaphysics: A critical translation with Kant's elucidations, selected notes, and related materials* (pp. 1–65). Bloomsbury.

Gava, G. (2018). Kant, Wolff, and the method of philosophy. *Oxford Studies in Early Modern Philosophy, 8*, 271–303.

Goubet, J.-F. (2018). Rationale psychologie. In R. Theis & A. Aichele (Eds.), *Handbuch Christian Wolff* (pp. 153–174). Springer VS.

Hatfield, G. (1995). Remaking the science of the mind: Psychology as natural science. In C. Fox, R. Porter, & R. Wokler (Eds.), *Inventing human science: Eighteenth-century domains* (pp. 184–231). University of California Press.

Hatfield, G. (2018). Baumgarten, Wolff, Descartes, and the origins of psychology. In C. Fugate & J. Hymers (Eds.), *Baumgarten and Kant on metaphysics* (pp. 61–77). Oxford University Press.

Heis, J. (2014). Kant (vs. Leibniz, Wolff, and Lambert) on real definitions in geometry. *Canadian Journal of Philosophy, 44*(5–6), 605–630.

Hinske, N. (1998). *Zwischen Aufklärung und Vernunftkritik: Studien zum Kantschen Logikcorpus*. Frommann-Holzboog.

Hinske, N. (1999). Wolffs empirische Psychologie und Kants pragmatische Anthropologie: Zur Diskussion über die Anfänge der Anthropologie im 18. Jahrhundert. *Aufklärung, 11*(1), 97–107.

Kant, I. (1911a). Kritik der reinen Vernunft. In Königlich Preussische Akademie der Wissenschaften (Ed.), *Kant's gesammelte Schriften* (Vol. III, 2nd ed.). Reimer. (Original work published 1787).

Kant, I. (1911b). Metaphysische Anfangsgründe der Naturwissenschaft. In Königlich Preussische Akademie der Wissenschaften (Ed.), *Kant's gesammelte Schriften* (Vol. IV). Reimer. (Original work published 1786).

Kant, I. (1911c). Prolegomena zu einer jeden künftigen Metaphysik, die als Wissenschaft wird auftreten können. In Königlich Preussische Akademie der Wissenschaften (Ed.), *Kant's gesammelte Schriften* (Vol. IV). Reimer. (Original work published 1784).

Kant, I. (1917). Anthropologie in pragmatischer Hinsicht. In Königlich Preussische Akademie der Wissenschaften (Ed.), *Kant's gesammelte Schriften* (Vol. VII). Reimer. (Original work published 1798).

Kant, I. (1942). Erste Einleitung in die *Kritik der Urteilskraft*. In Preussische Akademie der Wissenschaften (Ed.), *Kant's gesammelte Schriften* (Vol. XX). De Gruyter.

Kant, I. (1966a). Logik Blomberg. In Deutsche Akademie der Wissenschaften (Ed.), *Kant's gesammelte Schriften* (Vol. XXIV). De Gruyter.

Kant, I. (1966b). Logik Dohna-Wundlacken. In Deutsche Akademie der Wissenschaften (Ed.), *Kant's gesammelte Schriften* (Vol. XXIV). De Gruyter.

Kant, I. (1966c). Wiener Logik. In Deutsche Akademie der Wissenschaften (Ed.), *Kant's gesammelte Schriften* (Vol. XXIV). De Gruyter.

Kant, I. (1970a). Metaphysik Dohna. In Deutsche Akademie der Wissenschaften (Ed.), *Kant's gesammelte Schriften* (Vol. XXVIII). De Gruyter.

Kant, I. (1970b). Metaphysik L_1. In Deutsche Akademie der Wissenschaften (Ed.), *Kant's gesammelte Schriften* (Vol. XXVIII). De Gruyter.

Kant, I. (1970c). Metaphysik Volckmann. In Deutsche Akademie der Wissenschaften (Ed.), *Kant's gesammelte Schriften* (Vol. XXVIII). De Gruyter.

Kant, I. (1983). Metaphysik Mrongovius. In Deutsche Akademie der Wissenschaften (Ed.), *Kant's gesammelte Schriften* (Vol. XXIX). De Gruyter.

Kant, I. (1998). *Critique of pure reason* (P. Guyer, & A. Wood, Trans.). Cambridge University Press. (Original work published 1781/1787).

Kant, I. (2002). Metaphysical foundations of natural science. (M. Friedman, Trans.). In H. Allison & P. Heath (Eds.), *Theoretical philosophy after 1781*. Cambridge University Press. (Original work published 1786).

Kim, S. B. (1994). *Die Entstehung der Kantischen Anthropologie und ihre Beziehung zur empirischen Psychologie der Wolffschen Schule*. Peter Lang.

Meier, G. F. (1755–1759). *Metaphysik* (4 Vols.). Johann Justinus Gebauer.

Mensch, J. (2019). From anthropology to rational psychology in Kant's *Lectures on metaphysics*. In C. Fugate (Ed.), *Kant's lectures on metaphysics: A critical guide* (pp. 194–213). Cambridge University Press.

Nowitzki, H.-P. (2003). *Der wohltemperierte Mensch: Aufklärungsanthropologien im Widerstreit*. De Gruyter.

Ramul, K. (1960). The problem of measurement in the psychology of the eighteenth century. *American Psychologist, 15*(4), 256–265.

Richards, R. (1980). Christian Wolff's prolegomena to empirical and rational psychology: Translation and commentary. *Proceedings of the American Philosophical Society, 124*(3), 227–239.

Rumore, P. (2018). Empirical psychology. In R. Theis & A. Aichele (Eds.), *Handbuch Christian Wolff* (pp. 175–196). Springer VS.

Shabel, L. (1998). Kant on the 'symbolic construction' of mathematical concepts. *Studies in History and Philosophy of Science, 29*(4), 589–621.

Sturm, T. (2009). *Kant und die Wissenschaften vom Menschen*. Mentis.

Sutherland, D. (2010). Philosophy, geometry, and logic in Leibniz, Wolff, and early Kant. In M. Domski & M. Dickson (Eds.), *Discourse on a new method: Reinvigorating the marriage of history and philosophy of science* (pp. 155–192). Open Court.

van den Berg, H. (2014). *Kant on proper science: Biology in the critical philosophy and the opus postumum*. Springer.

Vidal, F. (2011). *The sciences of the soul*. (S. Brown, Trans.). Chicago University Press.

Wolff, C. (1965). Vernünfftige Gedancken von den Kräften des menschlichen Verstandes und Ihrem richtigen Gebrauche in Erkänntnis der Wahrheit. In H. W. Arndt (Ed.), *Christian Wolff: Gesammelte Werke* (Deutsche Logik) (Vol. 1, Pt. I, 14th ed.). Olms. (Original work published 1754).

Wolff, C. (1968). Psychologia Empirica, methodo scientifica pertractata, qua ea, quae de anima humana indubia experientiae fide constant, continentur et ad solidam universae philosophiae practicae ac theologiae naturalis tractationem via sternitur. In J. École (Ed.), *Christian Wolff: Gesammelte Werke* (Psychologia Empirica) (Vol. 5, Pt. II). Olms. (Original work published 1738).

Wolff, C. (1972). Psychologia rationalis, methodo scientifica pertractata, qua ea, quae de anima humana indubia experientiae fide innotescunt, per essentiam et naturam animae explicantur, et ad intimiorem naturae ejusque autoris cognitionem profutura proponuntur. In J. École (Ed.), *Christian Wolff: Gesammelte Werke* (Psychologia rationalis) (Vol. 6, Pt. II, 2nd ed.). Olms. (Original work published 1740).

Wolff, C. (1973). Kurtzer Unterricht von der Mathematischen Methode oder Lehr-Art. In J. E. Hofmann (Ed.), *Christian Wolff: Gesammelte Werke* (Anfangs-Gründe aller Mathematischen Wissenschaften) (Vol. 12, Pt. I). Olms. (Original work published 1750).

Wolff, C. (1979). Philosophia practica universalis methodo scientifica pertractata. In J. École (Ed.), *Christian Wolff: Gesammelte Werke* (Philosophica Practica Universalis) (Vol. 11, Pt. II). Olms. (Original work published 1739).

Wolff, C. (1983a). Discursus Praeliminaris de Philosophia in Genera. In J. École (Ed.), *Christian Wolff: Gesammelte Werke* (Philosophia Rationalis, sive Logica) (Vol. 1, Pt. II). Olms. (Original work published 1740).

Wolff, C. (1983b). Vernünfftige Gedancken von Gott, der Welt und der Seele des Menschen, auch allen Dingen überhaupt. In C. Corr (Ed.), *Christian Wolff: Gesammelte Werke* (Deutsche Metaphysik) (Vol. 2, Pt. I, 11th ed.). Olms. (Original work published 1751).

Chapter 16
Hegel and Wolff's Psychologies

Werner Ludwig Euler

16.1 Introduction

There are two reasons for asking what it is that makes it necessary for Hegel to renew the question of philosophical psychology of modernity and why it should be of interest to the reader. The first relates to the critical philosophy of Kant, since the standard view is that Kant is the destroyer of the foundations of rational psychology. The second is that Hegel emphatically declares the end of the "former metaphysics." The question concerning the need for renewal is all the more insistent because, according to Hegel, Kant's critical work had released psychology from misleading questions about metaphysics, such as those about the immortality of the soul or its "seat" in the body. In order to give adequate answers to these questions, it would seem to be indispensable to mark out the general historical and systematic context and then to identify the multiple theoretical relations that are effective therein. Although he praises the merits of Kant's critical revision of rational psychology, Hegel reconstructs its framework himself; in connection with his analytic examinations concerning empirical psychology and rational psychology, he engages with its concrete content, the general form or method of its description in Wolff, and the epistemological theories of Aristotle and Kant.

Hegel rejects both the tradition of empirical psychology and the treatment of rational psychology because he thinks that they are not suitable to comprehend the essence of its object, that is, *spirit*. Instead of those fundamental doctrines, he recommends the study of the books of Aristotle on the soul. The reason for that preference consists, for him, in the fact that Aristotle treated the subject in a speculative manner, whereas the rational psychology of Wolff was merely a "philosophy of understanding" or an abstract metaphysics of understanding. Rational psychology, or "pneumatology," asks about the attributes of *substantiality* (identity), *simplicity*, and *immortality* of the spirit

W. L. Euler (✉)
Federal University for Latin American Integration, Foz do Iguaçu, Paraná, Brazil
e-mail: werner.euler@unila.edu.br

S. d. F. Araujo et al. (eds.), *The Force of an Idea*, Studies in History and Philosophy of Science 50, https://doi.org/10.1007/978-3-030-74435-9_16

or the soul, as well as the *community* of soul and body, and applies these categories to its object, which is a "thing," as the underlying ground to which those determinations have to be related as predicates. Hegel intends to replace the thing of the soul (*Seelending*), under the influence of Aristotle, by substance in the sense of a *substantial form* according to an end. In this process, he starts from two different meanings of the concept of a "thing" (Wolff, 1992, p. 127), and in his attempt, he stands within the tradition of the metaphysics of substance in Leibniz, who rediscovered the *substantial forms* of Aristotle and took that expression, together with *entelecheia*, as a basis of his metaphysics. In this respect, the question arises, for Hegel as well as for traditional metaphysics, as to the extent to which the soul can be treated as a "thing."

It is mainly in the "Paralogisms" chapter in the Transcendental Dialectic division of his *Critique of Pure Reason* that Kant attempted to prove the invalidity and the subreption through erroneous forms of inference of the rational attribution of the four classic ideas as determinations of the soul (Kant, 1787/1998). By analyzing these paralogisms and by criticizing the Kantian conclusion, which led to the inadmissibility of the use of the four categories for the determination of the soul, and especially by questioning whether its substantiality or thingness was avoided as a result of Kant's critique, Hegel reopens the question of the validity of the determinations of the essence of the soul. This does not mean, however, that he intends to return to rational psychology. Rather, it means that he tries to give (methodologically as well as with regard to the content) a new direction for the resolution of the problem connected with the question of the determination of the essence of the soul. To follow this philosophical program is an interesting and worthwhile task.

Hence, questions are raised as to the *How* and the *Why* of the transformation of the determinations of the essence of the soul, and thereafter an inquiry into Hegel's own conception of psychology in relation to and in contrast with the older doctrine of the soul.

The following remarks (Sect. 16.2) provide a short summary of the conceptual determinations that are relevant as background for the development of philosophical psychology and, moreover, for an adequate comprehension of Hegel's main objections to the classical knowledge of the soul. In Sect. 16.3, I present in greater detail the particular position of psychology within Wolff's metaphysics. Section 16.4 marks a very important step in the tradition of philosophical psychology, that is, the critical argumentation of Kant against the former metaphysical approaches. Those arguments serve as preparation for the "dissolution" (*Auflösung*) of and modifications to the problems of the soul by Hegel, which I discuss in Sect. 16.5 and the related subsections.

16.2 The Tradition of Philosophical Psychology

16.2.1 Aristotle

In all psychological inquiries before Hegel's philosophical investigations into the topic, the main concept was that of the *soul*. In Hegel, however, this subject belongs to "Anthropology," which is the title of the first part of the *Philosophy of Subjective Spirit* within his encyclopedic system. Its third part, "Psychology" (§§.440–482), which is based on the "Anthropology" as its material ground, deals with the "spirit." Hegel places the theory of consciousness between the "Anthropology" and "Psychology," in a chapter called "Phenomenology."[1]

Aristotle is supposed to be the forefather of psychology as the science of the soul. His theory, developed in *De anima* and *Parva naturalia*,[2] received substantial attention in modern philosophy, especially in Hegel.[3] While the modern mind-body problem was unknown to Aristotle, his theory of the soul was nonetheless full of unresolved difficulties. For Hegel, the mind-body problem is treated as an apparent (i.e., seeming, empty) question only, based on prejudices and therefore a nonproblem.

In Aristotle, the soul is essentially and primarily related to the corpus (Aristotle, 1995, 430a25, p. 171).[4] That relationship is rather complicated. The soul is in a threefold manner cause and principle of the body and of life. As in all natural beings, the soul is distinctive in itself; with regard to matter, it is the possibility of everything; and as cause and activity, it causes everything (Aristotle, 1995, 430a10–430a25). As cause, it is reason and principle in relation to matter. This principle alone is immortal (eternal). The suffering part of reason, however, is transitory.

The soul is also compared with *life* as the formal existence of living beings. As such, it is a purpose in itself (Aristotle, 1995, 412a27–412b6).[5] It is the first entelechy of an organic body (Aristotle, 1995, 412b5–412b6). According to the concept, it is substance (*ousia*) (Aristotle, 1995, 412b10–412b11), that is, the substance of the natural, living body. It is only through the soul that the corpus *is* with respect to its possibility.

However, matter is not the determined object or *thing*; rather, this is the figure or form (*eidos*). While the matter is only the possibility, the form is entelechy (realization). The natural, living body is the underlying ground (substratum) and matter. The soul, however, is necessarily substance as the formal cause of a natural corpus. The question whether soul and body are unified (the same) is irrelevant because the soul is the substance. It is inseparable from the natural body, whose form and concept it is, just like, for example, eyesight in relation to the eye: "Not the body which has lost the soul, but the body which has it, is the one which is in potential, so that it has the capacity to live" (Aristotle, 1995, 412b25–412b26, p. 64).[6] In other words, taken for itself, that is, isolated from the soul, the body is not even a body; taken as matter only, it has no existence. In fact, the soul is the purpose (end) for which a certain organ or faculty of a living being is there (e.g., eyesight), but it is not the

[1] With regard to the division of the *Philosophy of Subjective Spirit* into three parts, cf. Hespe (1991, pp. 490–521).

[2] See, for example, Aristotle (1995, 415b17–416a, p. 81): The soul is in a threefold manner cause and principle of the body and of life.

[3] A new discussion about the relationship between Hegel and Aristotle concerning psychology can be found in Corti et al. (2020).

[4] Cf. Wolff (1992, p. 36).

[5] Cf. Wolff (1992, p. 45).

[6] My own translation.

organ itself (e.g., the eye).[7] On the other hand, the soul is also dependent on the body and its organs. It is connected with it from the very beginning; it does not join it from outside, occasionally and temporarily, in order to animate the matter, which existed independently for itself. According to this conception, the community of the soul and the body is always presupposed, as when the soul is thought, for instance, to be the cause of the motion of the body.

The principle, which causes the local movement of living beings, is reason (*nous*), because it always happens for the sake of an end, that is, the representation (*phantasia*) or the striving (*orexis*) (Aristotle, 1995, 433a10–433b30). Practical reason and striving involve a purpose. The purpose is their principle. The self-movement of a living being is the effect of a faculty of striving. This faculty cannot act without representation (as a kind of thinking).

Thus, it is obvious that as early as Aristotle, we are unable to find a simple and unambiguous answer to the four classic questions concerning the property (nature) of the soul, that is, whether the soul is substance, whether it is simple, whether it is immortal (immaterial), and how it refers to the body. It is, for example, not simply *immaterial*, but also, in a certain sense, material; it is not only *simple*, but also different in itself; and it is not substance as a *thing*, but substance of the body; and it is on the one hand *immortal*, while on the other, mortal.

In Hegel's view, we do not find in Aristotle something like a "metaphysics of the soul" (Hegel, 1986a, Vol. 19, p. 199). As far as the representation of a *thing* belongs to classical metaphysics, according to Hegel, the soul in Aristotle's thought is *not* a thing.

16.2.2 Descartes

Under the new Cartesian definition of substance as a thing that does not need any further thing for its support (Descartes, 1644/1905, p. 24), the conditions of the problem concerning the determination of the soul and its relation to the body change. Strictly speaking, this definition is valid only for one substance, namely, God. With regard to finite items like the body and the soul, the definition applies only with the restriction that they need the support of God (Descartes, 1644/1905, pp. 24–25). Insofar as soul and body are understood as substances, they are considered as things independent of each other (*res extensa, res cogitans*) (Descartes, 1641/1904, pp. 71–90). Thus, the question arises as to how and under what conditions they constitute a community within the living human organism. From Hegel's point of view, the solution to the problem in relation to metaphysics in Descartes, Spinoza, Leibniz, Malebranche, and Wolff is only thinkable under the condition that God is taken as the ground of unity that mediates the opposition between body and soul (spirit) (Descartes, 1641/1904, pp. 71–90; 1649/1909, pp. 351–357). With

[7] Cf. Wolff (1992, p. 74).

regard to the critique of Descartes, he agrees with Wolff on this point (Wolff, 1751/1983, §§.763ff.). One has only to add that the respective models that conceptualize the unity turn out differently (see more on this below, in Sect. 16.2.4).

Furthermore, in relation to the Cartesian account, it is significant that the actions of the soul, insofar as they concern its suffering, are restricted to mechanical or physiological movements of the body or matter (through the so-called living spirits). Thus, it turns out that Descartes is asking for the location of the soul in the brain (Descartes, 1641/1904, pp. 71–90; 1644/1905, pp. 319–320; 1649/1909, pp. 351–355).[8] The *substantial forms* are rejected as being incomprehensible or merely different states of things (Descartes, 1644/1905, pp. 321–322). The soul is an immaterial substance, a thinking being (*cogito*) that exists differently from and independently of the body. With regard to their essences, soul and body are heterogeneous.

Unlike the body and in contrast to the textbook tradition, the spirit is inseparable. Through mediation by the *living spirits*, which are nothing else but subtle atoms of matter, and the "pores" of the brain, the body and the soul enter into interaction (Descartes, 1649/1909, pp. 354–355).

16.2.3 Leibniz

In Hegel's eyes, Leibniz's thinking belongs entirely to the tradition of metaphysics in the succession of Descartes' philosophy, representing also the doctrine of the thingness of the soul (Hegel, 1830/1992, §.389, p. 47; 1986a, Vol. 20, pp. 233–255.).[9] However, one should consider that Leibniz explicitly endorses the Aristotelian meaning of the *substantial forms* and the concept of entelechy (Liske, 2000, pp. 95ff.)[10] in order to explain and ground the spirituality and final activity of the monads (as the new conception of substances), as well as the general dynamics of nature. The soul is a monad equipped with apperception and perception.[11] Only in rare occasions, and half-decidedly, does he also refer to the body as a *substance*. Strictly speaking, he abandons the Cartesian position of a body-soul dualism of substances and comes close instead to hylomorphism (Liske, 2000, pp. 95–104ff.). In general, according to Leibniz, bodies are only pure modes of expression, not original unities or substances in the proper sense. To this extent, the Leibnizian metaphysics of the soul does not fit the scheme of the metaphysical tradition of psychology sketched and criticized by Hegel.

---

[8] Cf. Euler (2002, 453–480).

[9] Cf. Wolff (1992, pp. 160ff.).

[10] Cf. Euler (2001).

[11] In this context, see my article in print (Euler, 2021).

16.2.4 Wolff

The metaphysics of Wolff is generally supposed to be a pure systematization of Leibnizian philosophy. In this sense, Hegel also uses the expression *Leibniz-Wolffian philosophy* (Hegel, 1986a, Vol. 20, p. 256). This characterization fails to do justice to the autonomous significance of Wolffian metaphysics, however. The traditional classification of empirical psychology and rational psychology, as well as of their subdivisions, is Wolff's creation. Empirical psychology essentially involves being guided on the basis of observations of the context of appearances to the concept of the soul. At the same time, this same concept, discovered in this way, inversely functions as the foundation of the observed contents of the soul. Simultaneously, the concept of the soul in empirical psychology is anticipated (Wolff, 1751/1983, §§.192–197), that is, considered in a manner whereby the soul is defined by the assistance of the concept of consciousness, and determined, together with it, as the presupposition of the soul's perceptions:

> Just in order that one knows what he has to perceive; it is to say that I understand by the *soul* that thing, which is conscious of itself and of other things outside of it, insofar as we become aware [conscious] of ourselves and of other things as external to us. (Wolff, 1751/1983, §.192, p. 107)[12]

Although Wolff criticizes, in this context, Descartes' determination of the nature of the soul, the understanding of the soul as a thing appears in his metaphysics in its most developed form (Wolff, 1751/1983, §§.193 and 197). Insofar as Wolff keeps his distance from Leibniz's conception of the substance as a *monad*, he can correctly be called a typical representative of the thingness of the soul. Thus, we find once more in his writings the four main categories as predicates of the soul, which I would like to present in the next section (identity, substantiality, §§.743, 747, 768; simplicity, §§.106, 742; immateriality, §.742; community, §§.760f., 765, 767). The relationship between consciousness and the soul turns out to be a systematic problem in Wolff's philosophy (§§.192–197), which I shall also examine closely in the following consideration.

[12] Translations from German or Latin texts into English are my own. In fact, it can be shown that the unity of empirical psychology and rational psychology in Wolff's *German Metaphysics* is constructed as a circle. According to my reading, the rational part dominates the empirical part. Recently, I read a paper at the Christian-Wolff-Museum (University of Halle) on the subject Christian Wolff: "Was kann ich wissen?" (Euler, 2019). In that study, I showed that the truth of the so-called proposition of consciousness is seemingly true not because of immediate, indubitable, sensual certainty, but because it is demonstrated by the construction of a syllogism already involved silently in that principle of consciousness. This fact can be developed by the application of instruments presented in Wolff's own theory of logic.

16.3 Position and Particularity of Psychology within the Metaphysics of Wolff

In Wolff's empirical psychology (Wolff, 1751/1983, Chap. 3), the soul as object of perception is determined by consciousness: the "thing" "which is conscious of itself and of other things besides (outside) of it" (Wolff, 1751/1983, §.192, p. 107). Nevertheless, the concept (the "essence") of the soul in rational psychology (Wolff, 1751/1983, Chap. 5) is its "force." Experience will "lead" us to this concept. The concept of force must be independent from experience insofar as it cannot be properly perceived for itself. Rather, that "which we perceive of it" (Wolff, 1751/1983, §.727, p. 453) will be founded (*gegründet*) in it. Experience has to confirm the determinations, which result from the concept itself in a rational way.

In order to state the nature and essence of the soul, Wolff refers to the determination of consciousness according to §.1 (§§.728ff.). To be conscious of oneself therefore means to be aware of the difference between oneself (as a thing) and *other* things, of which one is also conscious (§.730). From that distinction, it follows indirectly that the "I" is to be understood as a *thing*. The difference is mediated by perception and caused by the soul. It appears as soon as one becomes conscious of *other* things. By knowing this relation, it is simultaneously something that has been thought. This thought has to be in conformity with one's experience. Without anything to be thought, one has no consciousness at all (§.730). The difference between representations depends on thinking. Consequently, consciousness requires a rational act of thinking (§.733). Thinking must be added to representing (feeling) in order that consciousness can emerge and that the content of representation can be transformed into thought.

This act presupposes still other capacities of the soul, such as *memory* (§.735). The soul itself represents through thoughts the things about which it is thinking as outside of itself; that is "why it recognizes these things as different from itself" (§.740, p. 462; cf. §.730, §.45).

That the soul is "a simple thing" is proved indirectly by the argument that material things (bodies) are *composed* and not apt to think (§§.741f.; cf. §.106).[13] Hence, its *simplicity* results from its supposed *immateriality*, its thingness from its naturalness, that is, its corporeality. Since animals are not capable of thinking, even though they have a soul, it will be necessary to prove their simplicity in another way (§.897). Because all simple things "exist for themselves" (and not through and for others), the soul also exists *for itself* (§.743). Now, Wolff affirms that all things existing for themselves must possess a *force* as the cause of their alterations (feelings) (§.744). Therefore, the soul too must have only one homogeneous (numerically identical) force as the striving to do something (cf. §.117). Hence, its identity consists in its causality as a simple thing. In §§.755 and 756, he concludes from the previous considerations of rational psychology that the essence and nature of the soul consist in its *force* as the reason for all the changes that happen in the soul. In this respect,

[13] Cf. the qualities of a simple thing in Wolff (1751/1983, §§.106–107).

"nature" means that which makes a thing active or capable of causing an effect (§.756; cf. §.628). That effective force is called "nature" by Wolff "insofar as it is determined through the essence of a thing in its species" (§.628, p. 384). This force is active in the world and refers primarily to the body. Insofar as the soul participates in the global force and is equipped with a corresponding force, it is also *natural* and *material*, that is, a simple *thing* of nature (cf. §.629).

Hence, it is evident that the force of the soul is contradictory in itself; that is, the global force or the force in general is composed and natural. However, as the *particular* form of the soul, it is simple but not "supernatural" (Euler, 2011), and for that reason, it is *one* thing (cf. Wolff, 1751/1983, §.758). Thus, in rational psychology, it is necessary to demonstrate how—from a unique representational force of the world—all the alterations that empirical psychology discovers (§.754) in the soul will follow. Through this conception, Wolff believes that the so-called problem of the community of soul and body has been resolved (§.760).

Wolff does not deny the existence of the problem (§.760). It presupposes necessarily (as in Descartes) that soul and body exist and act always for themselves; that is, both of them are independent, in a certain sense unalterable, simply *given* things (§§.765, 768, 779–781). Wolff discusses three models for resolving the problem. The direct interaction between soul and body (*influxus physicus*) is rejected as groundless (§.761), like the direct influence of God (erroneously) ascribed to Descartes (but which has to do objectively with the occasionalism of Malebranche) (§§.763, 766). In opposition to it, Wolff explicitly follows the Leibnizian model of pre-established harmony (between body and soul) for the purpose of explaining their community (§§.765, 767). With regard to other aspects of Leibnizian metaphysics (such as the doctrine of monads or the adoption of Aristotelian substantial forms), Wolff is more skeptical. These aspects are either not adopted or done so only hesitantly.

The fact that Wolff (like Leibniz) attributes qualities of the soul to animals is not very surprising (§.789, 892). The unity of force is common to them (§§.892f.). The souls of human beings and animals are simple things (§§.921, 742, 789) and not something composite, like bodies (§.897). However, Wolff's opinion that the animal soul possesses consciousness and knowledge (§§.793f.) but not reason (§.869), although it does possess something similar (§.892)—for spirits alone have reason (§.905)—is a divergence from Leibniz's doctrine (cf. §.793f.). Implicitly, Wolff adopts parts of the natural physiology of the senses from Descartes (§.778). Of course, the soul is acting in concomitance with the brain, the nerves, and the fluid matter. Consequently, the harmony between body and soul is mediated not only by God but also by material, organic factors. This relationship is rationalized in the form of a syllogism (§§.814f.; cf. §§.841f.).

The peculiarity of Wolff's psychology, in terms of its content, seems to arise from the tension between its Cartesian and Leibnizian elements described above. From Hegel's point of view, however, this tension is due to its method, in that Wolff's metaphysics only makes use of categories of understanding.

Wolff criticizes Leibniz's concept of spirit and the theory of the monads because they lead to the consequence that matter should be understood as an "accumulation of spirits" (§.898, p. 558; cf. §§.599, 900).[14] On that basis, one could even attribute understanding and will to parts of matter (§.898), but Wolff treats this problem as purely linguistic. Objectively, he believes himself to be in accordance with Leibniz. "Spirit" is understood by Wolff to be that and only that being that has understanding and free will (§.896). It matches neither with simple things nor with animals.

16.4 Kant's Critique of Wolffian Psychology and Hegel's Critique of Kant

If one intends to illuminate more closely Hegel's critique of Wolffian psychology, it is necessary to refer, at least cursorily, to Kant's critique of the traditional doctrine of the soul, published in metaphysical textbooks, as we can see in the Paralogism chapter of the *Critique of Pure Reason* (Kant, 1787/1998). However, in the present article, we cannot discuss Kant's critical analysis of metaphysical psychology in detail.[15] Our interest is rather to show why Hegel could have the opinion that Kant's critique was incomplete or perhaps to a certain extent even unjustified. Without going into the matter of Kant's reasoning in detail, it is important to keep in mind for the present investigation that from the point of view of transcendental idealism, he rejects all of the four principal statements of rational psychology as not applicable to the soul.

Kant's critique is constructed systematically and follows the scheme of his categories with the concept of substance at the top (Kant, 1787/1998, B 402). It refers to the "rational doctrine of the soul," from which Kant retains only the title (B 400). Its unique object—the "I think" as pure "apperception"[16] and carrier of the concepts of understanding—as an "object of the inner senses" refers to the "I" and bears the name of "soul" in opposition to the body (B 400).

On account of a naturally false conclusion of reason (a paralogism), predicates are coordinated improperly with this object.[17] The difference between the soul as the "I" and the body is found out through the "I think." The phrase "I think" is the foundation of the science of psychology (B 400): "*I think* is thus the sole text of rational psychology, from which it is to develop its entire wisdom" (B401, p. 413).

Coming from the category of *substance*, which represents the "I" as a thinking being, "all the concepts of the pure doctrine of the soul" as "transcendental predicates" of the soul will follow, in line with the classification of the other categories (B 402f.). Kant enumerates (1) *immateriality* (in consequence of the substance); (2)

---

[14] This assertion could be easily illustrated with examples used by Leibniz, such as the pool full of fishes or the drop of water consisting of microorganisms. Cf. also Hegel's ironic reaction to the smell of coffee moving from a coffee pot (Hegel, 1985, L II, p. 10).

[15] Cf. for that purpose Ameriks (2000), Sturma (1998), and Klemme (1996).

[16] In this regard, see my article in print (Euler, 2021).

[17] Cf. Hegel (1830/1992, §.47 *addendum*, pp. 125f.).

incorruptibility (simplicity); (3) *personality* (numeric identity, unity); (4) *spirituality*, as a consequence from (1) to (3); and (5) *interaction* (relation of the soul to objects in space). In Wolff's *Psychologia rationalis*, these titles correspond with paragraphs 47, 643ff., 669, and 741–743 (Wolff, 1740/1972).

The decisive aspect by means of which Kant criticizes the former rational psychology and restricts its scientific weight and claims to truth lies in depriving the "I" (and its predicates) of substantial or existential meaning. The "I" or the soul is certainly a "thing" but a "thing in itself." For such kind of a thing, it must be necessarily true that it is unknowable (Kant, 1787/1998, B 464; cf. A 350f.). Consequently, if one maintains the primary qualities quoted above, as the content of rational psychology, the resulting claims will not possess the status of cognition. This means that one cannot connect these properties as predicates with the object (the "I") by the claim to determine that very object through that procedure. In this way, one has to declare the syllogism in §.6 of Wolff's *German Metaphysics*, coming from the Cartesian cogito and ending with the conclusion "I am, I exist," to be invalid. It becomes a *paralogism* in Kant's sense.

Hegel does not have any doubts about the correctness of the Kantian paraphrasing of the Wolffian model of rational psychology described above (Hegel, 1830/1992, §.47). The true reason for the inadmissibility of the relationship between the primary categories and the substance of the soul is *not* regarded by Hegel (unlike Kant) to be that they were not in concordance with perception but that the "content of the thought" had not been examined (Hegel, 1830/1992, §.47 *addendum*). That is the issue that Hegel himself makes his own task with reference to the rational doctrine of the soul, and with that he goes beyond Kant's critique (Wolff, 1992, p. 123).

Hegel agrees with Kant in concluding that the predicates discussed cannot be attributed to the soul as an object, but his reasoning differs from Kant's: the question concerns determinations of understanding, which are "bad for the soul," where "bad" means "incomplete." For the soul is "still something completely other […] than the mere simple, unalterable, etc." It is, for instance, certainly simple identity, but at the same time something that acts and differs from itself in itself. As a merely abstract simple without difference, it is the "dead" (Hegel, 1830/1992, §.47 *addendum*, p. 126).[18]

It would involve too much labor here to identify and verify Hegel's review word by word in the respective places of Kant's text. Through his special presentation, Hegel aims to make the "nature of the former metaphysics of the soul" and Kant's critique of it in a certain manner transparent (Hegel, 1830/1992, p. 193).

With regard to the one-sidedness of the Kantian representation of the "I," it stands, for Hegel—and this is the salient point—"on the same line" as the "categories of the former metaphysics" (simplicity, persistence, immateriality, and so on). Intellectually, these ultimate categories were even something still "higher." The higher validity probably consists in the purpose of the cognition of truth, that is, to

[18] With regard to Hegel's critique of Kant's critique of the metaphysics of the soul, cf. Wolff (1992, pp. 116–118, 121–124, 126, 152).

recognize therein if its objects are substances or phenomena (Hegel, 1816/1981, p. 196). Hegel criticizes the absence of this purpose in Kant, for whom the question of truth seemed to be a triviality.

16.5 Hegel's Dissolution of Previous Doctrines of the Soul

16.5.1 Critique of Rational Psychology and Empirical Psychology

Hegel's general account of metaphysics distinguishes historically and systematically between three major stages: antiquity, rationalism, and Kantian critique. In a particular sense, it also concerns the form (method) and the content of psychology as a doctrine of the soul and the spirit. In this case, a contrast appears between Aristotelian concepts on the one hand and those of later metaphysical psychology on the other. Aristotelian concepts are compared with common metaphysical terms, which are used "without conception" or truth.

A short description of the "metaphysics of spirit" (or of the "soul")—in the way that it is founded in pre-Kantian modern metaphysics, according to Hegel—indicates "substance, simplicity, immateriality" (Hegel, 1816/1981, p. 192).[19] The metaphysical reflection on those contents proceeded as follows: first, to include data of perception from empirical consciousness in the representation (Euler, 2004). On that basis, a rational search was conducted for the appropriate predicates, which, because of their nature, still belonged—strictly speaking—to *empirical psychology* (Wolff, 1751/1983, §§.727–730; 1740/1972, especially §§.3–5; Baumgarten, 1783/2004, §§.179–198). In this way, the real metaphysical considerations remained "only quite poor determinations of reflection" (such as substance, simplicity, immateriality, and immortality) (Wolff, 1992).

Hegel characterizes empirical psychology and rational psychology as distinct sciences that are—contrary to Wolff's intentions—unrelated one to another logically for the purpose of foundation. However, that divergence results, on the one hand, from the different objects and, on the other, from the different methods of scientific treatment. The *rational* doctrine of the soul, as a metaphysics of understanding, was searching for the "metaphysical nature of the *soul*" (i.e., the spirit as a *thing*) (Hegel, 1830/1992, §.34).[20] *Empirical* psychology, however, treated the "*concrete* spirit" on the basis of observation and experience. In this way, the metaphysical content (as universal) did not attain concrete determinations (qualities),

[19] These titles can be found in Wolff (1740/1972): §.47 (*Immateriality*), §.669 (*Incorruptibility*), §§.741–743 (*Personality*), §§.643ff. (*Spirituality*); cf. Kant (1787/1998, B 403).

[20] I usually refer to the third edition of his *Encyclopedia* (1830). Quotations are my own translation because standard English translations, such as those by Petry (1978) and Wallace and Miller (Hegel, 1971), are sometimes corrupted.

and inversely, the empirical part of the science of the soul complied "the usual metaphysics of understanding of forces, different actions etc." and excluded speculative forms of consideration. From that constellation necessarily followed—in Hegel's view—the reorientation in Aristotle (Hegel, 1830/1992, §.378). Wherever Hegel reproaches metaphysical psychology's "conceptionlessness" (*Begriffslosigkeit*), this happens with regard to his own comprehension of "the concept," which he elaborated mainly in his *Science of Logic*.[21] He discovers traces and preliminary steps in the logic of concepts in Aristotle.

In opposition to this, pre-Kantian metaphysics naturally employed in its psychological part concepts and definitions, especially as determinations of predicates of the soul; however, they function, according to Hegel—like the categories in Kant's *CPR*—as fixed categories of understanding. According to Hegel, rational psychology (*Rationelle Psychologie*) is nothing other than the abstract metaphysics of understanding (Hegel, 1830/1992, §.378).[22] It is typical for that kind of metaphysics to ask for the *substantiality*, *simplicity*, *identity*, and *immateriality* of the spirit in the same way as it asks for the properties (predicates) of a *thing*, where the properties result from the *application* of definite categories to the object to be determined (Hegel, 1830/1992, §.378 *addendum*). Such an understanding of concepts and their employment justifies the *determination* of the object to which it is related, but neither rational psychology nor empirical psychology is suited through such procedures to recognizing the concept of spirit as something *true*, *alive*, *organic*, and *systematic* (Hegel, 1830/1992, §.379 addendum). Metaphysical psychology has to be deprived of the listed predicates, that is, relieved of rational psychology, because it turns the spirit into a "dead being," and it must be relieved of empirical psychology too, insofar as it "is killing the living spirit by tearing it apart to a diversity of independent forces not produced and not held together by the concept" (Hegel, 1830/1992, §.379 *addendum*, p. 15). Thus, Hegel disputes the success of the inference (conclusion) of the empirical determinations of the soul from its essence as the ground, intended by Wolff. Wolff's reasoning seems to be unacceptable to him because, first, Wolff's employment of the term *ground* (reason) itself should not be managed without any conceptual analysis, and second, because the essence of the soul is interpreted as a *worldforce*, which is itself an empirical quality of the physical world of matter.[23]

[21] In the *addendum* to §.389 of the *Philosophy of Subjective Spirit* (Hegel, 1830/1992, p. 47), Hegel refers to the speculative way of considering the categories of understanding in his *Logic* and in the *Philosophy of the Spirit*. Accordingly, abstract qualities like *thing*, *substance*, and *simplicity* are applied to the soul. However, as they "change into their opposite," they prove the untruth of such determinations.

[22] Cf. Hegel (1986a, Vol. 20, p. 256).

[23] We can find Hegel's critique of empirical psychology and rational psychology in the following places (and elsewhere): Hegel (1830/1992, §§.26–36, especially §§.28 and 34, §.378, §.389; 1986a, Vol. 20, pp. 260–263). His description of the content and function of empirical psychology and rational psychology can be read in Hegel (1830/1992, §.34 *addendum*).

16.5.2 *Critique of Wolffian Psychology: What Does It Mean to Refer to the Soul as a "Thing?"*

Hegel's critique of Wolffian psychology refers to both empirical metaphysics and rational metaphysics. There are only a few places in his text where he opens the discussion with the help of explicit references. In comparison with empirical psychology and rational psychology in general, the Wolffian version is distinguished by the author's claim to employ a systematic division and derivation of its concepts. This illuminates and extends the view of the systematic correlations of the theory of the soul and the spirit. Only under such an approach do the four main problems of rational psychology that Hegel discusses critically—and which he repeatedly relates implicitly to Wolff—become obvious and practicable (Wolff, 1992, p. 115). Above all, the central aspect of Hegel's critique—the transformation of the spirit into a thing—emerges in Wolff's psychology, as has been shown in Sect. 16.3, in the most distinct way. The thingness is expressed by the meaning of the concept of "substance" used by Wolff.

Chapter 2 of the *German Metaphysics* ("On the First Reasons of Our Cognition and All Things in General") defines a thing as that what is possible: "Everything that can be, regardless of whether it may be real or not, we call a *thing*" (Wolff, 1751/1983, §.16, p. 9). Different finished qualities as fixed categories are related to that thing, such as quantity and space. Moreover, a distinction must be made between simple and composite things (§.82), with the latter arising from the composition of the former. Some predicates, like space and quantity, are not referable to simple things (§.81). The beginning (origin) of simple things is held to be incomprehensible (§.90). Wolff calls a thing a "substance" that exists for itself (*vor sich*) (and not for others) and contains "the source of its modifications within itself" (§.114, p. 59). This source is "force" (§.115, p. 60). In addition, we have the concept of "soul" (§.114). Hence, the soul is substance ("a thing existing for itself"), insofar as it possesses the force "through which it produces its thoughts one after another in an irremovable order" (§.114, pp. 59–60). Its products (concepts) are external restrictions of that force (through other existing things).

The declaration of the soul as a thing makes use of a certain understanding of self-consciousness, namely, in the sense of representation, not of apperception. As representation, it functions only like something in which certain properties inhere. It is that which Hegel calls "the thing with its manifold properties" (Hegel, 1816/1981, pp. 17–19). However, apperception is called the acting substance, which goes through the related qualities by thinking and comes back to itself. It is not a thing but the object, thus the objectivity of cognition. The acting consciousness is an operation through which the self (the "I") transfers the qualities met outside of thinking into its own form of universality. It is the same arrangement that Hegel calls "the concept" (cf. Hegel, 1816/1981, p. 17).

Furthermore, concerning the clarification of the meaning and the systematic position of the concept of a thing, one should also consider the chapter "The Existence" (*Science of Logic*, second Book, second Section), where Hegel, parsing

Kant's thing in itself, shows at the end (C. note) how this concept dissolves itself by means of its inner contradiction. Moreover, in an analogy with matter, he comments on the representation of forces and capacities of the soul, and how they are used in the former metaphysics, and says that since the soul is thought of as something consisting of such forces, which only exist for themselves, independent of any relation to others, it has the meaning of a thing (Hegel, 1812–1813/1978, pp. 339–340). The soul as a thing is thus represented at first sight as material and simple. However, as simple, it is presumed to be immaterial.

Hegel's own conception of the soul, which is foundational in his Anthropology—i.e., the first part of the *Philosophy of Subjective Spirit* in the *Encyclopedia*—will be comprehensible only if we keep in mind the relationships between spirit (as soul) and nature and between soul and spirit (Wolff, 1992, pp. 32–36). The soul is the lowest stage in the development of the spirit (Hegel, 1830/1992, §.387); "natural spirit" (*Naturgeist*) is just another name for the soul. In this regard, the first question with relation to the determinations of the soul refers neither to substantiality nor to simplicity but to immateriality, since this concerns its relationships to matter and to nature (Hegel, 1830/1992, §.389).

16.5.3 The Four Primary Determinations of the Soul: Hegel's New Investigation

Since Hegel resumes and examines critically the four classic foundational determinations of the soul, which were systematically given by Wolff and rejected by Kant, it may be supposed that he aimed to rehabilitate, in opposition to Kant, essential parts of the metaphysical doctrine of the soul. In truth, his approach to that determination is much more critical.[24] In one part, he preserves the titles but modifies the interpretation with regard to the context (Wolff, 1992, p. 124); in another part, he maintains that all four are irrelevant for the determination of the soul. For example, the quarrel about the immateriality of the soul makes no sense. It designates an unreal problem (Wolff, 1992, p. 115). We could perhaps talk of a modified adoption of the metaphysical categories of the soul. This would allow for a sharp distinction between Hegel's own account of the determination of the soul and the categories adopted from empirical psychology and rational psychology in Wolff, and would avoid through this analytical task any confusion between the conceptual inventories of both authors.

To what extent can the soul be called a "substance," according to Hegel? Two meanings, derived from Aristotle, should be distinguished: (1) substance as *substratum* refers to an underlying ground or supporter of properties, which are attributed to or lacking in an object; and (2) substance in the sense of *form* (*eidos*) or *concept*

[24] In this regard, cf. the ambivalent interpretation of §.389 of Hegel's *Encyclopedia* in Wolff (1992, pp. 115, 124).

(*logos*) is held to be the formal determination of something that is different in itself. The soul is a substance of the body insofar as it is the principle of the life, inner purpose (end in itself and self-organization), and activity of organic bodies (cf. Aristotle, 1995, 415a10–416b30).[25] In a similar way, Hegel often calls the concept of an object the "soul" of the respective item of cognition.

As a third variant, we have to mention substance as a *thing*, which comes not from Aristotelian philosophy but from Descartes. It was adopted by the thinkers of rational metaphysics, as it was by Wolff. Soul and body are, according to their view, things in themselves, independent of other existing substances. This way of thinking is rejected by Hegel, and as a result, he can refer positively to Kant's critique in the "Paralogisms" chapter in the *Critique of Pure Reason* (Kant, 1787/1998).

According to Hegel, substantiality in the first two meanings must be related to the soul. Hegel holds to a monism of substance: for him, the soul is the unique substance.[26] It is the *substratum* that contains the material basis (all the "stuff") for the development of the concept of spirit. It is also substance as *form* and *end* with reference to the body and to matter.

In its main aspect, however, the soul as substance is considered not in the sense of a substratum (foundation) of the body but rather in the sense of the *substantial form* of the body as a living organism (cf. Wolff, 1992, p. 127). It is expressed in the common language of logic (like the substance of Spinoza) as "the unity of thinking and being," that is, as the soul, still in an undifferentiated unity, which must proceed immanently through its own activity, developing from a "still simple difference to a real difference, reducing that difference to unity" (Hegel, 1986b, Vol. 10, §.389 *addendum*, p. 45). The "concept" that Hegel wants to reintroduce into psychology and that corresponds to the *substantial form* or the *logos* in Aristotle (1995, 415b14–415b15),[27] is "the substance of life" (Hegel, 1986b, Vol. 10, §.389, p. 44). It is the counterpart of the conceptionless apparatus as an aggregate of faculties or forces in rational metaphysics.

In a second regard, §.389 of Hegel's *Encyclopedia* discusses the substance of the soul in terms of a *substratum* or foundation. This meaning is attributed to it in relation to spirit, the determination of which has of course its material grounding in the soul (Hegel, 1986b, Vol. 10, §.389). However, as *substratum*, the soul cannot be the foundation (base) of the body and its organs (cf. Wolff, 1992, p. 131).

We have seen that a restriction to the determination of the essence of the soul also has its place in the representation of Wolff, insofar as it is maintained that the basic force is divisible into grades. The decisive difference consists in the fact that the limitation of the force occurs through an external impact from other things, although the force itself is of course something internal. In Hegel, however, the soul realizes itself due to an inner development of a purpose (self-organization), which is for its

[25] Cf. also Wolff (1992, p. 130).

[26] For the arguments in support of that position, I am referring once again to Wolff (1992, pp. 130f.).

[27] Cf. Wolff (1992, p. 128).

own part limited by the antagonisms occurring in the interchange with the natural environment that contribute to the determination of the life of an organism.

The soul can only be one (cf. Wolff, 1992, pp. 143f.). The different determinations belonging to it do not justify positing a plurality of substances. The natural soul as "stuff" (the feeling of the living organism) is and remains the identical foundation (the "substratum") of all resulting developments and determinations of spirit, insofar as they emerge from the emotions of the soul (cf. Wolff, 1992, pp. 144f.). However, they are also influenced by modifications, that is, transformations through the activities of spirit. Moreover, it can be only one as an identical unity of the form of the body connected with it since, without the identity of the soul as a determination of the form, it could not be something like a regulation center that coordinates the different neurophysical and sensual functions of the body.

The *simplicity* of the soul refers to sensation (§.447) and to ideal life (*Gefühlsleben*) (Hegel, 1986b, Vol. 10, §.389). It designates not simply the state of not being composite (in contrast to the body) but also the content of its own emotions, the "property" (*das Eigene*) as "immediate unity of the soul with its substance" that is not yet determined as consciousness (Hegel, 1986b, Vol. 10, §.400A, p. 98), the "simple inwardness" (*die einfache Innerlichkeit*) as a remaining individuality (Hegel, 1986b, Vol. 10, §.403A, p. 123). It is certainly uncontested that emotions are determined, but this determination is only an immediate one for the natural soul. That means, it is not a question of differences between objects that can be felt and the feeling itself of those objects—as in Aristotle between the hearing of a note [the sound] and the sound of the object, which we perceive (Aristotle, 1995, 425b11–427a15). Emotions and sensations are for the feeling subject not composed of many different (discrete) events; rather, they are considered from its subjective perspective, *immediately one* (cf. Wolff, 1992, p. 148), one "ideal life" of the individual (the natural soul, which is called "simple" in the sense that it has only one individual end in itself that does not, contrary to Aristotle, coincide with the physical (material) organization of the living being (cf. Wolff, 1992, pp. 45ff.) but refers to the life itself [as sensation]).

Life is an end in itself, the inner purpose of the individual. Consequently, life is regarded, first, from the point of view of a certain individual (animal or man). To that extent, it is "ideal." In other words, it is the individual's *own* life. Simultaneously, however, the end in itself is directly a universal (not only for a single person but for all individuals in general) and therefore *simple* (cf. Wolff, 1992, p. 49). In the sense of that ideality—that is, its independence from the material organism—the natural soul has already spiritualized itself at a lower stage.

From the simplicity and ideality of the soul as a feeling individual, *immateriality* seems to follow directly. However, one should restrain oneself from interpreting §.389 of Hegel's *Encyclopedia* to mean that *beside* the immateriality of the single soul, which exists for itself, it is *also still* "the universal immateriality of nature."[28] There are definite and important reasons why the fourth main assertion of rational

[28] With regard to the difficulties in relation to this interpretation, see the excellent study by Wolff (1992, pp. 43–45, 148–155).

psychology—that is, *immateriality* as a predicate of the substance of the soul—cannot be maintained, because Hegel holds that the problem involved in that assertion does not really exist.[29]

In the following subsection, I will explain why the problem of the immateriality of the soul simply disappears.

16.5.4 Two Realms of Problems: Immateriality and Community

According to Hegel, the empirical comprehension of the spirit as a *thing* is a (false) foundation for the dispute about questions concerning the immateriality of the soul and the community of soul and body.

In Hegel's view, the statement of the classic question about the soul's immateriality, as we find it in the psychology of Wolff (see Sect. 16.3), is posed incorrectly. This follows from the fact that the soul is, for the reasons already stated, the *unique* substance, and therefore *simple*. It is for precisely this reason that a division between material (corporeal) and immaterial (spiritual) substances is useless. Of course, we can continue to talk about immateriality, but we should do this only in the negative sense when referring to nature as a whole: that substance could be *not material*. In other words, the substance of *nature* as the total of all external things, especially of organisms, is what we call "immateriality." In *this* sense, the soul is immaterial not for its own purpose, but for nature (cf. Wolff, 1992, p. 149). But what about the inorganic things of nature, which are without life and sensation? How far can objects of nature—apparently having no inner purpose of the soul—be coordinated as a function of universal immateriality? This is thinkable only indirectly, insofar as natures that are not alive function as *external* ends of higher-order natures (living beings) and insofar as they are also material presuppositions of the life of the souls and therefore in a certain sense immaterial. Inversely, insofar as the soul is functionally bound to the body and its process of movement, it can also be regarded as corporeal (material) (cf. Wolff, 1992, p. 151).

Going beyond the conception of Kant, Hegel does not simply reject the *employment* of these categories (substance, identity, simplicity, immateriality) but tries to identify the conditions under which these predicates can be valid or invalid as "moments" (subordinated and qualified as significant determinations) of the determinations of the soul. The conditions that support the affirmation of the theses of immateriality, which result from §.389, had to be the truth or the substantiality of matter and spirit as *things*. However, Hegel claims to have demonstrated that both are inapplicable (and to that extent wrong): (organic) bodies cannot exist for themselves; the soul is *not* a thing. The concepts in question are untrue not because they could not be referred without contradiction as predicates to a subject but because

[29] On this topic, see the following section; cf. also Wolff (1992, pp. 151–155).

there are no real objects corresponding to them (Hegel, 1986b, Vol. 8, p. 100).[30] Accordingly, the interrogation of the *immateriality* of the soul is posed incorrectly and is therefore senseless and incapable of receiving a positive response. For the same reasons, the question of *materiality* would make no sense. In fact, the soul is neither immaterial nor material; rather, as a substance, it is necessarily connected with the matter of an organic body with reference to its form and function (cf. Wolff, 1992, p. 155).

The meaning of the problem of the *community* of soul and body follows, according to Wolff and the whole tradition of rational psychology and empirical psychology, from the existence of the soul and the body as independent substances on the one hand and from the necessity to unify them without contradiction on the other. Living animals can only exist (live) as their unity.

In ancient philosophy, the question of how the community of soul and body is to be explained did not arise because the modern comprehension of substance underlying that question was unknown. That question followed from the Cartesian conception of the body and soul as substances (things) that exist independently of each other and for themselves (cf. Wolff, 1992, p. 78). Hypothetically and skeptically, however, the idea had been considered by Aristotle (1995, 407b15–407b19).

The problem of community results from the assumption of the immateriality, extensionlessness, and simplicity of the soul as a *thing*. This assumption seems to exclude any relationship of community. Consequently, if community is impossible, it will be presupposed by metaphysics as an undeniable fact (Hegel, 1986b, Vol. 10, §.389 *addendum*). Indeed, according to Hegel, it is impossible to overcome such a fixed opposition by means of traditional metaphysical psychology (Hegel, 1986b, Vol. 10, §.389 *addendum*, p. 46).

The problem of community consists, strictly speaking, in a twofold problem, that is, (1) in relation to the shared location ("seat") of the soul within the body and (2) the functioning (the interaction) of body and soul.[31] These problems are not separated carefully enough from one another by rational psychology. The community of body and soul was always supposed as a fact. In his empirical psychology and rational psychology, Wolff gives different answers, both of which are discussed by Hegel (cf. Wolff, 1992, p. 80). According to §.768 of Wolff's *German Metaphysics* (Wolff, 1751/1983), soul and body can exist without any reference to their respective counterparts, but in relation to their nature or essence, they belong together. The harmony of soul and body is only possible if God has unified both originally.

To ground the unity (community) of body and soul in conformity with the principles of empirical psychology, Wolff is satisfied with natural observations and conclusions. There is no need for the concept of God in order to relate, finally (in the last paragraph of empirical psychology), the control over the body to the soul (Wolff, 1751/1983, §.539). The soul commands the body insofar as the arbitrary movements of the latter are determined through the free considerations of the soul

[30] Cf. Wolff (1992, pp. 152f.).

[31] On this matter, see Euler (2002, 2007).

(Wolff, 1751/1983, §.539). The unification of the soul and the body is based on the correspondence between thoughts of the soul and variations of the body. All we can perceive is that mental and bodily changes occur at the same time without causal relation (Hegel, 1830/1992, §§.529, 534): external things bring about modifications in the organs of the senses. In this way, emotions in the soul and thoughts about external things of which we have consciousness emerge at the same time (Hegel, 1830/1992, §§.219, 528). The relationship of simultaneity can be proved, according to Wolff, by stimulations of the body, such as pain, which is accompanied by a negative humor of the mind (Wolff, 1751/1983, §.531), but by passions, which are, on the one hand, bound to neurophysical processes in the body, and which, on the other, command the activity of the mind and distract from the faculty of truth (Wolff, 1751/1983, §.533; cf. §.444).

For certain undeniable reasons, Hegel does not accept the justification of the question of the community and claims its dissolution. According to him, there is no community between soul and body in the sense held by metaphysical psychology (the "usual physiological and psychological consideration") (Hegel, 1986b, Vol. 10, §.389 *addendum*), following the "metaphysics of understanding of forces." He criticizes empirical psychology for its erroneous comprehension of the opposition of soul and matter (an opposition that he, after all, holds to be necessary) (Hegel, 1986b, Vol. 10, §.389 *addendum*, p. 46). The fixing of this opposition makes any community between them and any explanation for it impossible. Thus, it would appear that the whole question has been declared unanswerable. Moreover, the division into forces and faculties of the soul gives no true unity (identity of the soul) but only an aggregate, because each of them acts for itself and is standing only in an external reciprocal interaction with others (cf. Wolff, 1992, p. 80). However, what seems to be necessary is, first, a renunciation of the conceptual inventory of metaphysics, which is presupposed by empirical psychology and, second, its replacement by a completely different ("speculative") manner of conceptual development and determination of the soul and its moments (among which some of the faculties, depending on their nature, could then be arranged again as activities, e.g., sensation and emotion).

16.5.5 The Idea of Hegel's Psychology

Finally, I would like to sketch Hegel's own idea of psychology. I recommend that it should be read and studied in connection and comparison with Wolff's metaphysics of the soul.

First, it should be remembered that the topic of what Hegel calls (in the title of the third section of the *Philosophy of Subjective Spirit*) "*Psychologie. Der Geist*" is the concept of "spirit" (Hegel, 1830/1992, §§.440–482). In the subsections—*a. Der theoretische Geist* (the theoretical spirit), *b. Der praktische Geist* (the practical spirit), and *c. der freie Geist* (the free spirit)—the object is what in the former metaphysics, particularly in the psychology of Wolff, contained the so-called doctrine of the faculties, although treated entirely differently (cf. Wolff, 1992, p. 29). What Hegel calls "the spirit" is not identical with the content of his concept of the "soul." Hegel's "Psychology" is not a doctrine of the soul, although it is closely connected with it. Furthermore, the spirit is not, as in Wolff, a motionless being, but activity (cf. Wolff, 1992, p. 118). It is a process "in its energy" in the sense that "the expressions of it are recognized definitely through its own intrinsicality" (Hegel, 1986b, Vol. 8, §.34 *addendum*). This is exactly the relationship that Wolff had in mind as the aim of his double-sided doctrine of the soul in his metaphysics. For Wolff, spirits differ from beings possessing a soul only in that they are suitable for distinct cognition at a higher degree (Wolff, 1751/1983, §.902).[32] "Spirit" in the strongest sense is, according to Wolff, that being capable of cognition in the "almost perfect degree," namely, God.

The context of Hegel's "Psychology" has his doctrine of the soul as its ground, that is, as matter (stuff) to its activity (analogously to the relation of the soul to the body) (Hegel, 1986b, Vol. 10, §.389).[33] To that extent, the spirit determines itself "to the truth of the soul and of consciousness" (Hegel, 1986b, Vol. 10, §.440). This means, first, that spirit is immanent in nature, but this happens only with regard to the spirit that is unfree or imprisoned (Hegel, 1986b, Vol. 10, §.381 *addendum*, p. 24). The soul, however, as "simple, spiritual substance," is "immediate spirit" (Hegel, 1986b, Vol. 10, §.440 *addendum*, p. 230).

However, the object of spirit is neither the "simple immediate totality" of the soul nor the knowledge of consciousness. Rather, spirit refers strictly to itself, producing its determinations through its own free activity (Hegel, 1986b, Vol. 10, §.440). The object (or the objects) of Hegelian "Psychology" is the free determination of spirit.

Hence, what follows from the change of the basic assertions given by the traditional doctrine of the soul? My answer is that they are incorporated into the conception of the spirit. In the "I," the spirit is simple but is not an overall simple like the soul; rather, it is "something distinguished in itself despite its simplicity" (Hegel, 1986b, Vol. 10, p. 21). Accordingly, the return of the distinctiveness of itself to unity could instead be called "Identity" (Hegel, 1986b, Vol. 10, §.441 *addendum*, p. 233).

It is the free activity of the spirit that creates its specific significance in relation to the soul and to consciousness. It is thus through this that the soul turns into something to be presupposed, which becomes necessary for the self-determination of spirit: "I is the infinite spirit itself, which *presupposes* itself as *soul* and as *consciousness*, and which makes itself *finite* in this way, but which also poses as surpassed (*aufgehoben*) that self-made presupposition, that is, that finiteness or that opposition, which is in itself surpassed, of consciousness against the soul, on the one hand, and the external object, on the other" (Hegel, 1986b, Vol. 10, §.441 *addendum*, p. 233). This determination of spirit marks the difference in relation to the psychology of Wolff. The spirit as the psychic element in men is productive (self-acting) and not only an existing faculty; psychic determinations are not

[32] Cf. Euler (2004, pp. 44–49).

[33] Cf. Wolff (1992, p. 131).

something given independently of it, not fixed categories of understanding, but as flexible as spirit itself, whose products they are. My spirit is neither the knowing of my own individual, mental properties (e.g., the properties of character) nor these properties itself, but the properties mediated through knowing and degraded to "moments" of my own personal spirituality. The simultaneous self-cognition and self-determination of spirit is its liberation from the dependence on nature. Its activities are not linked together in a series of mere forces or faculties but must be comprehended as the stages of development of this liberation.

This difference is not an external and contingent distinction of a simple method of treatment of a philosophical problem but of essential significance for the cognition of truth in relation to the contents of the science of psychology. With this new foundation of psychology, Hegel thus takes a very important step beyond traditional psychology in its Wolffian version and even the Kantian critique. This configuration of psychology is generally described by Hegel in the following way: "Psychology belongs, like Logic, to those sciences that have in more recent times profited least from the universal education (formation) of the spirit and the deeper concept of reason, and it is still in a very bad estate" (Hegel, 1986b, Vol. 10, §.444A, p. 228).

One should not go so far as to maintain that the "stuff of Hegelian 'Psychology'" is "essentially identical with the empirical psychology and rational psychology of the Wolffian school" (Hespe, 1991, p. 517). A more detailed investigation would also reveal considerable differences with respect to the traditional contents of spirit collected through metaphysics. However, one can surely start by saying that the metaphysical psychology in its most developed form—as in the empirical psychology and rational psychology of Wolff, along with the tradition of Aristotle—is the starting point of Hegel's critical revision, new foundation, and further promotion of the development of psychology as a science.

16.6 Conclusion

It has been shown that Wolff pretends to derive the main qualities of the soul from its original force, which is a concept contradictory in itself. In spite of this causal or energetic approach, he determines the soul as a thing, that is, in the tradition of the Cartesian *res* and not in the Aristotelian way as a principle of the body (Sects. 16.2 and 16.3). Kant, however, gave good reasons for destroying the four main categories of the metaphysics of the soul (Sect. 16.4). Hegel actualizes these four categories because he does not accept Kant's arguments for their elimination. However, this does not lead to a rehabilitation of either rational or empirical psychology. On the contrary, Hegel rejects both doctrines, and he does so by revising the very meaning of those determinations on the basis of Aristotelian assumptions. Thus, it comes about that simplicity, substantiality, and immateriality are to be understood as qualities of the soul in a totally new sense. The question of the unity (and community) of soul and body disappears completely or dissolves itself as a consequence of the reinterpretation of the other qualities of the soul. Understanding these Hegelian results is not easy. Doing so requires, on the one hand, a careful study and analysis of the questions discussed in traditional metaphysical psychology as well as the Kantian critique of them and, on the other hand, grasping Hegel's own theory of subjective spirit and his characteristically speculative method of treating the categories of that subject. The theory of the soul in Hegel's *Philosophy of Subjective Spirit* is limited to being only the universal ground for the development of the spirit as a universal active power, far removed from the concept of an isolated immaterial (supernatural) quality of individual thinking.

All these new investigations and insights made by Hegel required him to eliminate radically traditional issues of psychology, now revealed as being dependent on false presuppositions. With this altogether impressive result, Hegel succeeds in creating a new basis for the development of psychology as a science.[34]

References

Ameriks, K. (2000). *Kant's theory of mind: An analysis of the paralogisms of pure reason*. Clarendon Press.

Aristotle. (1995). *Über die Seele*. Felix Meiner.

Baumgarten, A. G. (2004). *Metaphysik*. Dietrich Scheglmann Reprints. (Original work published 1783).

Corti, L., Stein, S., & Plevrakis, E. (Eds.) (2020). Special issue devoted to Hegel and Aristotle on the metaphysics of mind. *Hegel-Bulletin, 41*(1).

Descartes, R. (1904). Meditationes de prima philosophia. In C. Adam & P. Tannery (Eds.), *Oeuvres de Descartes* (Vol. 7, pp. 1–90). Léopold Cerf. (Original work published 1641).

Descartes, R. (1905). Principia philosophiae. In C. Adam & P. Tannery (Eds.), *Oeuvres de Descartes* (Vol. 8.1). Léopold Cerf. (Original work published 1644).

Descartes, R. (1909). Les Passions de l'ame. In C. Adam & P. Tannery (Eds.), *Oeuvres de Descartes* (Vol. 11, pp. 291–497). Léopold Cerf. (Original work published 1649).

Euler, W. L. (2001). Substanz, Organismus, Leben. Das "Neue" im "Neuen System der Natur" von Leibniz. In H. Poser, C. Asmuth, U. Goldenbaum, & W. Li (Eds.), *Nihil sine ratione. Mensch, Natur und Technik im Wirken von G. W. Leibniz* (pp. 370–377). Gottfried-Wilhelm Leibniz Gesellschaft.

Euler, W. L. (2002). Die Suche nach dem "Seelenorgan." Kants philosophische Analyse einer anatomischen Entdeckung Soemmerrings. *Kant-Studien, 93*, 453–480.

Euler, W. L. (2004). Bewußtsein—Seele—Geist. Untersuchungen zur Transformation des Cartesischen "Cogito" in der Psychologie Christian Wolffs. In O.-P. Rudolph & J.-F. Goubet (Eds.), *Die Psychologie Christian Wolffs. Systematische und historische Untersuchungen* (pp. 11–50). Max Niemeyer Verlag.

Euler, W. L. (2007). *Commercium mentis et corporis*? Ernst Platners medizinische Anthropologie in der Kritik von Marcus Herz und Immanuel Kant. *Aufklärung, 19*, 21–68.

Euler, W. L. (2011). Über das Verhältnis des Natürlichen und Übernatürlichen und seine Konsequenzen für die Begründung göttlicher Wunder in der Metaphysik Christian Wolffs. *Aufklärung, 23*, 123–145.

Euler, W. L. (2019). *Christian Wolff: Was kann ich wissen?* Paper read at the University of Halle, December 2019. Unpublished manuscript.

[34] I would like to thank William Altman for his pleasant corrections and kind recommendations.

Euler, W. L. (2021). Apperzeption bei Leibniz und Kant. In G. Motta, D. Schulting, & U. Thiel (Eds.), *Kant's transcendental deduction and the theory of apperception: New interpretations*. de Gruyter.

Hegel, G. W. F. (1971). *Philosophy of mind*. (W. Wallace & A. V. Miller, Trans.). Oxford University Press.

Hegel, G. W. F. (1978). Wissenschaft der Logik. Erster Band. Die objektive Logik (1812/1813). In F. Hogemann & W. Jaeschke (Eds.), *G. W. F. Hegel: Gesammelte Werke* (Vol. 11). Nordrhein-Westfälische Akademie der Wissenschaften. (Original work published 1812/1813).

Hegel, G. W. F. (1981). Wissenschaft der Logik. Zweiter Band. Die subjektive Logik. In F. Hogemann & W. Jaeschke (Eds.), *G. W. F. Hegel: Gesammelte Werke* (Vol. 12). Nordrhein-Westfälische Akademie der Wissenschaften. (Original work published 1816).

Hegel, G. W. F. (1985). Wissenschaft der Logik. Erster Teil. Die objektive Logik. Erster Band. Die Lehre vom Sein. In F. Hogemann & W. Jaeschke (Eds.), *G. W. F. Hegel: Gesammelte Werke* (Vol. 21). Nordrhein-Westfälische Akademie der Wissenschaften. (Original work published 1832).

Hegel, G. W. F. (1986a). Vorlesungen über die Geschichte der Philosophie (3 Vols.). In E. Moldenhauer & K. M. Michel (Eds.), *G. W. F. Hegel: Werke in 20 Bände* (Vol. 18–20). Suhrkamp.

Hegel, G. W. F. (1986b). Enzyklopädie der Philosophischen Wissenschaften im Grundrisse (# Vols.). In E. Moldenhauer & K. M. Michel (Eds.), *G. W. F. Hegel: Werke in 20 Bände* (Vol. 8–10). Suhrkamp.

Hegel, G. W. F. (1992). Enzyklopädie der Philosophischen Wissenschaften im Grundrisse. In W. Bonsiepen, H.-C. Lucas, & U. Rameil (Eds.), *G. W. F. Hegel: Gesammelte Werke* (Vol. 20). Nordrhein-Westfälische Akademie der Wissenschaften. (Original work published 1830).

Hespe, F. (1991). System und Funktion der Philosophie des subjektiven Geistes. In F. Hespe & B. Tuschling (Eds.), *Psychologie und Anthropologie oder Philosophie des Geistes. Beitraege zu einer Hegel-Tagung in Marburg 1989* (pp. 490–521). Frommann-Holzboog.

Kant, I. (1998). *Critique of pure reason* (P. Guyer & A. W. Wood, Trans. & Ed.). Cambridge University Press. (Original work published 1781/1787).

Klemme, H. F. (1996). *Kants Philosophie des Subjekts. Systematische und entwicklungsgeschichtliche Untersuchungen zum Verhältnis von Selbstbewußtsein und Selbsterkenntnis*. Felix Meiner.

Liske, M. T. (2000). *Gottfried Wilhelm Leibniz*. C.H. Beck.

Petry, M. J. (Ed. & Trans.) (1978). *Hegel's philosophy of subjective spirit* (3 Vols). D. Reidel.

Sturma, D. (1998). Die Paralogismen der reinen Vernunft in der zweiten Auflage (B406-432; A381-405). In G. Mohr & M. Willaschek (Eds.), *Immanuel Kant. Kritik der reinen Vernunft* (pp. 391–411). Akademie Verlag.

Wolff, C. (1972). Psychologia rationalis methodo scientifica pertractata qua ea, quae de anima humana indubia experientiae fide innotescunt, per essentiam et naturam animae explicantur, et ad intimiorem naturae ejusque autoris cognitionem profutura proponuntur. Editio nova priori emendatior. In J. École (Ed.), *Christian Wolff: Gesammelte Werke* (Psychologia rationalis) (Vol. 6, Pt. II). Olms. (Original work published 1740).

Wolff, C. (1983). Vernünfftige Gedancken von Gott, der Welt und der Seele des Menschen, auch allen Dingen überhaupt. In J. École (Ed.), *Christian Wolff: Gesammelte Werke* (Deutsche Metaphysik) (Vol. 2, Pt. I). Olms. (Original work published 1751).

Wolff, M. (1992). *Das Koerper-Seele-Problem. Kommentar zu Hegel, Enzyklopaedie (1830), §.389*. Klostermann.

Chapter 17
"The Most Excellent Psychological Systematist": Wolff's Psychology in the Eyes of Wilhelm Wundt

Saulo de Freitas Araujo and Thiago Constâncio Ribeiro Pereira

17.1 Introduction

Christian Wolff's writings found a wide audience in the eighteenth century. His German treatises became hotly debated in German universities (Albrecht, 2018; Hammerstein, 1983), which led to his banishment from Prussia. Moreover, his Latin works found resonance in European universities, attracting many international students to Marburg and Halle (Carboncini, 2018; Röhling, 1983).

After Wolff's death, his psychological project was continued by various Enlightenment philosophers. Authors such as J. N. Tetens (1736–1803) attempted to integrate both rational and empirical psychology (Tetens, 1777, 2015), but a growing perception of their independence took over. For example, whereas M. Mendelssohn (1729–1786) focused on the development of rational psychology (e.g., Mendelssohn, 1767/2009), J. G. Krüger (1715–1759) and K. P. Moritz (1756–1793) helped to advance empirical psychology without its rational counterpart (e.g., Krüger, 1756; Moritz, 1783–1793). As a consequence, the unity of Wolff's psychology began to crumble, and a tension between empirical and rational psychology became the rule.

By the mid-nineteenth century, in the aftermath of Kant's influence, Wolff's psychology, although not completely forgotten, was no longer a hallmark to be adopted by the new psychological science that was emerging at the time. Rational or metaphysical psychology, as it was called back then, fell from grace among empirically inclined psychologists. The ideal of an empirical psychology free from metaphysics

S. d. F. Araujo (✉)
Department of Psychology, Federal University of Juiz de Fora,
Juiz de Fora, Minas Gerais, Brazil
e-mail: saulo.araujo@ufjf.edu.br

T. C. R. Pereira
Department of Psychology, Fluminense Federal University,
Volta Redonda, Rio de Janeiro, Brazil

S. d. F. Araujo et al. (eds.), *The Force of an Idea*, Studies in History and Philosophy of Science 50, https://doi.org/10.1007/978-3-030-74435-9_17

was becoming the order of the day (e.g., Beneke, 1820, pp. 1–9; Fortlage, 1855, pp. viii–ix; Waitz, 1846, pp. iii–vi). In addition, the references to Wolff's psychological writings became scattered and mostly superficial in the psychological literature.[1] Be that as it may, it is still possible to speak of a reception here, especially by German authors,[2] despite the fact that such a *Rezeptionsgeschichte* remains to be done.[3]

In this chapter, we will illustrate this situation using Wilhelm Wundt (1832–1920) as a case study. Wundt was one of the most cited and debated authors of the new physiological or experimental psychology in the second half of the nineteenth century. It is thus to be expected that his work would show some kind of dialogue with Wolff's central psychological ideas. In fact, Wundt's work is permeated by references to Wolff's psychology. For example, he praises Wolff as "the most excellent psychological systematist" (Wundt, 1874, p. 13) and "the most influential psychological systematist" (Wundt, 1887, I, p. 13).[4] At the same time, however, he often criticizes Wolff's psychological system for representing a kind of outdated thinking to be superseded by the emerging scientific psychology. Our goal, then, is to investigate how Wundt received and interpreted Wolff's psychology in the process of elaborating his own psychological project.

To pursue our goal, we will focus on four topics that provide the main lines of that reception: (1) the separation of empirical and rational psychologies, (2) the definition of basic psychological concepts, (3) the faculties of the soul, and (4) voluntarism as a response to intellectualism. In the end, we hope to contribute to reducing the gap in the historiography of psychology concerning Wolff's reception in the nineteenth-century psychology.

---

[1] In the second half of the nineteenth century, it was difficult to find a thorough discussion of Wolff's psychology. Citations of his work do not show a proper treatment of his ideas. Sometimes, one has the impression that he was more cited than read.

[2] It seems that Wolff's psychological program did not have a significant impact upon the establishment of French or North American scientific psychology. If one takes, for instance, Théodule Ribot (1839–1916) and William James (1842–1910) as representatives of the new scientific psychology in France and the USA, respectively, one will find but a few scattered references to Wolff's psychology (e.g., James, 1890/1981, pp. 199, 386; Ribot, 1870, p. 19).

[3] Traditional accounts of the history of German psychology recognize the influence of Wolff's psychological program in eighteenth-century German culture, but very little, if anything, is said about its reception in the nineteenth century (e.g., Dessoir, 1912, pp. 64–108; Klemm, 1911, pp. 60–65; Sommer, 1892, pp. 1–23). More recent accounts have not changed this pattern. For instance, Bell (2005, pp. 19–53) shows the pervasive influence of Wolff's psychology in German literature and philosophy of the eighteenth century, praising "a series of innovations grounded in Wolff's system" (p. 29). When it comes to the nineteenth century, however, the most one gets is the claim that Wolff's idea of *psycheometria* preempted "by more than a hundred years the development of psychometrics" (p. 19).

[4] The Wundt translations were made by the first author. The Wolff translations were made by both authors.

17.2 The Separation of Empirical and Rational Psychologies

In 1862, Wundt published his first psychological book, comprising several articles on sense perception that had been published previously in medical journals. However, he added to those articles an original theoretical essay on the meaning and scope of scientific psychology, which serves to unify them, thus inaugurating his early psychological program (Araujo, 2016). Here, Wolff appears for the first time to illustrate the historical development of psychology. According to Wundt:

> With *Aristotle*, psychology is divided in two parts, one deductive—in which the essence of the soul is developed from concepts—and another inductive, in which the properties of the soul that are given in experience are investigated. The same distinction reappears later with *Christian Wolff*, who attempts to determine in *rational* psychology the supernatural nature of the soul according to Leibniz's metaphysics and deals in *empirical* psychology with the several faculties of the soul as he believes them to be given through observation. (Wundt, 1862, p. xx)

In this passage, Wundt is not properly concerned with the originality of Wolff's psychological program. To him, the general idea of dividing psychology into two parts was not new and neither was the specific methodological commitment of each part (deduction versus induction): both could be traced back to Aristotle. The point, here, is Wundt's insistence on seeing Wolff's psychology in purely methodological terms. On the one hand, rational psychology involved deductive reasoning. On the other, empirical psychology implied induction from particular observations.

This methodological reading of Wolff is unsurprising, since Wundt's chief goal in his introductory essay was to discuss the traditional methods employed in psychology, which in his view were responsible for its theoretical stagnation. According to his diagnosis, those methods were insufficient and unsatisfactory, hence the need for methodological reform. Wundt's solution is radical:

> Since the chosen deductive method is to be rejected on principle, as we have demonstrated,[5] one will adopt only the inductive method, which empirical psychology has been following for a long time. However, it will be necessary to investigate whether induction cannot be used in psychological research in a much broader scope than has previously been the case. In my opinion, there are two ways in which this is possible: the first consists in extending the present observational methods, the second in selecting the experiment as a method of investigation. (p. xxiv)

At least two points deserve attention here. First, it is clear that the problem, for Wundt, laid in rational psychology because of its association with metaphysical reasoning, that is, the deduction of psychological theories and concepts from metaphysical hypotheses and postulates, leading to stagnation. The solution, as he saw it, was to abandon rational psychology *tout court* and improve empirical psychology. In other words, psychology would be transformed by a methodological reform

[5] Up to this point, Wundt had been discussing the limitations of the deductive method of metaphysical psychologists, who derived psychological theories from a priori metaphysical hypotheses, such as the immateriality or simplicity of the soul. For him, this submission of psychology to metaphysics, as represented by J. F. Herbart (1776–1841), for example, led the former to stagnation.

similar to that undergone by the natural sciences. In this sense, Wundt understood his scientific psychology as an extension and improvement of Wolff's empirical psychology, but not of rational psychology.[6] This is why he "deliberately assigned no place for the so-called *rational psychology*" (Wundt, 1887, I, p. 7, original italics) in his psychological project.

Second, Wundt wanted to extend the very notion of induction by using two strategies: widening the range of observation and defending the systematic use of experiments in psychology. In the first case, observation should go beyond the individual level to reach social and cultural phenomena as well, such as language, customs, homicide, and suicide.[7] In the second, the experimental method should be used systematically.[8] Both strategies make clear that there is an implicit acceptance by Wundt of the distinction between observation and experiment, the latter being only a particular form of the former.

This is not the whole story, however. For Wundt, Wolff's differentiation between empirical and rational psychology is based on a failure to appreciate the proper place of both psychology and philosophy in the system of sciences. In Wundt's words:

> In fact, psychology is as good an *empirical science* as physics and chemistry. The task of philosophy, however, can never be to replace the particular sciences; rather, it has first to take their secure results as its fundamentals. Thus, the elaborations of rational psychology are related to the real progress of our science, just as Schelling's or Hegel's natural philosophy is related to the development of modern science. Instead of the critically examined concepts of empirical science, those metaphysical elaborations are based on common, uncritical experience, the indefinite concepts of which are ordered in a dialectical schematism that possesses only a negative epistemic value, because it substitutes an empty pseudo-knowledge for real knowledge. (Wundt, 1887, I, p. 7)

In Wundt's psychological project, then, the separation between empirical psychology and metaphysics is much stronger than Wolff would accept. For Wundt, empirical psychology cannot be contaminated by a priori ontological assumptions (e.g., the soul is an immaterial substance). Otherwise, it would end up substituting ideas for reality. As much as metaphysical discussions are important in psychology, they must be empirically based and allowed only after the advancement of empirical findings, never before. In Wundt's eyes, Wolff's empirical psychology was not

[6] A year later, in the first edition of his *Vorlesungen über die Menschen-und Thierseele* (Lectures on the human and animal mind), Wundt repeated the same pattern of interpretation: he used Wolff's division of psychology to illustrate the insufficiency of traditional psychological methods and the need for reform (Wundt, 1863, p. 4). Again, he understood Wolff's psychology primarily in methodological terms.

[7] Initially, Wundt conceived of this supra-individual level of psychological analysis in two different ways, first, as social or moral statistics providing descriptive and correlational analyses of social phenomena (marriages, suicides, births, deaths, etc.), and second, as *Völkerpsychologie*, a psychological analysis of complex cultural products, such as language, myths, and customs. For a detailed account of this methodological reform, see Araujo (2016, Chap. 2).

[8] Wundt was well acquainted with the experimental tradition in both psychophysiology and psychophysics that had preceded him. His point, however, was that psychological experiments should not be restricted to those elementary forms. This is precisely what he would later develop in Leipzig.

empirical enough, to the extent that it depended from its beginning on Wolff's ontological assumptions about the ultimate nature of the soul. As a consequence, it could not be fully developed as genuine knowledge of psychological phenomena.

17.3 Basic Psychological Concepts

Although psychology should not begin with metaphysical hypotheses about the nature of the mind, it cannot proceed from scratch, that is, without initial concepts. According to Wundt, in order to enter the domain of psychological facts, psychologists need basic concepts and ideas, which he called "psychological preconceptions" (Wundt, 1874, p. 8).[9] The most general is the concept of spirit (*Geist*), which contains within it the concept of mind (*Seele*).[10] Here, Wolff's position appears as one of the precursors of this idea: spirit is the general concept under which the individual mind is included.[11] Wundt accentuates this relation between the spirit and the mind:

> The spirit denotes likewise the subject of inner experience, but its relations to a corporeal being are abstracted. The mind is the subject of inner experience with the conditions it carries by being empirically bound to an external being; the spirit is the same subject without reference to that binding. Hereafter, we will speak of the spirit and spiritual phenomena only when we do not want to put any weight on those moments of inner experience, by means of which the latter is dependent on our corporeal existence, that is, accessible through external experience. (Wundt, 1874, p. 12)

Two other concepts are crucial to Wundt's psychological program, namely, sensation (*Empfindung*) and representation (*Vorstellung*). Although he recognized that Wolff gave those words their psychological meaning,[12] he saw in Wolff's proposal a misunderstanding to be avoided. According to him, Wolff conceived of each sensation as a simple representation.[13] Wundt, by contrast, argued that representation and sensation are two different mental processes:

---

[9] Later, in his mature system, Wundt would speak not only of basic concepts but also of basic principles that guide psychological research, such as the principle of psychophysical parallelism and the principle of psychical causality (e.g., Wundt, 1911).

[10] In the German tradition, the word *Seele* can be translated either as soul or as mind, depending on the specific context in which it appears. In Wundt, the term refers to the subject matter of psychology, which is not conceived of as a metaphysical substance. In this sense, we prefer "mind" to "soul." For more details, see Araujo (2016).

[11] Wundt refers here to paragraph 643 of *Rational Psychology* (Wolff, 1734, §.643).

[12] As for sensation, Wundt says that "the word *Empfindung* has also been used, since Christian Wolff, as a translation of the Latin *sensus, sensatio*" (Wundt, 1902–1903, vol. 1, pp. 354–355). With regard to representation, he says that "Wolff first introduced the word [Vorstellung] into psychological terminology" (Wundt, 1902–1903, vol. 1, pp. 348–349).

[13] At this point, Wundt went awry in his referencing to Wolff's *Psychologia Empirica.* Instead of Part I, second section, first chapter, Wolff defines sensation only in §.65, which belongs to the second chapter of the second section in Part I (Wolff, 1732, §.65).

> Since the relation of representation to an object is only a secondary act, its original essence can only consist in the *connection of a plurality of sensations*. This connection always requires a special activity that turns the representation into a process that is different from sensation. Sensation is the most primordial content of consciousness, which no other mental act can precede. Representation, however, arises out of more simple processes, that is, sensations, which combine themselves according to certain psychological laws to form representations. (Wundt, 1874, p. 465, original italics)

The difference here is twofold. For Wundt, sensations and representations are mental acts or processes, but of different kinds. Whereas a sensation refers directly to its original object, a representation consists in a connection of sensations, thereby referring only indirectly to their corresponding objects. Representations, then, are acts of unification that require previous sensations in order to occur. The latter, however, cannot have the same nature as the former, to the extent that they lack precisely that unifying factor. Therefore, a sensation is not a single representation, as Wundt understood Wolff's position to be.

Next come the concepts of perception (*Perception*) and apperception (*Apperception*), which are part of Wundt's theory of consciousness. Recognizing that Leibniz had introduced the concept of apperception into philosophy, Wundt appealed to Wolff's *Psychologia Empirica* to reinforce the original meaning of the concept: "apperception is attributed to the soul insofar as it is conscious of its own perception" (Wolff, 1732, §.25, p. 17). In this way, apperception appears to be associated with self-consciousness. Here, Wundt did not reject Wolff's conception; rather, he wanted to extend it to include something that goes beyond self-consciousness, namely, the mere apprehension of a certain representation through attention (*Aufmerksamkeit*). Thus, whereas for Wolff attention was also a crucial aspect of consciousness, but distinct from apperception (Wolff, 1732, §.25), Wundt believed that it should be counted as a kind of apperception. This is the rationale for his famous visual metaphor of consciousness:

> As consciousness apprehends itself as active in the synthesis of sensations and in the association of representations, there arises its manifestation that we call *attention*. … If we say of representations given at a certain moment that they were in the visual field of consciousness, then we can describe that part of it, to which attention has turned, as the inner *focal point*. We will call *perception* the entrance of a representation in the inner visual field, and *apperception* its entrance in the focal point. (Wundt, 1874, pp. 717–718)

Next to representation, apperception, and attention, Wundt developed the concept of feeling (*Gefühl*) to refer to a class of mental processes that serve as "the subjective complement of objective sensations and representations" (Wundt, 1874, p. 462). In other words, feeling is a general category that subsumes a series of mental processes that indicate how consciousness "feels" or is affected by sensations and representations in contrasting terms (as different degrees of pleasure or displeasure). Every sensation is originally accompanied by a simple sensory feeling (e.g., a pleasant musical note), which can associate with other feelings to constitute more complex ones, such as aesthetic feelings (e.g., a pleasant melody). In its turn, each feeling can retroact immediately upon the course of representations, thus giving rise to what Wundt called an affect (*Affect*), which is constantly associated with bodily expressions (Wundt, 1874, pp. 800–807)[14]—for instance, when one bursts into tears from listening to a certain part of a symphony. Finally, a desire (*Begehren*) is another kind of feeling that consists in an expectation of "*future* impressions" (Wundt, 1874, p. 807).[15]

The point here is that Wundt criticized Wolff for not recognizing the independence of feeling from knowing, insofar as Wolff defined pleasure (*voluptas*) as a kind of knowledge, namely, the intuitive knowledge of any perfection.[16] On the contrary, Wundt said, feeling is a primary mental process that is given with sensation; it is already there from the beginning, not a secondary process derived from representations. By the same token, Wundt rejected Wolff's definition of affect (*affectus*) as a strong desire as well as his attempt to subordinate desires to intellectual phenomena, that is, to derive all desires from the knowledge of good and bad.[17] For Wundt, both affect and desire are feelings, but of different kinds; they belong to emotional life and cannot arise out of intellectual life.

17.4 Rejecting the Faculties of the Soul

In the beginning of his *Grundzüge*, Wundt recognized that philosophers had been trying for a long time to give an account of the faculties of the soul (*Seelenvermögen*), be it in terms of simple enumeration or of classification into general categories. In this context, Wundt praised Wolff as "the most excellent psychological systematist" (Wundt, 1874, p. 13). According to him, Wolff divided the faculties of the soul into two general categories, namely, the faculty of knowledge (*Erkenntnisvermögen*) and the faculty of desire (*Begehrungsvermögen*), each having a higher and a lower

[14] It is important to note that this is only the initial and still immature theory of emotional life that appears in the first edition of Wundt's *Grundzüge*. In his mature system, he would modify it. Be that as it may, this initial formulation is sufficient to show his opposition to what he understood Wolff's conception to be.

[15] We will see later (cf. Sect. 17.5) that Wundt postulated a third class of mental phenomena that occur together with feelings and representations: the will (*der Wille*). Accordingly, every mental process has three irreducible dimensions: the intellectual, the affective, and the volitional. Whereas the first is related to the apprehension of objects and the second to the corresponding subjectivity, the will refers to the class of phenomena that give direction to the mind, from simple impulses to complex choices that lead to action. The implications of Wundt's theory of the will will be seen later. For now, this is sufficient to clarify his opposition to Wolff.

[16] Wundt refers here to paragraph 511 of the *Psychologia Empirica*, in which Wolff offers the following definition: "pleasure is the intuition or the intuitive knowledge of the perfection of anything, whether true or apparent" (Wolff, 1732, §.511, p. 389).

[17] Wundt uses two passages from the *Psychologia Empirica* to substantiate his reading of Wolff. See Wolff, 1732, §§.603 and 509.

part.[18] In addition, Wolff tried to "deduce the various faculties from a single fundamental force, the representative force" (Wundt, 1874, p. 14).

Wundt saw two basic problems in Wolff's theory of mental faculties. First, their classification into four main classes lacked a leading systematic principle; instead, they "are strung together in a purely empirical way" (Wundt, 1874, p. 14).[19] Second, the very idea of force (*Kraft*) was problematic, to the extent that it "is transformed into a mythological being" (Wundt, 1874, p. 20). For him, both the concepts of mental faculties and force were substantialized, that is, compromised by an a priori ontological view of the mind according to which all inner phenomena should be understood as "expressions of a metaphysical substance or modifications of such a substance by external influences" (Wundt, 1874, p. 20). The solution, then, was to abandon both *Seelenvermögen* and *Kraft*.[20]

In Wundt's view, Wolff attempted to reconcile two incompatible philosophical systems. For this, Wundt accused him of a "superficial eclecticism" (Wundt, 1911, p. 156). On the one hand, he claimed, Wolff returned to Descartes's dualism, accepting the existence of a substantial soul, from which concept all mental phenomena could be deduced. On the other, he said, because Wolff had been strongly influenced by Locke, he had adopted the inductive method, analyzing mental phenomena through observation. The result of this eclecticism was that in both cases—empirical and rational psychology—class concepts (e.g., *Seelenvermögen*) took over "instead of a real interpretation" (Wundt, 1918, p. 231).[21] In other words, Wolff allegedly committed the mistake of conflating names with concrete psychological reality:

> From a psychological standpoint, concepts can serve only to order and differentiate the concrete facts of immediate experience according to general points of view. Such concepts are those of sensing, representing, feeling, desiring, and willing. However, since there are only particular sensations, representations, and wills, those collective concepts can *describe* the intended fact in general, but they cannot directly contribute to its knowledge. (Wundt, 1919, I, p. 139)

[18] Following Wundt, the lower part of knowledge consists of sense, imagination, and memory, while the higher part involves attention, reflection, and understanding. The lower part of desire comprises pleasure, desire, and affect, while the higher part is composed of the will and freedom (Wundt, 1874, p. x). In sum, the soul can be divided into four main classes, so that every mental phenomenon would have to fit into one of them. It is important to note that Wundt did not indicate the paragraphs in which Wolff proposed such a division of the faculties and its parts; instead, he referred to the 1738 edition of Wolff's *Psychologia Empirica* in general.

[19] Later, Wundt would criticize Wolff's theory, affirming that it consists "merely in a superficial classification of mental processes, in which general concepts—such as memory, imagination, sensibility, understanding, etc.—are treated as unitary fundamental forces of the mind" (Wundt, 1906, p. 4).

[20] In the last edition of his *Grundzüge*, Wundt clarified this point further: "they [the mental faculties] are not merely taken as class names for certain domains of inner experience, which in fact they are, but one considers them as forces by means of which the particular phenomena are produced" (Wundt, 1908–1911, vol. 1, p. 11).

[21] By real interpretation (*wirkliche Interpretation*), Wundt means concepts that refer to reality as given in experience, as opposed to abstract nouns.

In sum, Wundt had to reject Wolff's doctrine of the faculties of the soul because, in his eyes, it was contaminated by the idea of a substantial soul, which he could not accept, given his metaphysical restrictions. The solution, for him, was to find another way to characterize the mind and analyze mental phenomena, which leads us to the next section.

17.5 Voluntarism Against Intellectualism

Wundt called his mature psychological system voluntarism (*Voluntarismus*)[22] in opposition to intellectualism and faculty psychology (Wundt, 1911), which had had many representatives in the history of psychology. Roughly, he defined intellectualism as "those views that put the essence of the mental in the representation [*Vorstellung*]" (Wundt, 1919, vol. 1, p. 192). As a result, any philosopher or psychologist who defends the view that the intellectual dimension of the mind (representation, knowledge, reason) takes priority over all other dimensions (e.g., feeling and will) is an intellectualist. Now, in Wundt's eyes, as Wolff conceived of the soul as "a representing being" (Wundt, 1919, vol. 1, p. 367) and tried "to deduce all faculties of the soul from a single fundamental force" (Wundt, 1874, p. 14)—the *vis repraesentativa* or force of representation—he too must be counted among the intellectualists.[23]

Wundt himself was committed to intellectualism in his early years, during which time he defended a logical conception of the mind, meaning that all mental processes are logical processes (judgments, inferences, etc.). After perceiving what he considered to be a mistake, he struggled for many years to find a new way to depict the mind, thus proposing a new conception of psychology (Araujo, 2016). In this sense, voluntarism is his final answer not only to Wolff but also to many other intellectualists of his time.[24]

If Wundt chose the term "voluntaristic" to define his psychology, this means, above all, that the will (*der Wille*)—not the representation (*die Vorstellung*)—occupies pride of place. But such a characterization requires further clarification. We believe that Wundt's psychological voluntarism can be summarized in a few theses, as he himself admitted (Wundt, 1911). First, psychology deals only with concrete, immediate experience. Second, only concrete activities or processes are given immediately and can be classified into three main domains: feeling, representing, and willing. Since we are somehow conscious of what is given (albeit in varying degrees), it follows, third, that all mental activities are given in consciousness or that

[22] Wundt used the term *voluntaristiche Psychologie* to characterize his psychological theory for the first time in 1895 (Wundt, 1895, p. 166).

[23] In his mature discussions on intellectualism, Wundt does not mention Wolff as a typical representative. Instead, his focus is on Leibniz and Herbart. However, there can be no doubt that Wolff is implied in those discussions.

[24] For a detailed analysis of the development of Wundt's voluntarism, see Araujo (2016).

there is no such thing as an unconscious mind. Fourth, paying attention to what is given leads one to conclude that all mental processes have simultaneously three dimensions, which only psychological analysis can separate by abstraction. Wundt concluded, fifth, that feeling, representing, and willing are basic properties of consciousness, which means that all three dimensions are present to us at any moment, although we can select or focus on one of them, thus highlighting specific aspects of our mental processes. As he sees it, "representing, feeling, willing, etc., are not really separate" (Wundt, 1911, p. 157). None of this explains, however, why he chose the volitional dimension of the mind to characterize his psychological system. Why not sentimentalism, for example?

Wundt wanted to accentuate the original volitional nature of all mental processes, which had been lost in previous psychological systems. By accentuating the will, he intended to highlight the dynamic nature of the mind—mind as process, event, activity—and reject a common assumption of intellectualism, namely, that representations could be conceived of "as more or less permanent objects" (p. 157). For him, the will symbolizes much better than any other psychological process the constant activity of the mind.

Wundt's voluntarism represents not only an alternative to intellectualism but also a rejection of Wolff's doctrine of the faculties of the soul and its metaphysical assumptions in particular—the representative nature of the soul, its substantiality, and immateriality—which implies the rejection of Wolff's psychological project in general. In this sense, Wolff's psychology represented for Wundt an outdated psychological system to be overcome by the new scientific psychology. This does not mean, however, that Wundt always understood Wolff's positions correctly, as we will see in the next section.

17.6 Problems with Wundt's Interpretation

In order to assess Wundt's reception of Wolff, it is important to acknowledge the specific context. Above all, one should not forget that Wundt read Wolff's writings more than a century after their appearance, in the aftermath of Kant and German Idealism. Thus, it is natural to expect that his particular goals and assumptions led him to some misunderstandings regarding Wolff's ideas. Here, we will follow the order of our exposition to address those potential problems.

17.6.1 Psychology's Division

We have seen that Wundt understood Wolff's division of psychology in methodological terms. Accordingly, empirical psychology and rational psychology were associated with induction and deduction, respectively. However, this represents a misapprehension of Wolff's perspective in at least three respects: the methods employed by Wolff in his empirical and rational psychology, Wolff's justification for the division of psychology, and the nature of rational psychology.

First, both parts of Wolff's psychology involve from the outset a blend of inductive and deductive reasoning. For instance, the very first psychological finding, according to which we are conscious of ourselves and other things outside us (Wolff, 1751/2003, §.1; 1738/1968, §.11), is an inductive generalization from a particular experience, and the immediate conclusion that "we exist" is deduced with the help of the major premise "whoever is conscious of himself and other things, really exists" (Wolff, 1738/1968, §§.13–14; 1751/2003, §.1). Moreover, Wolff claimed that deductive demonstrations are the gold standard for all scientific knowledge (Wolff, 1738/1968, §§.15–17; 1751/2003, §.8), be it in empirical or rational psychology.

Second, Wundt seems to have ignored that Wolff's division of psychological science was motivated mainly by epistemic and practical reasons, not by methodological ones (in terms of deduction and induction):

> Since this [rational psychology] is a new enterprise, which is contrary to common opinion, and since new things are at first only reluctantly admitted by most people, the chief reason for my separation of rational and empirical psychology was to prevent psychological claims from being indistinctly rejected. In fact, the theory and practice of morals and even of politics depend on psychological principles and are derived from the latter by us, who take the demonstrative method into account. Practical philosophy is of the greatest importance, and we do not wish to build things of the greatest importance upon principles that are contested. For this reason, we build the truths of practical philosophy only upon principles that are clearly established by experience in empirical psychology. (Wolff, 1740/1983a, §.112, p. 51; cf. Wolff, 1733/1973, §.89, pp. 251–252)

In other words, rational psychology contains the more controversial psychological topics such as the essence and nature of the soul, the explanation of the body-soul relation, and the immortality of the soul. Empirical psychology concerns only the most obvious and secure psychological knowledge. Thus, there are no methodological issues at stake, only epistemic and practical ones.[25]

This leads us to the third point. Understanding rational psychology as a purely deductive metaphysical discipline ignores some of its relevant features. For instance, rational psychology presents a quite detailed physiological consideration of the bodily counterpart of the soul's faculties (e.g., Wolff, 1740/1972, §§.111–153), which is not a deductive chain based on the supernatural nature of the soul. It is a conjectural but empirically driven treatment of the material side of the soul's activities, which could well belong to empirical psychology. However, its presence in rational psychology can be justified by the controversial nature of the physiological science of that period, including the debates between mechanism and vitalism.[26] In

[25] For a full account of the main theses regarding Wolff's division of psychology in the literature, see Pereira (2017).

[26] In the eighteenth century, neurophysiological knowledge underwent a great expansion, which led to new empirical discoveries and various controversies: "It [the eighteenth century] saw the classic and time-honored ideas of neurophysiology—animal spirits moving in hollow nerve conduits to and from the ventricles of the brain—being gradually replaced by ideas more in accord with ana-

order to protect empirical psychology from unnecessary controversies, it makes more sense to deal with that kind of speculative "physiological psychology" within rational psychology. This not only reinforces our thesis that Wolff's division of psychology was not methodologically driven in the first place but also shows the novelty of his rational psychology: it was not merely the old pneumatology disguised by a new name—it contained innovative themes of the period that would pervade psychological science in the nineteenth century.[27]

Why, then, one might ask, should rational psychology have no place in Wundt's eyes? Could not the new *physiologische Psychologie* of the nineteenth century be understood in continuity with Wolff's neurophysiological remarks? Or has Wundt completely overlooked the physiological part of Wolff's rational psychology? We will leave these questions open here.

The second point in Wundt's critique concerned the presence of metaphysical assumptions behind Wolff's empirical psychology. For Wundt, however important metaphysical discussions may be, they should come only at the end, after the empirical findings, and never before. In fact, such a strong separation of empirical science and metaphysics has no place in Wolff's system, but this does not mean that Wolff's psychology suffered from the problems identified by Wundt.

In the first place, Wolff's empirical psychology does not depend on a substance conception of the soul, as Wundt claimed. Even if we consider the precedence of ontology in relation to empirical psychology, and the fact that the idea of the soul as a simple thing endowed with a representative force is already there (e.g., Wolff, 1751/2003, §.114, §.128; cf. Wolff, 1736/2001, part II, section II, chs. 1–2), we should not ignore that it plays a decisive role only in rational psychology, and not in empirical psychology. Wolff repeatedly emphasized the complete independence of empirical psychology in relation to the demonstrations of rational psychology (e.g., 1738/1968, *Preface*, pp. 16*–17*; 1733/1973, §.104; 1740/1983a, §§.111–112; 1751/2003, §.191, §.727). Moreover, this became one of the key subjects of his *Schutzschriften* (defense writings), such as the *Anmerckungen zur Deutschen Metaphysik* (Annotations to the German Metaphysics) (e.g., 1740/1983b, Ad §.191, Ad §.527 & seqq., Ad §.727).[28]

In the second place, all psychological knowledge, including the appreciation of the most abstract metaphysical hypotheses in rational psychology, is empirically

---

tomical reality. It also saw an enormous increase in interest in the nervous system as the source of many of the ills of both body and mind, along with new therapies" (Whitaker et al., 2007, p. 3). For a general overview of this transformation and the ensuing debates, see Brazier (1984) and Duchesneau (2012).

[27] For a detailed analysis of the peculiar character of Wolff's rational psychology, see the contribution from Corey Dyck to this volume.

[28] Wundt was not the first to formulate this accusation. It was already present in Wolff's time, causing him to make some specific defenses on this point in his *Schutzschriften*, which followed his *Deutsche Metaphysik*, and especially in his *Anmerckungen zur Deutschen Metaphysik*. For more details about the controversies regarding Wolff's psychology, see the contributions from Thiago Pereira and Saulo Araujo as well as from Ursula Goldenbaum to this volume.

based. An emendation to the first paragraph in Chap. 5 of the *Deutsche Metaphysik* (German Metaphysics) provides an illustration of this methodological requirement:

> What has been mentioned before [empirical psychology] about the soul from experience is the touchstone of what is taught here [rational psychology] about its nature and essence, and about those effects grounded in it. However, what is taught here by no means is a touchstone of what experience teaches us. (Wolff, 1751/2003, §.727, pp. 453–454)

In fact, Wolff's main criteria for the evaluation of metaphysical assumptions—including philosophical hypotheses—were their correspondence to experience and accordance with established truths (Wolff, 1740/1972, §§.532–533; 1740/1983a, §.127). It seems, then, that Wolff was not as far from Wundt as the latter believed. Contrary to Wundt's indictment, Wolff's rational psychology was in constant dialogue with human experience.

17.6.2 Conceptual Definitions

Wundt recognized Wolff's originality and systematic efforts to build a psychological terminology. Nevertheless, by defining his basic concepts, Wundt's main goal was to establish the novelty of his own psychological program. In this sense, the references to Wolff's conceptual definitions play an instrumental rather than an analytical role: that is, Wolff appears mostly as a contrast case to Wundt. As a result, the lack of a careful conceptual analysis led Wundt to some superficial and sometimes misplaced interpretations of Wolff's basic psychological concepts.

For instance, consider the relationship between representation and sensation. For Wolff, the general faculty of knowledge (*facultas cognoscendi*) encompasses the cognitive acts of the soul, which implies that the soul can be conscious of something. However, the soul can be conscious of things in different ways (and for several reasons), which receive different names that correspond to the soul's specific *faculties*—sensation (*sensatio*), perception (*perceptio*), imagination (*imaginatio*), and so on. For instance, perception is defined as "a mental act by means of which the mind represents for itself any object whatsoever" (Wolff, 1738/1968, §.24, p. 17), whereas sensation is "a perception that can be intelligibly explained by the modifications that occur in a certain organ of our body" (Wolff, 1738/1968, §.65, p. 38). Thus, even though sensations imply representations, Wolff nowhere claims that sensations are simple representations, regardless of Wundt's understanding.

As for the nonintellectual dimension of mental life, there are striking dissimilarities between Wundt and Wolff. For the latter, the general faculty of desire (*facultas appetendi*) embraces all appetitive phenomena, which arise from cognitions: pleasure (*voluptas*), sensual desire (*appetitus sensitivus*), affect (*affectus*), and rational desire (*apetitus rationalis*) or will (*voluntas*).[29] For instance, Wolff defines appetite

[29] Wolff conceives of the will in two different senses: as an empirical act of the soul (*volitio*) and as faculty or potency (*voluntas*). See, for example, Wolff (1738/1968, §.882). However, he does not

or desire in general (*appetitus*) as "the inclination of the soul toward an object grounded in the good it perceives in it" (Wolff, 1738/1968, §. 579, p. 440).[30] If a desire involves a confused idea of the good, then it is a sensual desire (§.580); if it involves a distinct representation of the good, it is called a rational desire or will (§.880). What Wundt calls feeling (*Gefühl*) in general corresponds only partially to what Wolff calls appetite or desire (*appetitus*) in general. Besides, the will is a kind of desire (*appetitus rationalis*) in Wolff's theory, whereas it is a distinct and autonomous mental process in Wundt's psychology. Most importantly, however, Wundt defends the original independence of feelings (simple feeling, affect, etc.) and cognitive phenomena (sensation, perception, etc.), whereas Wolff derives feelings from intellectual phenomena (such as the perception of good and bad).

It is important to understand the meaning of that derivation. For Wolff, all natural phenomena follow the Leibnizian principle of continuity, according to which nature does not make leaps: "Appetite originates from cognition, but not by jumps" (Wolff, 1738/1968, §.509, p. 387). Explaining how appetites emerge out of our cognitions is a way to avoid "jumps" in psychological phenomena and to guarantee their continuity (Wolff, 1738/1968, §.509; cf. Wolff, 1733/1973, §.94; 1740/1983b, Ad §.404, Ad §.873). Accordingly, desires presuppose knowledge of the thing desired because one cannot desire what one does not know: "*ignoti nulla cupido*" (no desire for the unknown) (Wolff, 1738/1968, §.588, p. 445).

There is a second aspect of Wolff's standpoint that seems to have escaped Wundt's attention. The demonstration of how appetites derive from cognitions is fundamental for Wolff's defense of the freedom of the soul (*animae libertas*).[31] Here, the key factors are the notions of motive (*motivum*), tendency or inclination (*conatus*), and spontaneity (*spontaneitas*). As we saw above, an appetite or desire is a tendency or inclination of the soul toward a certain object due to the good it has perceived in it (Wolff, 1738/1968, §.579). Such a perception, which can be a distinct or indistinct representation of the good in a thing, is the motive of the appetite or desire (Wolff, 1738/1968, §.887, §.890). It happens that such motives do not determine the future states of the soul because they are contingent, not necessary (Wolff, 1738/1968, §§.940). It is the soul that has an intrinsic principle of self-determination, which allows it to choose among different inclinations or volitions: the principle of spontaneity (Wolff, 1738/1968, §§.933). This is to say that our volitions are motivated, but not determined, by previous perceptions or representations. As strong as

always follow his own distinction.

[30] It should be noted that in his *Psychologia Rationalis*, Wolff attributed to the soul "an inclination to change a present perception," which he called *percepturitio* (Wolff, 1740/1972, §.481, p. 396). According to him, every perception brings in itself such a tendency or inclination (*conatus*).

[31] This was already explicit in his German writings (e.g., Wolff, 1726, Ad §.873). In this sense, one can say that Wolff's reinforcement of the same point in his Latin writings represents one more step in his attempt to defend himself against the old charge of determinism, which had contributed to his expulsion from Halle. For more details, see the contribution from Ursula Goldenbaum to this volume, as well as Biller (2018) and Pečar et al. (2015).

those inclinations may be, the soul has "the capacity to choose spontaneously among the various possibilities" (Wolff, 1738/1968, §.941, p. 706). Thus, it is free.

However important this derivation of appetites from cognitions may be, it is crucial to mention that it does not imply an identity between appetitive and cognitive phenomena. They are different acts of the soul that express the same fundamental force in different ways.[32] Had Wundt understood this point better, he might have suspended his criticism.

17.6.3 The Faculties of the Soul

Wundt's assessment of Wolff's theory of mental faculties is based on three charges: (a) lack of a systematic principle, (b) commitment to a substantial view of the soul, and (c) confusion between names and psychological reality. Here, again, the deeper layers of Wolff's system seem to have escaped Wundt's attention.

It is not true that the faculties of the soul in Wolff's empirical psychology "are strung together in a purely empirical way" (Wundt, 1874, p. 14). There are systematic principles behind them. Above all, Wolff followed the principles of his philosophical method, which display the reasons for the connection and organization of all philosophical notions (see Wolff, 1740/1983a, §§.115–140). In the case of empirical psychology, they amount to perceiving the soul's changes, establishing their distinct concepts, and extracting principles therefrom (Wolff, 1751/2003, §.191; cf. Wolff, 1738/1968, §§.1–3). In this sense, the faculties of the soul are "strung together" not by chance, but by logical and methodological principles, as well as by empirical evidence.

In rational psychology too, one finds a principle for the unification of the faculties: they are different kinds of manifestation or activity of a single force—the representative force—which is the essence of the soul (Wolff, 1740/1972, §§.54–66). In other words, the soul's faculties are expressions (or different limitations) of that force. As a consequence, far from being a mere "superficial classification of mental processes" (Wundt, 1906, p. 4), Wolff's account of mental faculties follows his basic principles.

As for Wolff's commitment to a substantial conception of the soul, Wundt is only partially correct, since he ignored some important distinctions. In fact, the soul is, for Wolff, a substance (Wolff, 1740/1972, §.53). However, mental faculties are not substances but active capacities or potentialities of the soul (Wolff, 1738/1968, §.29; 1740/1972, §.54, §.81). They refer to classes of phenomena that are given in our common experience. Moreover, in his German writings, Wolff had already warned against this confusion. For him, "*Facultates* as special *Substantias* ... are

[32] To illustrate this distinction, Wolff uses an interesting analogy: "who would say that ignition and melting are the same just because both effects take place due to a force of fire?" (Wolff, 1740/1983b, Ad §.873, p. 533).

fictional concepts," which have no place in metaphysics (Wolff, 1740/1983b, Ad §.873, p. 534).[33]

Finally, Wundt accused Wolff of conflating names with reality. Here, it should be noted that Wolff differentiated between nominal and real definitions (Wolff, 1740/1983c, §§.191–197). Nominal definitions give the distinctive features through which a thing can be recognized and differentiated from others; real definitions show how a definite thing is possible (Wolff, 1740/1983c, §.197). For instance, empirical psychology provides a nominal definition of the soul: "that being in us which is conscious of itself and of other things outside us" (Wolff, 1738/1968, §.20, p. 15). However, this nominal definition does not say anything about how the soul is actually possible. This happens only in rational psychology, when the essence of the soul is defined and its possibilities demonstrated (Wolff, 1740/1972, §.66). A real definition, or the notion of how a certain being is possible, involves a deeper understanding of that being, implying a closer connection with psychological reality. The question remains open, however, as to how much of Wundt's critique applies to Wolff's psychological concepts.

17.6.4 Voluntarism and Intellectualism

Wundt proposed the name "voluntarism" to identify his psychological system. For him, the volitional dimension of the mind best characterizes the ultimate nature of psychological phenomena: dynamic, active, and process-like. In so doing, he wanted to criticize and reject intellectualism, which gives priority to the intellectual dimension of the mind. Thus, Wolff appears in Wundt's works as one of the representatives of intellectualism.

It is important to keep in mind the precise sense in which Wolff derives the volitional (or appetitive) faculties from the intellectual ones. For example, desiring something presupposes the perception or consciousness of the desired thing and is manifested through the soul's *tendency* toward that thing (Wolff, 1738/1968, §.509, §.588, §.929; 1733/1973, §.94; 1740/1983b, Ad §.404, Ad §.873, Ad §.878). However, this does not seem to be far from Wundt's claim that all mental activities are given in consciousness and that desires involve the expectation of future impressions. Thus, consciousness is *logically*, but not *chronologically* or *empirically*, prior to volition. In Wolff's psychology, sensation, representation, and feelings are simultaneous and continuous dimensions of mental life, but consciousness is a logical condition of the empirical *knowledge* of any psychological phenomena. That should at least attenuate the ascription of intellectualism to Wolff.

It is also true that Wolff saw the activity of representing as the very essence of the soul. Rational psychology unfolds a detailed demonstration of how the soul's

[33] For a more in-depth discussion of this complex problem in Wolff and Wolffianism, see the contribution from Stefan Heßbrügger-Walter to this volume.

faculties derive from that essence, and this irremediably connects Wolff to the intellectualist tradition. Wundt was right about that. One can speak here of a kind of "ontological intellectualism." However, by considering the practical and moral aspects of Wolff's philosophy, one concludes that this ontological intellectualism does not prevent Wolff from considering the nonintellectual dimension of mental life. For instance, empirical psychology is endowed with the key task of providing secure principles for practical philosophy (e.g., Wolff, 1738/1968, *Preface*, §.6, §.8; 1740/1983a, §.112; 1740/1983b, §.1, Ad §.191; 1751/2003, §.191). As practical philosophy is "the science of directing the appetitive faculty to choose good and escape evil" (Wolff, 1740/1983a, §.62, p. 31), the investigation of the appetitive faculties plays a major role there (Wolff, 1740/1983b, Ad §.406 & seqq., p. 213; cf. Wolff, 1738/1968, §.945). In fact, given the importance attached to the study of the volitional phenomena of the soul, and the practical ends of Wolff's psychology, one can see that he does not fit so perfectly the image of a full-blown intellectualist.

17.7 Conclusion

The reception of Wolff's psychology by Wundt occurred more than a century after the original publications, in the aftermath of Kant and German Idealism. In this context, the very idea of a new scientific psychology, methodologically aligned with the natural sciences, began to gain force. It is hardly surprising, then, that Wundt conceived of his psychological program along those same lines. More specifically, he called for a reform of psychology, based on two axes: a methodological improvement and a rejection of a priori metaphysical assumptions.

This helps us understand how he interpreted Wolff in general: in terms of a methodological and metaphysical critique. Despite recognizing Wolff's originality and systematic efforts, Wundt used Wolffian psychology to contrast with his own ideas. Instead of seeing continuities, he emphasized discontinuities. In so doing, however, he sometimes lost touch with Wolff's original ideas to the point of misunderstanding them, as happened in the case of the division of psychology in two parts.

There is no doubt that Wundt's psychological program helped expand the field of psychology as a whole, not only by promoting the widespread use of the experimental method but also by defending the necessity of nonexperimental investigations. However, had he analyzed Wolff's writings from a different perspective, he would certainly have found some points of contact too. Like Wundt, Wolff tried to offer a unified conception of the mind based on solid conceptual and theoretical grounds, which Wundt seems not to have been able to grasp—at least, not in its deeper sense.

We realize that a single case study does not allow us to generalize for the whole period under investigation. However, there can be no doubt that Wundt was a major figure at that time. In fact, it would be interesting to expand the field of analysis to include other authors who represent scientific psychology in the nineteenth century. It might well be the case that Wundt, rather than being an isolated case, represents a general tendency among the promoters of the emerging scientific psychology. Be that as it may, the reception of Wolff's psychology in the nineteenth century remains an open field for investigation.

References

Albrecht, M. (2018). Wolff an den deutschsprachigen Universitäten. In R. Theis & A. Aichele (Eds.), *Handbuch Christian Wolff* (pp. 418–465). Springer.

Araujo, S. F. (2016). *Wundt and the philosophical foundations of psychology: A reappraisal*. Springer.

Bell, M. (2005). *The German tradition of psychology in literature and thought 1700-1840*. Cambridge University Press.

Beneke, F. E. (1820). *Erfahrungsseelenlehre als Grundlage alles Wissens*. Mittler.

Biller, G. (2018). Biographie und Bibliographie. In R. Theis & A. Aichele (Eds.), *Handbuch Christian Wolff* (pp. 5–31). Springer.

Brazier, M. (1984). *A history of neurophysiology in the 17th and 18th centuries: From concept to experiment*. Raven Press.

Carboncini, S. (2018). Wolffsrezeption in Europa. In R. Theis & A. Aichele (Eds.), *Handbuch Christian Wolff* (pp. 467–495). Springer.

Dessoir, M. (1912). *Outlines of the history of psychology*. (D. Fisher, Trans.). The Macmillan Company. (Original work published 1911).

Duchesneau, F. (2012). *La physiologie des Lumières. Empirisme, modèles et théories. [The physiology of enlightenment. Empiricism, models, and theories]*. Classiques Garnier.

Fortlage, K. (1855). *System der Psychologie als empirischer Wissenschaft aus der Beobachtung des innern Sinnes* (Vol. 1). Brockhaus.

Hammerstein, N. (1983). Christian Wolff und die Universitäten. Zur Wirkungsgeschichte des Wolffianismus im 18. Jahrhundert. In W. Schneiders (Ed.), *Christian Wolff 1679–1754. Interpretationen zu seiner Philosophie und deren Wirkung* (pp. 267–277). Meiner.

James, W. (1981). *The principles of psychology* (3 vols.). Harvard University Press. (Original work published 1890).

Klemm, O. (1911). *Geschichte der Psychologie*. Teubner.

Krüger, J. G. (1756). *Versuch einer Experimental-Seelenlehre*. C. H. Hemmerde.

Mendelssohn, M. (2009). Phaedon: Oder über die Unsterblichkeit der Seele in drey Gesprächen. In C. Schulte, A. Kennecke, & G. Jurewicz (Eds.), *Mendelssohns Ausgewählte Werke: Schriften zu Aufklärung und Judentum* (Vol. 1, pp. 341–427). WBG. (Original work published 1767).

Moritz, K. P. (Ed.). (1783–1793). *ΓΝΩΘΙ ΣΑΥΤΟΝ oder Magazin zur Erfahrungsseelenkunde als ein Lesebuch für Gelehrte und Ungelehrte*. Mylius.

Pečar, A., Zaunstöck, H., & Müller-Bahlke, T. (Eds.). (2015). *Die Causa Christian Wolff. Ein epochemachender Skandal und seine Hintergründe*. Franckesche Stiftungen.

Pereira, T. C. R. (2017). *A Unidade da Psicologia no Pensamento de Christian Wolff*. Unpublished doctoral dissertation, Universidade Federal de Juiz de Fora, Juiz de Fora, Brazil.

Ribot, T. (1870). *La psychologie anglaise contemporaine [Contemporary English psychology]*. Ladrange.

Röhling, H. (1983). Bemerkungen zur Wirkung Christian Wolffs in Ost- und Südosteuropa. In W. Schneiders (Ed.), *Christian Wolff 1679–1754. Interpretationen zu seiner Philosophie und deren Wirkung* (pp. 267–277). Meiner.

Sommer, R. (1892). *Geschichte der deutschen Psychologie und Aesthetik von Wolff-Baumgarten bis Kant-Schiller*. Stahel.

Tetens, J. N. (1777). *Philosophische Versuche über die menschliche Natur und ihre Entwickelung [Philosophical essays on human nature and its development]*. Weidmann.

Tetens, J. N. (2015). *Metaphysik [Metaphysics]*. Meiner.

Waitz, T. (1846). *Grundlegung der Psychologie*. Perthes.

Whitaker, H., Smith, C., & Finger, S. (2007). Introduction. In H. Whitaker, C. Smith, & S. Finger (Eds.), *Brain, mind, and medicine: Neuroscience in the 18th century* (pp. 3–4). Springer.

Wolff, C. (1726). *Ausführliche Nachricht von seinen eigenen Schrifften, die er in deutscher Sprache von den verschiedenen Theilen der Welt-Weißheit heraus gegeben, auf Verlangen ans Licht gestellet*. Franckfurt am Main: Andrea und Hort.

Wolff, C. (1732). *Psychologia Empirica, methodo scientifica pertractata, qua ea, quae de anima humana indubia experientiae fide constant, continentur et ad solidam universae philosophiae practicae ac theologiae naturalis tractationem via sternitur*. Renger.

Wolff, C. (1734). *Psychologia rationalis, methodo scientifica pertractata, qua ea, quae de anima humana indubia experientiae fide innotescunt, per essentiam et naturam animae explicantur, et ad intimiorem naturae ejusque autoris cognitionem profutura proponuntur*. Renger.

Wolff, C. (1968). Psychologia Empirica, methodo scientifica pertractata, qua ea, quae de anima humana indubia experientiae fide constant, continentur et ad solidam universae philosophiae practicae ac theologiae naturalis tractationem via sternitur. In J. École (Ed.), *Christian Wolff: Gesammelte Werke* (Psychologia Empirica) (Vol. 5, Pt. II). Olms. (Original work published 1738).

Wolff, C. (1972). Psychologia rationalis, methodo scientifica pertractata, qua ea, quae de anima humana indubia experientiae fide innotescunt, per essentiam et naturam animae explicantur, et ad intimiorem naturae ejusque autoris cognitionem profutura proponuntur. In J. École (Ed.), *Christian Wolff: Gesammelte Werke* (Psychologia rationalis) (Vol. 6, Pt. II, 2nd ed.). Olms. (Original work published 1740).

Wolff, C. (1973). Ausführliche Nachricht von seinen eigenen Schrifften, die er in deutscher Sprache heraus gegeben. In H. W. Arndt (Ed.), *Christian Wolff: Gesammelte Werke* (Ausführliche Nachricht) (Vol. 9, Pt. I). Olms. (Original work published 1733).

Wolff, C. (1983a). Philosophia rationalis sive Logica, methodo scientifica pertractata et ad usum scientiarum atque vitae aptata. Praemittitur Discursus praeliminaris de philosophia in genere. In J. École (Ed.), *Christian Wolff: Gesammelte Werke* (Discursus praemlinaris) (Vol. 1.1, Pt. II, 3rd ed.). Olms. (Original work published 1740).

Wolff, C. (1983b). Der vernünfftigen Gedancken von Gott, der Welt und der Seele des Menschen, auch allen Dingen überhaupt, anderer Theil, bestehend in ausführlichen Anmerckungen, und zu besserem Verstande und bequemerem Gebrauche derselben heraus gegeben. In C. A. Corr (Ed.), *Christian Wolff: Gesammelte Werke* (Anmerckungen zur Deutschen Metaphysik) (Vol. 3, Pt. I). Olms. (Original work published 1740).

Wolff, C. (1983c). Philosophia rationalis sive Logica, methodo scientifica pertractata et ad usum scientiarum atque vitae aptata. Praemittitur Discursus praeliminaris de philosophia in genere. In J. École (Ed.), *Christian Wolff: Gesammelte Werke* (Philosophia rationalis sive Logica) (Vol. 1.2, Pt. II, 3rd ed.). Olms. (Original work published 1740).

Wolff, C. (2001). Philosophia prima, sive Ontologia, methodo scientifica pertractata, qua omnis cognitionis humanae principia continentur. In J. École (Ed.), *Christian Wolff: Gesammelte Werke* (Ontologia) (Vol. 3, Pt. II, 2nd ed.). Olms. (Original work published 1736).

Wolff, C. (2003). Vernünfftige Gedancken von Gott, der Welt und der Seele des Menschen, auch allen Dingen überhaupt. In C. Corr (Ed.), *Christian Wolff: Gesammelte Werke* (Deutsche Metaphysik) (Vol. 2, Pt. I). Olms. (Original work published 1751).

Wundt, W. (1862). *Beiträge zur Theorie der Sinneswahrnehmung*. C. F. Winter.

Wundt, W. (1863). *Vorlesungen über die Menschen- und Thierseele* (2 vols.). Leipzig: Voß.Wundt, W. (1874). *Grundzüge der physiologischen Psychologie [Fundamentals of physiological psychology]*. Engelmann.

Wundt, W. (1887). *Grundzüge der physiologischen Psychologie [Fundamentals of physiological psychology]* (2 vols., 3rd ed.). Engelmann.

Wundt, W. (1895). *Logik [Logic]* (Vol. 2, 2nd ed.). Enke.

Wundt, W. (1902–1903). *Grundzüge der physiologischen Psychologie [Fundamentals of physiological psychology]* (3 vols., 5th ed.). Engelmann.

Wundt, W. (1906). *Vorlesungen über die Menschen- und Tierseele [Lectures on the human and animal mind]* (4th ed.). Voss.

Wundt, W. (1908–1911). *Grundzüge der physiologischen Psychologie [Fundamentals of physiological psychology]* (3 vols, 6th ed.). Engelmann.

Wundt, W. (1911). Über die definition der Psychologie [The definition of psychology]. In W. Wundt (Ed.), *Kleine Schriften* (Vol. 2, pp. 113–166). Engelmann.

Wundt, W. (1918). *Einleitung in die Philosophie [Introduction to philosophy].* Kröner.

Wundt, W. (1919). *System der Philosophie [System of philosophy]* (2 vols., 4th ed.). Kröner.

Index

S. d. F. Araujo et al. (eds.), *The Force of an Idea*, Studies in History and Philosophy of Science 50, https://doi.org/10.1007/978-3-030-74435-9

Q

R

Z

CPSIA information can be obtained
at www.ICGtesting.com
Printed in the USA
BVHW010219210721
612511BV00002B/31

9 783030 744342